CHAPMAN & HALL/CRC FINANCIAL MATHEMATICS SERIES

Numerical Methods for Finance

CHAPMAN & HALL/CRC
Financial Mathematics Series

Aims and scope:
The field of financial mathematics forms an ever-expanding slice of the financial sector. This series aims to capture new developments and summarize what is known over the whole spectrum of this field. It will include a broad range of textbooks, reference works and handbooks that are meant to appeal to both academics and practitioners. The inclusion of numerical code and concrete real-world examples is highly encouraged.

Published Titles

American-Style Derivatives; Valuation and Computation, *Jerome Detemple*

Financial Modelling with Jump Processes, *Rama Cont and Peter Tankov*

An Introduction to Credit Risk Modeling, *Christian Bluhm, Ludger Overbeck, and Christoph Wagner*

Portfolio Optimization and Performance Analysis, *Jean-Luc Prigent*

Robust Libor Modelling and Pricing of Derivative Products, *John Schoenmakers*

Structured Credit Portfolio Analysis, Baskets & CDOs, *Christian Bluhm and Ludger Overbeck*

Numerical Methods for Finance, *John A. D. Appleby, David C. Edelman, and John J. H. Miller*

Proposals for the series should be submitted to one of the series editors above or directly to:
CRC Press, Taylor and Francis Group
24-25 Blades Court
Deodar Road
London SW15 2NU
UK

CHAPMAN & HALL/CRC FINANCIAL MATHEMATICS SERIES

Numerical Methods for Finance

Edited by

John A. D. Appleby

David C. Edelman

John J. H. Miller

CRC Press
Taylor & Francis Group
Boca Raton London New York

CRC Press is an imprint of the
Taylor & Francis Group, an **informa** business
A CHAPMAN & HALL BOOK

CRC Press
Taylor & Francis Group
6000 Broken Sound Parkway NW, Suite 300
Boca Raton, FL 33487-2742

First issued in paperback 2019

ISBN-13: 978-1-58488-925-0 (hbk)
ISBN-13: 978-0-367-38859-1 (pbk)

Library of Congress Cataloging-in-Publication Data

Miller, John (John James Henry), 1937-
Numerical methods for finance / John Miller and David Edelman.
p. cm. -- (Financial mathematics series)
Papers presented at a conference.
Includes bibliographical references and index.
ISBN-13: 978-1-58488-925-0 (alk. paper)
ISBN-10: 1-58488-925-X (alk. paper)
1. Finance--Mathematical models--Congresses. I. Edelman, David. II. Title. III. Series.

HG106.M55 2007
332.01'5195--dc22 2007014372

Visit the Taylor & Francis Web site at
http://www.taylorandfrancis.com

and the CRC Press Web site at
http://www.crcpress.com

Contents

Preface

This volume contains a refereed selection of papers, which were first presented at the international conference on Numerical Methods for Finance held in Dublin, Ireland in June 2006 and were then submitted for publication. The refereeing procedure was carried out by members of the International Steering Committee, the Local Organizing Committee and the Editors.

The aim of the conference was to attract leading researchers, both practitioners and academics, to discuss new and relevant numerical methods for the solution of practical problems in finance.

The conference was held under the auspices of the Institute for Numerical Computation and Analysis, a non-profit company limited by guarantee; see http://www.incaireland.org for more details.

It is a pleasure for us to thank the members of the International Steering Committee:

Elie Ayache (ITO33, Paris, France)
Phelim Boyle (University of Waterloo, Ontario, Canada)
Rama Cont (Ecole Polytechnique, Palaiseau, France)
Paul Glasserman (Columbia University, New York, USA)
Sam Howison (University of Oxford, UK)
John J. H. Miller (INCA, Dublin, Ireland)
Harald Niederreiter (National University of Singapore)
Eckhard Platen (University of Technology Sydney, Australia)
Wil Schilders (Philips, Eindhoven, Netherlands)
Hans Schumacher (Tilburg University, Netherlands)
Ruediger Seydel (University of Cologne, Germany)
Ton Vorst (ABN-AMRO, Amsterdam, Netherlands)
Paul Wilmott (Wilmott Associates, London, UK)
Lixin Wu (University of Science & Technology, Hong Kong, China)

and the members of the Local Organizing Committee:

John A. D. Appleby (Dublin City University)
Nikolai Dokuchaev (University of Limerick)

David C. Edelman (Smurfit Business School, Dublin)
Peter Gorman (Chartered Accountant, Dublin)
Bernard Hanzon (University College Cork)
Frank Monks (Nexgen Capital, Dublin)
Frank Oertel (University College Cork)
Shane Whelan (University College Dublin)

In addition, we wish to thank our sponsors, without their enthusiasm and practical help, this conference would not have succeeded.

The Editors
John A. D. Appleby
David C. Edelman
John J. H. Miller
Dublin, Ireland

List of Contributors

Carlo Acerbi
Abaxbank
Corso Monforte 34, 20122 Milano

P.L. De Angelis
University of Naples Parthenope
Via Medina
40 -80133 Naples, Italy

S.J. Berridge
Man Investments
Sugar Quay,
Lower Thames St
London EC3R6DU

David Blake
Pensions Institute
Cass Business School
City University
106 Bunhill Row, London, EC1Y 8TZ, United Kingdom.
Kevin Dowd Centre for Risk & Insurance Studies
Nottingham University Business School
Jubilee Campus, Nottingham, NG8 1BB, United Kingdom

Thomas Breuer
PPE Research Centre
FH Vorarlberg,
Hochschulstrasse 1, A-6850
Dornbirn

Damiano Brigo
Credit Models -Banca IMI Corso Matteotti 6
20121 Milano, Italy

Andrew J.G. Cairns
Maxwell Institute
Edinburgh
Actuarial Mathematics and Statistics
Heriot-Watt University
Edinburgh, EH14 4AS,
United Kingdom

S. Corsaro
University of Naples Parthenope
Via Medina
40 -80133 Naples, Italy

Mark Cummins
Dept. Accounting & Finance
Kemmy Business School
University of Limerick

Stefania Corsaro
University of Naples Parthenope
Via Medina
40 -80133 Naples, Italy

Vladimír Dobiáš
University College Dublin

Kevin Dowd
Nottingham University Business School
University of Nottingham
Nottingham, NG8, IBB

David C. Edelman
University College Dublin

Eymen Errais
Department of Management Science & Engineering
Stanford University
Stanford, CA 94305-4026

Hakan Er
Department of Business Administration
Akdeniz University, Turkey

Alma Garcia
Department of Computer Science
University of Essex,
United Kingdom

Kay Giesecke
Department of Management Science & Engineering
Stanford University
Stanford, CA 94305-4026

Michael B. Giles
Professor of Scientific Computing
Oxford University Computing Laboratory
Oxford University

Thomas Gerstner
Institut für Numerische Simulation
Universität Bonn, Germany

Lisa R. Goldberg
MSCI Barra, Inc.
2100 Milvia Street
Berkeley, CA 94704-1113

Markus Holtz
Institut für Numerische Simulation
Universität Bonn, Germany

Martin Jandačka
PPE Research Centre
FH Vorarlberg,
Hochschulstrasse 1, A-6850 Dornbirn

Ralf Korn
Fachbereich Mathematik
TU Kaiserslautern, Germany

Sheri Markose
Department of Economics
University of Essex,
United Kingdom

Zelda Marino
University of Naples Parthenope
Via Medina
40 -80133 Naples, Italy

Olaf Menkens
School of Mathematical Sciences
Dublin City University
Glasnevin, Dublin 9, Ireland

Bernard Murphy
Dept. Accounting & Finance
Kemmy Business School
University of Limerick

Conall O'Sullivan
University College Dublin

Andrea Pallavicini
Credit Models -Banca IMI Corso
Matteotti 6
20121 Milano, Italy

Francesca Perla
University of Naples Parthenope
Via Medina
40 -80133 Naples, Italy

Francesco Sandrini
Pioneer Investments

J.M. Schumacher
Department of Econometrics and
Operations Research
Center for Economic Research
(CentER)
Tilburg University
PO Box 90153
5000 LE Tilburg,
The Netherlands

Edward Tsang
Department of Computer Science,
University of Essex
United Kingdom

About the Editors

John A. D. Appleby is a senior lecturer of stochastic analysis and financial mathematics in the School of Mathematical Sciences in Dublin City University (DCU). His research interests lies in the qualitative theory of stochastic and deterministic dynamical systems, both in continuous and discrete time. In particular, his research focuses on highly nonlinear equations, on equations which involve delay and memory, and on applications of these equations to modeling financial markets. He has published around 45 refereed journal articles in these areas since receiving his PhD in Mathematical Sciences from DCU in 1999. Dr Appleby is the academic director of undergraduate and postgraduate degree programs in DCU in Financial and Actuarial Mathematics, and in Actuarial Science, and is an examiner for the Society of Actuaries in Ireland.

David C. Edelman is currently on the faculty of the Michael Smurfit School of Business at University College Dublin in Finance, following previous positions including Sydney University (Australia) and Columbia University (USA). David is a specialist in Quantitative and Computational Finance, Mathematical Statistics, Machine Learning, and Information Theory. He has published over 50 refereed articles in these areas after receiving his Bachelors, Masters, and PhD from MIT and Columbia.

John J. H. Miller is Director of INCA, the Institute for Numerical Computation and Analysis, in Dublin, Ireland and a Research Fellow in the Research Institute of the Royal College of Surgeons in Ireland. Prior to 2000, he was in the Department of Mathematics, Trinity College, Dublin. He received his Sc.D. from the University of Dublin, his PhD in mathematics from the Massachusetts Institute of Technology and two bachelor degrees from Trinity College Dublin.

Sponsors

DEPFA BANK

CHAPTER 1

Coherent Measures of Risk into Everyday Market Practice

Carlo Acerbi
Abaxbank, Milan, Italy

Contents

1.1 MOTIVATIONS

This chapter presents a guided tour of the recent (sometimes very technical) literature on coherent risk measures (CRMs). Our purpose is to overview the theory of CRMs from the perspective of practical risk-management applications. We have tried to single out those results of the theory that help in understanding which CRMs can be considered as realistic candidate alternatives to value at risk (VaR) in the financial risk-management practice. This has also been the spirit of the author's research line in recent years [1, 4–6] (see Acerbi [2] for a review).

1.2 COHERENCY AXIOMS AND THE SHORTCOMINGS OF VAR

In 1997, a seminal paper by Artzner et al. [7, 8] introduced the concept of coherent measure of risk by imposing, via an axiomatic framework, specific mathematical conditions that enforce some basic principles that a sensible risk measure should always satisfy. This cornerstone of financial mathematics was welcomed by many as the first serious attempt to give a precise definition of financial risk itself, via a deductive approach. Among the four celebrated axioms of coherency, a special role has always been played by the so-called *subadditivity axiom*

$$\rho(X+Y) \leq \rho(X) + \rho(Y) \tag{1.2.1}$$

where $\rho(\cdot)$ represents a measure of risk acting on portfolios' profit-loss random variable (r.v.s) (X, Y) on a chosen time horizon. The reason why this condition has been long debated is probably due to the fact that VaR—the most popular risk measure for capital adequacy purposes—turned out to be not subadditive and consequently not coherent. As a matter of fact, since inception, the development of the theory of CRMs has run in parallel with the debate on whether and how VaR should be abandoned by the risk-management community.

The subadditivity axiom encodes the risk-diversification principle. The quantity

$$H(X, Y; \rho) = \rho(X) + \rho(Y) - \rho(X+Y) \tag{1.2.2}$$

is the *hedging benefit* or, in capital adequacy terms, the *capital relief* associated with the merging of portfolios X and Y. This quantity will be larger when the two portfolios contain many bets on the same risk driver, but of opposite direction, which therefore hedge each other in the merging. It will be zero in the limiting case when the two portfolios bet on the same directional move of every common risk factor. However, the problem with nonsubadditive risk measures such as VaR is that there happen to be cases in which the hedging benefit turns out to be negative, which is simply nonsensical from a risk-theoretical perspective.

Specific examples of subadditivity violations of VaR are available in the literature [5, 8], although these may appear to be fictitious and unrealistic. It may be surprising to learn, however, that examples of subadditivity violations of VaR can also be be built with very inoffensive distribution functions. An example is known [3] where the two marginal distributions of X and Y are both standard normals, leading to the conclusion that it is never sufficient to study the marginals to ward off a VaR violation of subadditivity, because the trigger of such events is a copula property.

Other examples of subadditivity violation of VaR (see Acerbi [2], examples 2.15 and 4.4) allow us to display the connection between the coherence of a risk measure and the *convexity of risk surfaces*. By risk surface, we mean the function $\vec{w} \mapsto \rho(\Pi(\vec{w}))$, which maps the vector of weights $\vec{w}$ of the portfolio $\Pi(\vec{w}) = \sum_i w_i X_i$ onto the risk $\rho(\Pi(\vec{w}))$ of the portfolio. The problem of ρ-portfolio optimization amounts to the global search of minima on the surface. An elementary consequence of coherency is the convexity of risk surfaces

$$\rho \text{ coherent} \Rightarrow \rho(\Pi(\vec{w})) \text{ convex.} \tag{1.2.3}$$

This immediate result tells us that risk optimization—if we carefully define our variables—is an intrinsically convex problem. This bears enormous practical consequences, because the border between convex and nonconvex optimization delimits solvable and unsolvable problems when things are complex enough, whatever supercomputer you may have. In the examples (see Acerbi [2]), VaR exhibits nonconvex risk surfaces, infested with local minima, that can easily be recognized to be just artifacts of the chosen (noncoherent) risk measure. In the same examples, thanks to convexity, a CRM displays, on the contrary, a single global minimum, which can be immediately recognized as the correct optimal portfolio, from symmetry arguments.

The lesson we learn is that, by adopting a noncoherent measure as a decision-making tool for asset allocation, we are choosing to face formidable (and often unsolvable) computational problems related to the minimization of risk surfaces plagued by a plethora of risk-nonsensical local minima. As a matter of fact, we are persuaded that no bank in the world has actually ever performed a *true* VaR minimization in its portfolios, if we exclude multivariate Gaussian frameworks à la Riskmetrics, where VaR is actually just a disguised version of standard deviation and hence convex.

Nowadays, sacrificing the huge computational advantage of convex optimization for the sake of VaR fanaticism is pure masochism.

1.3 THE OBJECTIVIST PARADIGM

The general representation of CRMs is well known [8,9]. Any CRM $\rho_{\mathcal{F}}$ is in one-to-one correspondence with a family $\mathcal{F}$ of probability measures $\mathbb{P}$.

The formula is strikingly simple

$$\rho_{\mathcal{F}}(X) = \sup_{\mathbb{P} \in \mathcal{F}} \mathbb{E}^{\mathbb{P}}[-X]. \tag{1.3.1}$$

But this representation is of little help for a risk manager, as it provides too much freedom. More importantly, it generates a sort of philosophical impasse, as it assumes an intrinsically *subjectivist* point of view that is opposite to the typical risk manager's philosophy, which is *objectivist*. The formula defines the CRM $\rho_{\mathcal{F}}$ as the worst case expected loss of the portfolio in a family $\mathcal{F}$ of "parallel universes" $\mathbb{P}$.

Objectivists are statisticians who believe that a unique, real probability measure of future events must necessarily exist somewhere, and their principal aim is to try to estimate it empirically. Subjectivists, in contrast are intransigent statisticians who posit that even if this real probability measure existed, it would be unknowable. They simply reject this concept and think of probability measures as mere mathematical instruments. Equation (1.3.1) is manifestly subjectivist, as it is based on families of probability measures.

Risk managers are objectivists, and the algorithm they use to assess the capital adequacy via VaR is intrinsically objectivist. We can in fact split this process into two clearly distinct steps:

1. Model the probability distribution of your portfolio
2. Compute VaR on this distribution

An overwhelmingly larger part of the computational effort (data mining, multivariate risk-factors distribution modeling, asset pricing, etc.) is done in step 1, which has no relation with VaR and is just an objectivist project. The computation of VaR, given the distribution, is typically a single last code line. Hence, in this scheme, replacing VaR with any other CRM is immediate, but it is clear that, for this purpose, it is necessary to identify those CRMs that fit the objectivist paradigm.

If we look for something better than VaR, we cannot forget that, despite its shortcomings, this risk measure brought into risk management practice a real revolution thanks to some features that were innovative at the time of its advent and that nobody today would be willing to give up.

- Universality (VaR applies to risks of any nature)
- Globality (VaR condenses multiple risks to a single figure)
- Probability (VaR contains probabilistic information on the measured risks)

- Right units of measure (VaR is simply expressed in terms of "lost money")

The last two features explain why VaR is worshipped by any firm's boss, whose daily refrain is: "How much money do we risk and with what probability?" Remember that risk sensitivities (aka "greeks," namely partial derivatives of the portfolio value to a specific risk factor) do not share any of the above features, and you will immediately understand why VaR became so popular. As a matter of fact, a bank's greeks-based risk report is immensely more cumbersome and less communicative than a VaR-based one.

If we look more closely at the features that made the success of VaR, we notice that they have nothing to do with VaR itself in particular, but rather with the objectivist paradigm above. In other words, if in step 2 above, we replace VaR with any sensible risk measure defined as a monetary statistic of the portfolio distribution, we automatically preserve these features. That is why looking for CRMs that fit the objectivist paradigm is so crucial.

In our opinion, the real lasting heritage of VaR in the development of the theory and practice of risk management is precisely the very fact that it served to introduce, for the first time, the objectivist paradigm into the market practice. Risk managers started to plot the distribution of their portfolios and learned to fear its left tail thanks to the lesson of VaR.

1.4 ESTIMABILITY

The property that characterizes the subset of those CRMs that fit the objectivist paradigm is *law invariance*, first studied in this context by Kusuoka [11]. A measure of risk ρ is said to be law invariant (LI) if it is a functional of the portfolio's distribution function $F_X(\cdot)$ only. The concept of law invariance therefore can be defined only with reference to a single chosen probability space

$$\rho \text{ law invariant} \quad \Leftrightarrow \quad \rho(X) = \rho[F_X(\cdot)] \tag{1.4.1}$$

or equivalently

$$\rho \text{ law invariant} \quad \Leftrightarrow \quad [F_X(\cdot) = F_Y(\cdot) \quad \Rightarrow \quad \rho(X) = \rho(Y)]. \tag{1.4.2}$$

It is easy to realize that law invariance means *estimability* from empirical data.

THEOREM 1.4.1

$$\rho \text{ law invariant} \Leftrightarrow \rho \text{ estimable} \tag{1.4.3}$$

PROOF ($\Leftarrow$): suppose ρ to be estimable and let X and Y be r.v.s with identical probability distribution (i.i.d.) function. Consider N i.i.d. realizations $\{x_i\}_{i=1,\ldots,N}$ and $\{y_i\}_{i=1,\ldots,N}$ and an estimator $\hat{\rho}$. We will have

$$\hat{\rho}(\{x_i\}) \stackrel{N\to\infty}{\longrightarrow} \rho(X)$$

$$\hat{\rho}(\{y_i\}) \stackrel{N\to\infty}{\longrightarrow} \rho(Y).$$

But for large N, the samples $\{x_i\}$ and $\{y_i\}$ are indistinguishable, hence $\rho(X) = \rho(Y)$

PROOF ($\Rightarrow$): suppose ρ to be LI. Then a (canonical) estimator is defined by

$$\hat{\rho}(\{x_i\}) \equiv \rho(\hat{F}_X(\{x_i\})) \tag{1.4.4}$$

where $\hat{F}_X(\{x_i\})$ represents an empirical distribution estimated from the data $\{x_i\}$.

It is then clear that for a CRM to be measurable on a single given probability distribution, it must be also LI. That is why, unless an unlikely subjectivistic revolution takes place in the market, risk managers will always turn their attention just to the subset of LI CRMs for any practical application. Law invariance, in other words, is a sort of unavoidable "fifth axiom" for practitioners.

Popular examples of LI CRMs include, for instance, α-expected shortfall (ES_α) (aka CVaR, AVaR, etc.) [5, 13]

$$ES_\alpha(X) = -\frac{1}{\alpha}\int_0^\alpha F_X^{\leftarrow}(p)\,dp \qquad \alpha \in (0\%, 100\%) \tag{1.4.5}$$

namely the "average loss of the portfolio in the worst α cases" or the family of CRMs based on one-sided moments [10]

$$\rho_{p,a}(X) = -\mathbb{E}[X] + a\|(X - \mathbb{X})^-\|_p \qquad a \in [0,1],\ p \geq 1 \tag{1.4.6}$$

among which we recognize *semivariance* (when $a = 1$, $p = 2$).

1.5 THE DIVERSIFICATION PRINCIPLE REVISITED

There is one aspect of the diversification principle that subadditivity does not capture. It is related to the limiting case when we sum two portfolios X and Y that are *comonotonic*. This means that we can write $X = f(Z)$ and $Y = g(Z)$, where f and g are monotonic functions driven by the same random risk factor Z. Such portfolios always go up and down together in all cases, and hence they provide no mutual hedge at all, namely no diversification. For comonotonic random variables, people speak also of "perfect dependence" because it turns out that the dependence structure of such variables is in fact the same (*copula maxima*) that links any random variable X to itself.

The diversification principle tells us that, for a measure of risk ρ, the hedging benefit $H(X, Y; \rho)$ should be exactly zero when X and Y are comonotonic. This property of ρ is termed *comonotonic additivity* (CA)

$$\rho \text{ comonotonic additive } \Leftrightarrow [X, Y \text{ comonotonic} \\ \Rightarrow \rho(X+Y) = \rho(X) + \rho(Y)]. \quad (1.5.1)$$

Subadditivity does not imply CA. There are in fact CRMs that are not comonotonic additive, such as equation (1.4.6), for instance.

We think that the diversification principle is well embodied only in the combination of both subadditivity and CA. Each one separately is not enough. To understand this fact, the clearest explanation we know is to show that, in the absence of each of these conditions, there exists a specific *cheating strategy* (CS) allowing a risk manager to reduce the capital requirement of a portfolio without reducing at all the potential risks.

CS_1, lack of subadditivity: split your portfolio into suitable subportfolios and compute capital adequacy on each one

CS_2, lack of comonotonic additivity: merge your portfolio with the one of new comonotonic partners and compute capital adequacy on the global portfolio

CA is therefore a natural further condition to the list of properties of a good risk measure. It becomes a sort of "sixth axiom," because it is a distinct condition from LI when imposed on a CRM. There exist CRMs that satisfy LI and not CA and vice versa.

The above arguments support the interest to describe the class of CRMs that also satisfy both LI and CA (LI CA CRMs).

1.6 SPECTRAL MEASURES OF RISK

The class of LI CA CRMs was first described exhaustively by Kusuoka [11]. It has a general representation

$$\rho_{\mu}(X) = \int_0^1 d\mu(p)\, ES_p(X) \qquad d\mu \text{ any measure on } [0,1]. \tag{1.6.1}$$

The same class was defined as *spectral measures of risk* independently by Acerbi [1] with an equivalent representation

$$\rho_{\phi}(X) = -\int_0^1 \phi(p)\, F_X^{\leftarrow}(p)\, dp \tag{1.6.2}$$

where the function $\phi : [0,1] \mapsto \mathbb{R}$, named the *risk spectrum*, satisfies the coherence conditions

1. $\phi(p) \geq 0$
2. $\int_0^1 \phi(p)\, dp = 1$
3. $\phi(p_1) \geq \phi(p_2)$ if $p_1 \leq p_2$

Despite the complicated formula, a spectral measure ρ_{ϕ} is nothing but the ϕ-weighted average of all outcomes of the portfolio, from the worst ($p = 0$) to the best ($p = 1$). This is the most general form that a LI CA CRM can assume. The only residual freedom is in the choice of the weighting function ϕ within the above conditions.

Condition 3 is related to subadditivity. It just says that, in general, worse cases must be given a larger weight when we measure risk, and this seems actually very reasonable. This is also where VaR fails, as it measures the severity of the loss associated with the quantile threshold, forgetting to give a weight to the losses in the tail beyond it. Expected shortfall ES_{α} is a special case of spectral measure of risk whose risk spectrum is a constant function with domain $[0, \alpha]$.

Spectral measures of risk turned out to be strictly related to the class of *distortion risk measures* introduced in actuarial math in 1996 in a different language by Wang [15].

1.7 ESTIMATORS OF SPECTRAL MEASURES

It is easy to provide estimators of spectral measures. Given N i.i.d. scenario outcomes $\{x_i^{(k)}\}_{i=1,\dots,N}$ for the vector of the market's variables (possibly assets) $\vec{X} = X^{(k)}$, and given any portfolio function of them $Y = Y(\vec{X})$, we can just exploit law invariance and use the $\Leftarrow$ proof of theorem 1.4.1 to obtain canonical estimators.

$$\hat{\rho}_{\phi}^{(N)}(Y) = \rho_{\phi}[\hat{F}_Y^{(N)}(\{\vec{x}_i\})] \tag{1.7.1}$$

where we have defined the empirical marginal distribution function of Y

$$\hat{F}_Y^{(N)}(\{\vec{x}_i\})(\vec{t}) \equiv \frac{1}{N}\sum_{i=1}^{N} \theta\,(t - y_i) \tag{1.7.2}$$

which is nothing but the cumulative empirical histogram of the outcomes $y_i \equiv Y(\vec{x}_i)$. Equation (1.7.1) can in fact be solved to give simply

$$\hat{\rho}_{\phi}^{(N)}(Y) = -\sum_{i=1}^{N} y_{i:N}\,\bar{\phi}_i \tag{1.7.3}$$

where we have adopted the notation of the *ordered statistics* $y_{i:N}$ (i.e., the sorted version of the vector y_i) and defined the weights

$$\bar{\phi}_i \equiv \int_{(i-1)/N}^{i/N} \phi(p)\,dp. \tag{1.7.4}$$

In equation (1.7.3), we see in a very transparent language that a spectral measure is nothing but a weighted average of all the cases of a portfolio sorted from the worst to the best.

In the case of ES_α, the estimator in equation (1.7.3) specializes to

$$\widehat{ES}_{\alpha}^{(N)}(Y) = -\frac{1}{[N\alpha]}\left(\sum_{i=1}^{[N\alpha]} y_{i:N} + (N\alpha - [N\alpha])\; y_{[N\alpha]+1:N}\right). \tag{1.7.5}$$

All of these estimators can be easily implemented in a simple spreadsheet or in any programming language.

We note that these estimators not only converge for large N to the estimated measure, but also preserve coherency at every finite N by construction.

1.8 OPTIMIZATION OF CRMS: EXPLOITING CONVEXITY

As we have already stressed, CRM surfaces are convex. Nonetheless, setting up an optimization program using, say, the estimator of ES in equation (1.7.5) requires some clever trick. In fact, suppose you want to minimize the $ES_\alpha(Y_{\vec{w}})$

of a portfolio $Y_{\vec{w}} = \sum_k w_k X^{(k)}$ choosing the optimal weights w_k under some given constraints. A naive minimization procedure using equation (1.7.5) will simply fail, because the ordered statistics $y_{i:N} = \{\texttt{sort}_{j=1,\ldots,N}[\sum_k w_k x_j^{(k)}]\}_i$ reshuffle discretely when the parameters $\vec{w}$ change continuously. In other words, $\widehat{ES}_\alpha^{(N)}(Y_{\vec{w}})$ is not analytical in the weights $\vec{w}$, and this creates big troubles for any standard minimization routine. Moreover, the `sort` algorithm is very slow and memory consuming on large samples.

The problem of finding efficient optimization routines for ES_α was elegantly solved by Pflug [12] and Uryasev and Rockafellar [13, 14], who mapped it onto the equivalent problem of finding the minima of the functional

$$\Gamma_\alpha^{(N)}(Y_{\vec{w}}, \psi) = -\psi + \frac{1}{N\alpha}\sum_{i=1}^{N}(\psi - y_i)^+ \tag{1.8.1}$$

with $y_i = \sum_k w_k x_i^{(k)}$. Despite the presence of an additional parameter ψ, Γ is a much simpler objective function to minimize, thanks to the manifest analyticity and convexity with regard to the weights $\vec{w}$. Notice, in particular, that equation (1.8.1) is free from ordered statistics.

The main result of [13] is that $\widehat{ES}_\alpha(Y_{\vec{w}})$ (as a function of $\vec{w}$) and $\Gamma_\alpha^{(N)}(Y_{\vec{w}}, \psi)$ (as a function of both $\vec{w}$ and ψ) attain their minima on the same argument weights $\vec{w}$. So, we can find ES_α-optimal portfolios by minimizing Γ_α instead, which is dramatically easier. Furthermore, it can be shown [13] that this convex nonlinear program can be mapped onto an equivalent *linear* program, at the price of introducing further additional parameters. It is in this linearized version that the most efficient routines are obtained, making it possible to set up an optimization procedure for portfolios of essentially any size and complexity.

It is difficult to overestimate the importance of this result. It allows us to fully exploit the advantages of convex optimization with ES_α and opens the way to efficient routines for large and complex portfolios, under any distributional assumptions. With this ES_α methodology, risk managers finally have the ability to solve problems that they could only dream of solving using VaR.

This methodology was extended from ES_α to any spectral measure ρ_ϕ (i.e., any LI CA CRM) by Acerbi [4]. Also, in the general case, the main problem to tackle is the presence of sorting routines induced in equation (1.7.3) by the ordered statistics. In parallel to the above result, one introduces the functional

$$\Gamma_\phi^{(N)}(Y_{\vec{w}}, \vec{\psi}) = \sum_{j=1}^{N-1}(\bar{\phi}_{j+1} - \bar{\phi}_j)\left\{ j\,\psi_j - \sum_{i=1}^{N}(\psi_j - y_i)^+ \right\} - \bar{\phi}_N \sum_{i=1}^{N} y_i \tag{1.8.2}$$

whose minimum (as a function of both $\vec{w}$ and $\vec{\psi}$) can be shown to occur at the same argument $\vec{w}$ that minimizes $\hat{\rho}_{\phi}^{(N)}(Y_{\vec{w}})$. Therefore, we see that in generalizing from ES_{α} to any spectral measure ρ_{ϕ}, the only difference is that the additional parameter ψ has become a vector of additional parameters $\vec{\psi}$. For this extended methodology, it is also possible to map the nonlinear convex problem onto an equivalent linear one. This also extends the efficiency of the available ES_{α} optimization routines to any other LI CA CRMs. See Acerbi [2,4] for more details.

1.9 CONCLUSIONS

We have discussed why, in our opinion, the class of CRMs is too large under the perspective of practical risk-management applications. If the practice of risk management remains intrinsically objectivistic, the additional constraint of law invariance will always be implicitly assumed by the market. A further restriction is provided by a closer look at the risk-diversification principle, which naturally introduces the condition of comonotonic additivity.

The subset of CRMs that possess both LI and CA coincides with the class of spectral measures. This class lends itself to immediate transparent representation, to straightforward estimation, and—adopting nontrivial tricks—to powerful optimization techniques that exploit the convexity of the risk minimization programs and allow risk managers, probably for the first time, to face the problem of finding optimal portfolios with virtually no restrictions of size, complexity, and distributional assumptions.

REFERENCES

1. Acerbi, C. (2002). Spectral measures of risk: a coherent representation of subjective risk aversion. *Journal of Banking and Finance*, 26: 1505–1518.
2. Acerbi, C. (2003). Coherent representations of subjective risk aversion. In *Risk Measures for the XXI Century*, ed. G. Szego. Wiley, New York.
3. Acerbi, C. (2007). To be published in *Quatitative Finance*..
4. Acerbi, C., Simonetti, P. (2002). Portfolio Optimization with Spectral Measures of Risk. Abaxbank preprint, available on www.gloriamundi.org.
5. Acerbi, C., Tasche, D. (2002). On the coherence of expected shortfall. *Journal of Banking and Finance*, 26: 1487–1503.
6. Acerbi, C., Tasche, D. (2002). Expected shortfall: a natural coherent alternative to value at risk. *Economic Notes*, 31 (2): 379–388.
7. Artzner, P., Delbaen, F., Eber, J.-M., Heath, D. (1997). Thinking coherently. *Risk*, 10: (11).

8. Artzner, P., Delbaen, F., Eber, J.-M., Heath, D. (1999). Coherent measures of risk. *Math. Fin.*, 9 (3): 203–228.
9. Delbaen, F. (2000). Coherent Risk Measures on General Probability Spaces. Preprint, ETH, Zürich.
10. Fischer, T. (2001). Examples of Coherent Risk Measures Depending on One–Sided Moments. Working paper, Darmstadt University of Technology.
11. Kusuoka, S. (2001). On law invariant coherent risk measures. *Adv. Math. Econ.*, 3: 83–95.
12. Pflug, G. (2000). Some remarks on the value-at-risk and the conditional value-at-risk. In *Probabilistic Constrained Optimization: Methodology and Applications* ed. Uryasev. S., Kluwer Academic Publishers, Dordrecht.
13. Rockafellar, R.T., Uryasev, S. (2000). Optimization of conditional value-at-risk. *Journal of Risk*, 2 (3) 21–41.
14. Rockafellar, R.T., Uryasev, S. (2001). Conditional value-at-risk for general loss distributions. *Journal of Banking and Finance*, 26/7: 1443–1471, 2002.
15. Wang, S. (1996). Premium calculation by transforming the layer premium density. *Astin Bulletin*, 26: 71–92.

CHAPTER 2

Pricing High-Dimensional American Options Using Local Consistency Conditions

S.J. Berridge
Man Investments, London, United Kingdom
J.M. Schumacher
Tilburg University, Tilburg, The Netherlands

Contents

Abstract: We investigate a new method for pricing high-dimensional American options. The method is of finite difference type, in that we obtain solutions on a constant grid of representative states through time. We alleviate the well-known problems associated with applying standard finite difference techniques in high-dimensional spaces by using an irregular grid, as can be generated for example by a Monte Carlo or quasi–Monte Carlo method. The use of such a grid calls for an alternative method for discretizing the convection-diffusion operator in the pricing partial differential equation; this is done by considering the grid points as states of an approximating continuous-time Markov chain, and constructing transition intensities by appealing to local consistency conditions in the spirit of Kushner and Dupuis [22]. The actual computation of the transition intensities is done by means of linear programming, which is a time-consuming process but one that can be easily parallelized. Once the transition matrix has been constructed, prices can be computed quickly. The method is tested on geometric average options in up to ten dimensions. Accurate results are obtained, in particular when use is made of a simple bias control technique.

Keywords: American options, high-dimensional problems, free boundary problems, optimal stopping, variational inequalities, numerical methods, unstructured mesh, Markov chain approximation

2.1 INTRODUCTION

The pricing of American options has been extensively discussed in recent years (cf. Detemple [12] for a survey), and in particular much attention has been paid to the computational challenges that arise in the high-dimensional case. The term "high dimensional" in this context refers to situations in which the number of stochastic factors to be taken into account is at least three or four. State–space dimensions in this range occur quite frequently, in particular in models that involve multiple assets, interest rates, and inflation rates.

Standard finite-difference methods that work well in low-dimensional problems quickly become unmanageable in high-dimensional cases; on the other hand, standard Monte Carlo methods cannot be applied as such, due to the optimization problem that is embedded in American options. Many recent papers have been devoted to finding ways of adapting the Monte Carlo method for American and Bermudan option pricing; see for instance [6, 7, 16, 20, 23, 26, 28, 30, 31]. A survey of Monte Carlo methods for American option pricing is provided by Glasserman [14, chap. 8].

As has been pointed out by Glasserman [14], a unifying framework for simulation-based approaches to American option pricing is provided by the stochastic mesh method that was developed by Broadie and Glasserman [6]. The scope of this framework is extended even more if the term "stochastic" in "stochastic mesh" is interpreted broadly to include also quasi–Monte Carlo approaches, as proposed by Boyle et al. [4, 5]. The stochastic mesh method has as its main ingredients: firstly, a time-indexed family of meshes (i.e., collections of points in the state space that have been generated to match a given distribution), and secondly, for each pair of adjacent meshes, a collection of mesh weights. The mesh weights are used in a backward recursion to compute conditional expectations $E[f_{i+1}(x_{i+1}) \mid x_i]$. Broadie and Glasserman [6] have advocated the use of independent sample paths to generate the meshes, combined with likelihood-ratio weights corresponding to the average conditional density associated with the mesh points at time t_i. They also show that other choices of the mesh weights can result in large variances. The least-squares Monte Carlo method [23] can be interpreted as a stochastic mesh method with implicitly defined mesh weights [14, chap. 8]. The weights implied by the least-squares method do not coincide with the likelihood-ratio weights; still, the method converges, as both the number of sample paths and the number of basis functions tend to infinity in appropriate proportions [8, 28]. This shows that alternatives to likelihood-ratio weights can be feasible.

The dynamic programming problem that arises in American option pricing can be written in terms of a partial differential equation, and especially in dimensions one and two, the finite-difference method (FDM) is an effective way of computing solutions. The FDM employs a grid at each time step and computes conditional expectations by applying suitable weights; in that sense, it can be viewed as a member of the family of stochastic mesh methods, interpreted in the wide sense. The mesh is typically a regular grid on a certain finite region; interpreted stochastically, such a mesh would correspond to a uniform distribution on the truncated state space. The mesh weights are usually derived from Taylor expansions up to a certain order. The actual weights depend on the chosen order as well as on the form of time discretization that is being used, such as explicit, fully implicit, or Crank–Nicolson. They are not constructed as likelihood-ratio weights, nor can they necessarily be interpreted as such at a given level of time discretization. The finite-difference method is well known to converge if both the space-discretization step and the time-discretization step tend to zero in appropriate proportions, depending on the degree of implicitness [19]. While the convergence result holds true in any state–space dimension, the computational feasibility of the standard finite-difference method is very strongly affected by the fact that the number of grid points (for a fixed number of grid points per dimension) is exponential in the dimension parameter. This is a well-known problem in dynamic programming, usually referred to as the "curse of dimensionality."

The method that we propose in this chapter is based on a blend of finite-difference and Monte Carlo techniques. From the Monte Carlo method, and in particular from its realization as the stochastic mesh method, we take the idea of employing an irregular mesh (produced by a Monte Carlo or quasi–Monte Carlo technique) that is, to a certain extent, representative of the densities corresponding to a given process. In this way we gain flexibility and avoid the unnatural sharp cutoff of the usual regular finite-difference grids. We stay in line with basic finite-difference methods, in that we use the same grid at every time step. This is in contrast with the methods based on forward generation of Monte Carlo paths starting from a given initial state, which produce meshes that are different at different points in time. Although the method that we propose allows the use of different grids at different time points, we use only a single grid in this chapter, both to simplify the presentation and to provide the sharpest possible test of the applicability of the proposed method.

We use the term "irregular" here in the sense of "nonuniform," and in this way there may be a superficial similarity between the method we propose and local mesh-refinement techniques in the context of the finite-element method. The use of nonuniform meshes can be very effective in finite-element computations, but the construction of such meshes in high-dimensional spaces is unwieldy. Our approach in this chapter is closer to the finite-difference method than to the finite-element method, even though we do not use finite differences in the strict sense of the word, and the irregularity of the grid is guided by general importance-sampling considerations rather than by the specific form of a payoff function.

By using irregular grids we gain flexibility, but there is a price to pay. In the standard finite-difference method based on regular grids, simple formulas for weights can be derived from Taylor expansions. Such simple rules are no longer available if we use irregular grids, and so we must look for alternatives to the classical finite-difference formulas. We propose here a method based on Markov chain approximations. In a discrete-time context, transformations from a continuous-state (vector autoregressive) model to a Markov chain model have been constructed, for instance by Tauchen [29]. Financial models are often formulated in continuous time, typically by means of stochastic differential equations (SDE). In this chapter we construct continuous-time Markov chain approximations starting from a given SDE. Even though time will eventually have to be discretized in the numerical procedure, we find it convenient to build a continuous-time-approximating Markov chain, because in this way we preserve freedom in choosing a time-stepping method and, in particular, we are able to use implicit methods. When implicit time stepping is used, the mesh weights are defined implicitly rather than explicitly, as is also the case in the Longstaff–Schwartz method [23].

The benefit of using a single mesh is that we have natural candidates for the discrete states in an approximating Markov chain, namely the grid points in this mesh. To define transition intensities between these points, we work in the spirit of Kushner and Dupuis [22] by matching the first two moments of the conditional densities given by the original SDE. This leads to a collection of linear programming (LP) problems, one for each grid point. Implementation details are given in section 2.3, where we also discuss the interpretation and treatment of grid points whose associated LP problems are infeasible. The idea of using the local consistency criterion is an innovation with respect to an earlier paper [2] that used irregular grids in combination with a root extraction method for determining transition intensities. An important advantage of the local consistency method is that it has certain guaranteed stability properties, as discussed in section 2.4. Moreover, we propose an implementation that guarantees that we obtain a sparse generator matrix.

The proposal of this chapter can be viewed essentially as an attempt to make the finite-difference method work in high dimensions by incorporating some Monte Carlo elements. As in the standard finite-difference method, the result of the computation is the option price as a function of the value of the underlying, as opposed to the standard Monte Carlo method, which focuses just on the option price for a single given value of the underlying. We can therefore obtain partial derivatives at virtually no extra computational cost. An important advantage of the method that we propose is that the number of tuning parameters is small. The popular Longstaff–Schwartz method [23] is often sensitive to the choice of basis functions, and it seems difficult to give general guidelines on the selection of these functions. In contrast, the method proposed here is based on two main parameters, namely the number of mesh points used (replacing the space-discretization step in the standard finite-difference method) and the time-discretization step. The method does not require much specific knowledge about the option to be priced, in contrast to methods that optimize over a parametric collection of exercise boundaries or hedging strategies to find upper or lower bounds; of course this may be an advantage or a disadvantage, depending on whether or not such specific knowledge is available. Our proposed computational procedure is such that essentially the same code can be used for problems in different dimensions, in stark contrast to the standard finite-difference method. The computational results presented below suggest that this freedom of dimension is achieved at the price of a fairly small deterioration of convergence speed.

The chapter continues in section 2.2 with a formulation of the problem of interest. Section 2.3 presents the proposed methodology. A stability analysis is presented in section 2.4. We discuss grid specification and boundary conditions in section 2.5, and the results of experiments are described in section 2.6. Finally, section 2.7 concludes the chapter.

2.2 FORMULATION

We consider an arbitrage-free market described in terms of a state variable $X(s) \in \mathbb{R}^d$ for $s \in [t, T]$, which under the risk-neutral measure follows a Markov diffusion process

$$dX(s) = \mu(X(s), s)\, ds + \sigma(X(s), s)\, dW(s) \tag{2.2.1}$$

with initial condition $X(t) = x_t$. For simplicity, we assume a constant interest rate r. In this market we consider a derivative product on $X(s)$ with immediate exercise value $\psi(X(s), s)$ at time s and value $V(s) = v(X(s), s)$ for some pricing function $v(x, s)$. The process $V(s)$ satisfies

$$dV(s) = \mu_V(X(s), s)\, ds + \sigma_V(X(s), s)\, dW(s) \tag{2.2.2}$$

where μ_V and σ_V can be expressed in terms of μ and σ by means of Itô's lemma. The terminal value is given by $V(\cdot, T) = \psi(\cdot, T)$, and intermediate values satisfy $V(\cdot, s) \geq \psi(\cdot, s)$, $s \in [t, T]$.

The value of the derivative product can be expressed as a supremum over stopping times

$$v(x_t, t) = \sup_{\tau \in \mathcal{T}} \mathbb{E}_t^{\mathbb{Q}}(e^{-r(\tau - t)}\psi(X(\tau))) \tag{2.2.3}$$

where $\mathcal{T}$ is the set of stopping times on $[t, T]$ with respect to the natural filtration; the expectation is taken with respect to the risk-neutral measure $\mathbb{Q}$;

and the initial value is $X(t) = x_t$. Standard dynamic programming arguments (see for instance Jaillet et al. [19]) allow reformulation of the problem in terms of the Black–Scholes operator

$$\mathcal{L} = \frac{1}{2}\operatorname{tr}\sigma\sigma'\frac{\partial^2}{\partial x^2} + \mu\frac{\partial}{\partial x} - r. \tag{2.2.4}$$

The option value can be found by solving the differential complementarity problem

$$\begin{cases} \dfrac{\partial v}{\partial t} + \mathcal{L}v \leq 0 \\ v - \psi \geq 0 \\ \left(\dfrac{\partial v}{\partial t} + \mathcal{L}v\right)(v - \psi) = 0 \end{cases} \tag{2.2.5}$$

for $(x, s) \in \mathbb{R}^d \times [t, T]$ with the terminal condition $v(\cdot, T) \equiv \psi(\cdot, T)$. The solution of this problem divides the time–state space into two complementary regions: the continuation region, where it is optimal to hold the option, and the stopping region, where it is optimal to exercise. In the continuation region, the first line of equation (2.2.5) is satisfied with equality, and the option is not exercised. In the stopping region, the second line of equation (2.2.5) is active, and the stopping rule calls for the option to be exercised.

We assume below that the risk-neutral process, equation (2.2.1), is time homogeneous, i.e., the drift and volatility parameters μ and σ do not depend on time explicitly. Our basic state process is therefore given by a stochastic differential equation of the form

$$dX(s) = \mu(X(s))\,ds + \sigma(X(s))\,dW(s) \tag{2.2.6}$$

where $X(s)$ is d-dimensional, and $W(s)$ is a k-dimensional standard Wiener process. The time-homogeneity assumption is in principle not necessary for the proposed method, but it does simplify both the analysis and the implementation of the method considerably.

2.3 OUTLINE OF THE METHOD

The proposed method essentially consists of the following steps:

1. Construction of an irregular grid in the state space
2. Construction of a continuous-time Markov chain, using the grid points as states

3. Discretization of time by selecting a time-stepping method
4. Resolution of the resulting sequence of linear complementarity problems

We discuss these steps now in more detail.

2.3.1 Step 1: State–Space Discretization

As noted before, the method discussed in this chapter calls for the selection of a single grid in the state space $\mathbb{R}^d$. The main issues are: how to choose the grid density, and how to construct a grid that is representative of the selected density.

Importance-sampling considerations tell us that the most efficient grid density is given by the density of the process itself. The process density, however, is time dependent as well as state dependent, and so a compromise has to be made if one is to work with a single grid. As outlined in Evans and Swartz [13], the rate of convergence for importance sampling of normal densities using normal importance-sampling functions is most damaged when the variance of the importance-sampling function is less than that of the true density. Conversely, convergence rates are not greatly affected when the variance of the importance-sampling function is greater than that of the true density. The situation we should try to avoid is that the process has a significant probability of lying in the "tails" of the grid density. Berridge and Schumacher [2] used a root method to construct transition probabilities, and the process considered was a five-dimensional Brownian motion with drift; a grid covariance of 1.5 times the process covariance at expiry was found to give the best convergence rate when tested against grids with covariances of 1.0 and 2.0 times the covariance at expiry.

A grid with a given density can be formed by crude Monte Carlo in combination with a suitable grid transformation. However, the literature on Monte Carlo (MC) and quasi–Monte Carlo (QMC) integration indicates that better results can be obtained by using low-discrepancy (Niederreiter [24]) or low-distortion (Pagès [25]) methods. Two-dimensional plots, as shown in Figure 2.3.1, do indeed suggest that the latter methods provide nicer samplings. In the numerical experiments, we have used both low-discrepancy (Sobol) grids and low-distortion grids.

2.3.2 Step 2: Markov Chain Approximation

Suppose now that a grid

$$\mathcal{X} = \{x_1, \ldots, x_n\} \subset \mathbb{R}^n \tag{2.3.1}$$

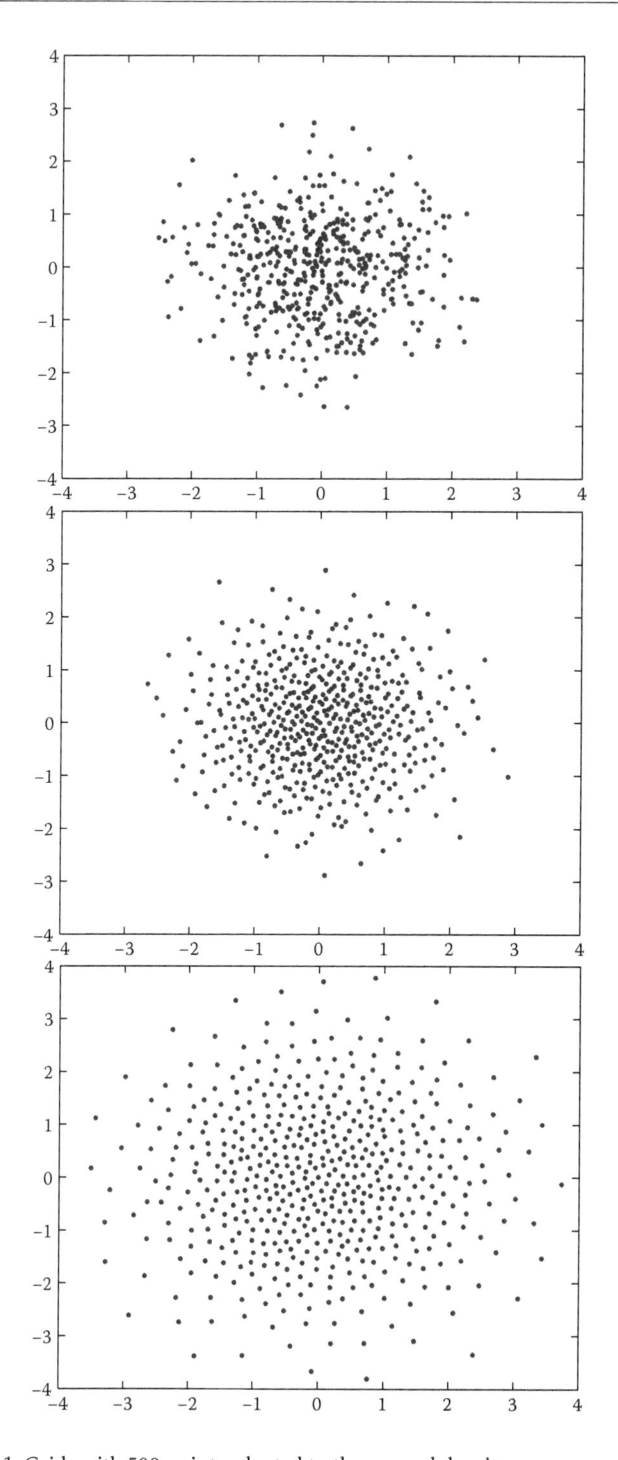

FIGURE 2.3.1 Grids with 500 points adapted to the normal density.

is given. The next step is to construct a continuous-time Markov chain approximation to the process given by equation (2.2.6), with states that correspond to the points in a given irregular grid. First, consider the more standard discrete-time approximation. Let δt denote a time step. A Markov transition matrix P is said to be locally consistent with the given process [22] if for each state $i = 1, \ldots, n$ we have[1]

$$\Sigma(x_i)\delta t = \sum_{j=1}^{n}(x_j - x_i - \mu(x_i)\delta t)(x_j - x_i - \mu(x_i)\delta t)' p_{ij} \tag{2.3.2a}$$

$$\mu(x_i)\delta t = \sum_{j=1}^{n}(x_j - x_i) p_{ij} \tag{2.3.2b}$$

$$1 = \sum_{j=1}^{n} p_{ij} \tag{2.3.2c}$$

$$p_{ij} \geq 0 \tag{2.3.2d}$$

where p_{ij} is the (i, j)th entry of P. For each state i, the above problem is a linear programming feasibility problem with n variables $p_{i1}, \ldots, p_{in}$, which are subject to $\frac{1}{2}d(d+3)+1$ equality constraints and n inequality constraints. In a typical application, the number of equality constraints is much smaller than the number of variables.

To arrive at a continuous-time approximation method, we note that transition intensities relate to transition probabilities via

$$a_{ij} = \lim_{\delta t \downarrow 0} \frac{1}{\delta t}(p_{ij}(\delta t) - \delta_{ij}) \tag{2.3.3}$$

where δ_{ij} is the Kronecker delta. The numbers a_{ij} can be collected into a matrix A, which is known as the transition intensity matrix or the infinitesimal generator matrix. Taking limits in equation (2.3.2) leads to the conditions

$$\Sigma(x_i) = \sum_{j \neq i}(x_j - x_i)(x_j - x_i)' a_{ij} \tag{2.3.4a}$$

$$\mu(x_i) = \sum_{j \neq i}(x_j - x_i) a_{ij} \tag{2.3.4b}$$

$$a_{ii} = -\sum_{j \neq i}^{n} a_{ij} \tag{2.3.4c}$$

$$a_{ij} \geq 0 \quad (j \neq i). \tag{2.3.4d}$$

[1] The formulation in Kushner and Dupuis [22] is more general in that it allows $o(\delta t)$ terms to be added on the right-hand side of the first two conditions.

Again we have a linear-programming feasibility problem at each grid point x_i; the number of variables and the number of equality constraints are both reduced by 1 with respect to the discrete-time formulation. In particular, the number of equality constraints is

$$\eta_d = \frac{1}{2}d(d+3). \tag{2.3.5}$$

In applications, the number of grid points n is typically on the order 10^4 or 10^5. In the actual implementation, we drastically reduced the number of variables by taking into consideration only variables a_{ij}, such that the corresponding grid points x_j are close to the given point x_i. In this way, we simplify the feasibility problem and moreover obtain a sparse infinitesimal generator matrix.

To obtain specific solutions, it is useful to convert the feasibility problem to an optimization problem by adding an objective function. The objective function that we have used in experiments is $\sum_{j\neq i} f_j a_{ij}$, where f_j is equal to k^3 if x_j is the kth nearest neighbor of x_i. Minimization of this objective function expresses a preference for using points close to x_i to satisfy the conditions of equation (2.3.4).

It follows from results in linear programming (see, for instance, Schrijver [27, Cor. 7.11]) that the solution of the linear program described above is in general a corner solution using as many zero variables as possible. The number of nonzero transition intensities per point is then the minimum number, namely η_d. These points are not necessarily the η_d nearest neighbors of x_i, as these may not satisfy the feasibility conditions; they rather form the closest possible feasible set, as measured by the objective function.

There is clearly a similarity between the method that we propose here and the method of lines as discussed for instance by Hundsdorfer and Verwer [18]. We cannot use the standard finite-difference formulas in the space direction, however, due to the fact that we have an irregular grid. Instead, we use transition intensities obtained from local consistency conditions. The resulting sign constraints are sufficient for stability, as discussed in section 2.4 below, but are not necessary. The use of a regular grid in the method of lines brings alternative ways of ensuring stability, but it also restricts the method to low-dimensional applications. In the case of an irregular grid, we do not know of any systematic way to ensure stability other than via the sign constraints resulting from the Markov chain approximation.

2.3.3 Step 3: Time Discretization

In the previous step, we have discretized in space but not in time. In this way we have obtained a finite-dimensional differential variational inequality instead of the partial differential variational inequality shown in equation (2.2.5). The finite-dimensional version can be written in the form

$$0 \leq v(t) - \psi(t) \perp -\frac{dv}{dt}(t) + (rI - A)v(t) \geq 0 \tag{2.3.6}$$

where A denotes the infinitesimal generator matrix, and the "perp" symbol is used to express the complementarity relations. This follows a notation that is standard in the theory of linear complementarity problems [9]. We abuse notation to some extent in that we keep the notation v for the pricing function, although this is now a vector-valued function of continuous time providing a space-discretized version of the actual pricing function $v(x,t)$.

The usual time stepping methods for linear ordinary differential equations (ODEs) can all be used, but the complementarity conditions have to be taken into account. From equation (2.3.6), a general θ-scheme is given by

$$0 \leq v_k - \psi_k \perp \frac{v_k - v_{k+1}}{\delta t} + (rI - A)(\theta v_k + (1-\theta)v_{k+1}) \geq 0 \tag{2.3.7}$$

for $k = K-1, \ldots, 0$, with terminal condition $V_K = \psi_K$. The choice $\theta = 0$ corresponds to the *explicit* scheme; the *fully implicit* scheme is obtained by taking $\theta = 1$, and the Crank–Nicolson scheme by taking $\theta = \frac{1}{2}$. Each scheme leads to a series of problems of the form

$$0 \leq v_k - \psi_k \perp M_L v_k - M_R v_{k+1} \geq 0. \tag{2.3.8}$$

The final step consists of solving this sequence of linear complementarity problems (LCPs).

2.3.4 Step 4: Solving the Sequence of LCPs

There are many methods available for solving LCPs, including the projected successive overrelaxation (PSOR) method proposed in Cryer [10]. Another possible candidate is linear programming, which is used for example by Dempster and Hutton [11] to solve the one-dimensional American option pricing problem. Several possible approaches are discussed and compared by Huang and Pang [17]. The solution is particularly simple when the explicit method is used, since in this case the matrix M_L multiplying the unknown at the right hand side of equation (2.3.8) is the identity and we can immediately write down the solution:

$$v_k = \max(\psi_k, (I + \delta t(rI - A))v_{k+1}). \qquad (2.3.9)$$

Nevertheless, as in the European case, one may prefer the Crank–Nicolson method because of its better convergence properties with respect to the time step, or the implicit method because of its unconditional stability, as discussed below.

2.4 STABILITY ANALYSIS

In the theory of the numerical solution of convection-diffusion equations, it is well known that, under certain conditions, time-stepping algorithms can be unstable and produce solutions that are very far from the truth. Similar effects may arise in the context of differential variational inequalities of the convection–diffusion type. In this case, the problem that is solved at each step is a linear complementarity problem (LCP) rather than a linear system; thus the stability problem is more difficult to analyze and in particular cannot be formulated in terms of eigenvalues. We present a stability analysis here that makes essential use of the fact that our proposed numerical scheme is based on a continuous-time Markov chain interpretation. The payoff function (value of immediate exercise) is allowed to be time dependent and is assumed to be bounded.

It should be noted that stability is necessary for convergence to the true solution, but this alone is not generally sufficient. Conditions for convergence of the method proposed here are discussed by Berridge [3, Chap. 6].

We remind the reader of some matrix classes that will be used in the following analysis. For a full treatment of matrix classes, we refer the reader to Berman and Plemmons [1] or Cottle, Pang, and Stone [9].

DEFINITION 2.4.1

A real square matrix is said to be a Z-matrix if its off-diagonal entries are nonpositive.

DEFINITION 2.4.2

A real square matrix is said to be a P-matrix if all of its principal minors are positive.

2.4.1 The Explicit Method

The explicit version of the complementarity problems presented in equation (2.3.7) is

$$0 \le v_k - \psi_k \perp v_k - M_R v_{k+1} \ge 0 \tag{2.4.1}$$

for $k = K-1, \ldots, 0$, where the matrix M_R is given by

$$M_R = (1 - r\delta t)I + A\delta t. \tag{2.4.2}$$

In the explicit case, the complementarity problems presented in equation (2.4.1) are readily solved. The solution is given by

$$v_k = \max(\psi_k, M_R v_{k+1}) \tag{2.4.3}$$

where max denotes the componentwise maximum. A stability condition for the sequence of vectors generated in this way is given below.

LEMMA 2.4.1
Suppose that the matrix A satisfies equations (2.3.4c) and (2.3.4d). Let $r \ge 0$, and let M_R be defined by equation (2.4.2). Under the stability condition

$$\delta t \le \frac{1}{\|A - rI\|_\Delta} \tag{2.4.4}$$

where $\|A\|_\Delta = \max_i |a_{ii}|$, the solution at $k = 0$ of the explicit system of complementarity problems, shown in equation (2.4.1), with M_R defined by equation (2.4.2), satisfies

$$\|v_0\|_\infty \le \max_{k=0,\ldots,K} (1 - r\delta t)^k \|\psi_k\|_\infty. \tag{2.4.5}$$

PROOF The conditions in equations (2.3.4c), (2.3.4d), and (2.4.4) imply that M_R is elementwise nonnegative, so that for any two vectors x and y with $x \le y$ (componentwise inequality) we have $M_R x \le M_R y$. Furthermore, the row sums of M_R are equal to $1 - r\delta t$. Therefore, writing $\mathbb{1}$ for the vector all of whose entries are equal to 1, we have $M_R v \le M_R \|v\|_\infty \mathbb{1} = (1 - r\delta t)\|v\|_\infty \mathbb{1}$ for any vector v. It follows that $\|M_R v\|_\infty (1 - r\delta t) \le \|v\|_\infty$ for all v. From equation (2.4.3) we have

$$\begin{aligned} \|v_k\|_\infty &\le \max(\|\psi_k\|_\infty, \|M_R v_{k+1}\|_\infty) \\ &\le \max(\|\psi_k\|_\infty, (1 - r\delta t)\|v_{k+1}\|_\infty). \end{aligned}$$

Applying this inequality recursively with $v_K \equiv \psi_K$ produces the required result.

The stability bound on δt is presented in terms of the maximum absolute value of the entries of the diagonal of the matrix A, which under the conditions of equations (2.3.4c) and (2.3.4d) is the same as the maximum absolute value of all entries of A. It is shown below that the bound can be estimated in terms of a ratio between the time step and the square of a "space step."

LEMMA 2.4.2
Suppose that A satisfies the conditions described in equations (2.3.4a), (2.3.4c), and (2.3.4d). Let ε be such that

$$\|x_i - x_j\|^2 \geq \varepsilon^2 \quad \forall i, j \ s.t. j \neq i, \ a_{ij} \neq 0. \tag{2.4.6}$$

Then the norm appearing in lemma 2.4.1 is bounded as follows:

$$\|A - rI\|_\Delta \leq \frac{1}{\varepsilon^2} \max_i \operatorname{tr} \Sigma(x_i) + r. \tag{2.4.7}$$

PROOF Applying the trace operator to both sides of equation (2.3.4a), we find

$$\begin{aligned} \operatorname{tr} \Sigma(x_i) &= \sum_{j \neq i} a_{ij} \operatorname{tr} (x_j - x_i)(x_j - x_i)' \\ &= \sum_{j \neq i} a_{ij} \|x_j - x_i\|^2 \\ &\geq \varepsilon^2 \sum_{j \neq i} a_{ij}. \end{aligned}$$

It follows that $\max_i |a_{ii}| \leq \varepsilon^{-2} \max_i \operatorname{tr} \Sigma(x_i)$, and this immediately implies the inequality (2.4.7).

Lemmas 2.4.2 and 2.4.1 can be combined to give a stability condition in terms of the ratio between the time step and the minimum point separation.

LEMMA 2.4.3
Under the conditions of lemma 2.4.2, the stability condition in lemma 2.4.1 holds provided that

$$\frac{\delta t}{\varepsilon^2} \leq \frac{1 - r\delta t}{\max_i \operatorname{tr} \Sigma(x_i)}. \tag{2.4.8}$$

The lemma indicates that, if the number of grid points is increased so that the distances between grid points are reduced by a factor c, then for stability reasons, the number of time steps should be increased by a factor c^2. This quadratic relation is typical for the explicit method.

2.4.2 The Fully Implicit Method

The stability condition in the explicit case can be rather restrictive, especially in a low dimension where the point separation decreases more rapidly with grid size. Because implicit methods often exhibit greater stability, we now investigate their properties.

The complementarity problems shown in equation (2.3.7) for the fully implicit case are rewritten as

$$0 \leq v_k - \psi_k \perp M_L v_k - v_{k+1} \geq 0 \tag{2.4.9}$$

for $k = K - 1, \ldots, 0$, where the matrix M_L is given by

$$M_L = (1 + r\delta t)I - A\delta t. \tag{2.4.10}$$

LEMMA 2.4.4
Suppose that A satisfies equations (2.3.4c) and (2.3.4d). Let $r \geq 0$, and let M_L be defined by equation (2.4.10). Under these conditions, the complementarity problems in equation (2.4.9) have unique solutions for all $k = K - 1, \ldots, 0$, and the solution at time index $k = 0$ of the sequence of problems in equation (2.4.9) satisfies

$$\|v_0\|_\infty \leq \max_{k=0,\ldots,K} (1 + r\delta t)^k \|\psi_k\|_\infty. \tag{2.4.11}$$

PROOF Under the condition of equation (2.3.4d), the matrix M_L defined in equation (2.4.10) is a Z-matrix. By equation (2.3.4c), we also have $M_L \mathbb{1} = (1 + r\delta t)\mathbb{1}$, which shows, moreover, that M_L is semipositive,[2] and so we can conclude that M_L is a P-matrix [1, Thm. 6.2.3]. It follows that the complementarity problems in equation (2.4.9) all have unique solutions.

By the fact that M_L is a Z-matrix, the unique solution of equation (2.4.9) can be characterized as the least element (in the sense of componentwise inequality) of the set of "feasible vectors" of the problem, i.e., the vectors that satisfy the inequality constraints in equation (2.4.9) but not necessarily the complementarity conditions (cf. for instance [9, Thm. 3.11.6] or [1, Thm. 10.4.11]).

[2] A square real matrix M is said to be *semipositive* if there exists a vector x such that $x_i > 0$ and $(Mx)_i > 0$ for all i.

Now consider the vector

$$v = \max((1 + r\delta t)^{-1}\|v_{k+1}\|, \|\psi_k\|_\infty)\mathbb{1}.$$

We claim that v is feasible for equation (2.4.9). Concerning the first inequality in equation (2.4.9), this is obvious. For the second, recalling that A has zero row sums, we have,

$$\begin{aligned} M_L v - v_{k+1} &= (1 + r\delta t)v - v_{k+1} \\ &= \max\left(\|v_{k+1}\|, (1 + r\delta t)\|\psi_k\|_\infty\right)\mathbb{1} - v_{k+1} \\ &\geq 0. \end{aligned}$$

Therefore, the true solution v_k satisfies $v_k \leq v$, and consequently we have

$$\begin{aligned} \|v_k\|_\infty &\leq \|v\|_\infty \\ &= \max((1 + r\delta t)^{-1}\|v_{k+1}\|, \|\psi_k\|_\infty). \end{aligned}$$

Applying this inequality recursively with $v_K \equiv \psi_K$ produces the desired result.

The lemma indicates that the fully implicit method is unconditionally stable; again, this is a familiar result.

2.4.3 The θ Method

The θ case lies "between" the explicit and implicit problems. The complementarity problems shown in equation (2.3.7) for the θ case are rewritten as

$$0 \leq v_k - \psi_k \perp M_L v_k - M_R v_{k+1} \geq 0 \tag{2.4.12}$$

for $k = K - 1, \ldots, 0$, where the matrices M_L and M_R are given by

$$M_L = (1 + r\theta\delta t)I - A\theta\delta t \tag{2.4.13a}$$

$$M_R = (1 - r(1 - \theta)\delta t)I + A(1 - \theta)\delta t. \tag{2.4.13b}$$

By combining the arguments used for the explicit and for the implicit case, we obtain the following results.

LEMMA 2.4.5
Suppose A satisfies equations (2.3.4c) and (2.3.4d). Let $r \geq 0$ and let M_L and M_R be given by equation (2.4.13a). Under the stability condition

$$\delta t \leq \frac{1}{(1-\theta)\|A - rI\|_\Delta} \tag{2.4.14}$$

where $\|A\|_\Delta = \max_i |a_{ii}|$, the solution at time index $k = 0$ of the θ-case sequence of complementarity problems in equation (2.4.12), satisfies

$$\|v_0\|_\infty \leq \max_{k=0,\dots,K} \left(\frac{1 - r(1-\theta)\delta t}{1 + r\theta\delta t}\right)^k \|\psi_k\|_\infty. \tag{2.4.15}$$

LEMMA 2.4.6
Under the conditions of lemma 2.4.2, the stability condition in lemma 2.4.5 holds provided that

$$\frac{\delta t}{\varepsilon^2} \leq \frac{1 - r(1-\theta)\delta t}{(1-\theta)\max_i \operatorname{tr} \Sigma(x_i)}. \tag{2.4.16}$$

We do get a stability condition for any $\theta < 1$, be it that the condition for positive θ is weaker than the one in the explicit case. This is in line with the conditions found in the literature [15, 19] for finite-difference methods applied to American options.

2.5 BOUNDARY POINTS

It is clear that the problem expressed in equation (2.3.4) may be infeasible for some i. In such a case, we say that x_i is an implied boundary point; otherwise we call x_i an implied interior point. Given nondegenerate Σ and a well-adapted grid, one expects that the implied boundary points are indeed found at the extremities of the grid, and that the implied interior points lie away from the extremities. The location of infeasible points is illustrated in Figure 2.5.1 for a 500-point low-distortion grid in two dimensions. In this case the number of infeasible points is 21, about 4% of the total, and these points are indeed farthest from the grid center.

The partial differential equations of option pricing are typically formulated on infinite domains, so that truncation is necessary for the application of finite-difference methods. An artificial boundary is thus created. It can be seen as an advantage of the irregular grid method that the transition to the boundary is smoother than in the case of standard finite-difference methods based on

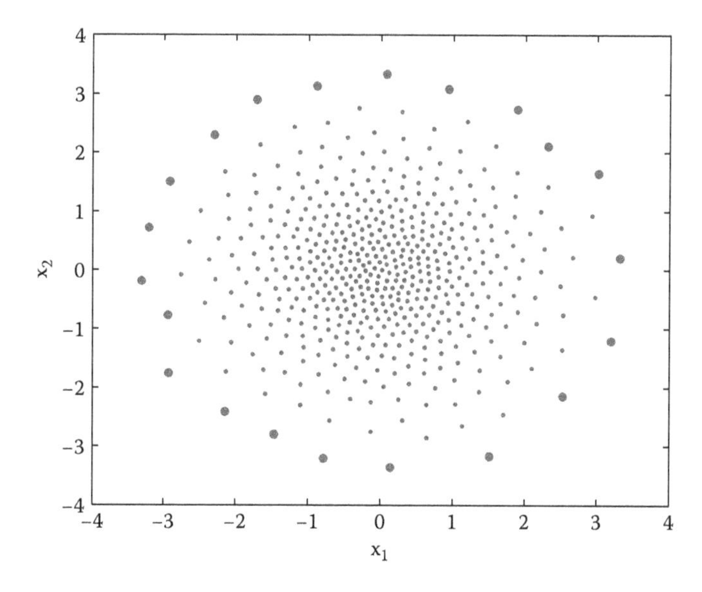

FIGURE 2.5.1 Interior points (small) and boundary points (large) on a normal low-distortion grid for $d = 2$, $n = 500$.

regular grids. It is important for the accuracy of our method that the criterion that we use for identification of boundary points, namely infeasibility of the linear program shown in equation (2.3.4), is correct in the sense that infeasible points are indeed located in regions where the process of interest has a low density. A formal proof of such a property does not seem easy to give, however, in view of the intricate interplay that exists in general between grid generation, process parameters, and feasibility determination. In the experiments that we carried out, as reported below in section 2.6, we did find that infeasible points occur only at a sufficiently large distance from the grid center. We call this distance the boundary radius.

To get an impression of the way in which the boundary radius depends on grid size and state–space dimension, one can reason as follows. Consider a grid of n points derived from a d-variate standard normal distribution. Let x_0 be a given point in d-dimensional space. The expected number of grid points in the shifted half-space tangent at x_0 to the ball with radius $\|x_0\|$ (i.e., the region of points x in the state space that satisfy $x^T x_0 \geq \|x_0\|^2$) is $\Phi(\|x_0\|)n$, where Φ is the cumulative normal distribution function. The minimum number of neighbors needed to satisfy the consistency conditions is η_d, given by equation (2.3.5). If we assume (rather optimistically, perhaps) that boundary points only start to appear when the expected number of grid points in the shifted

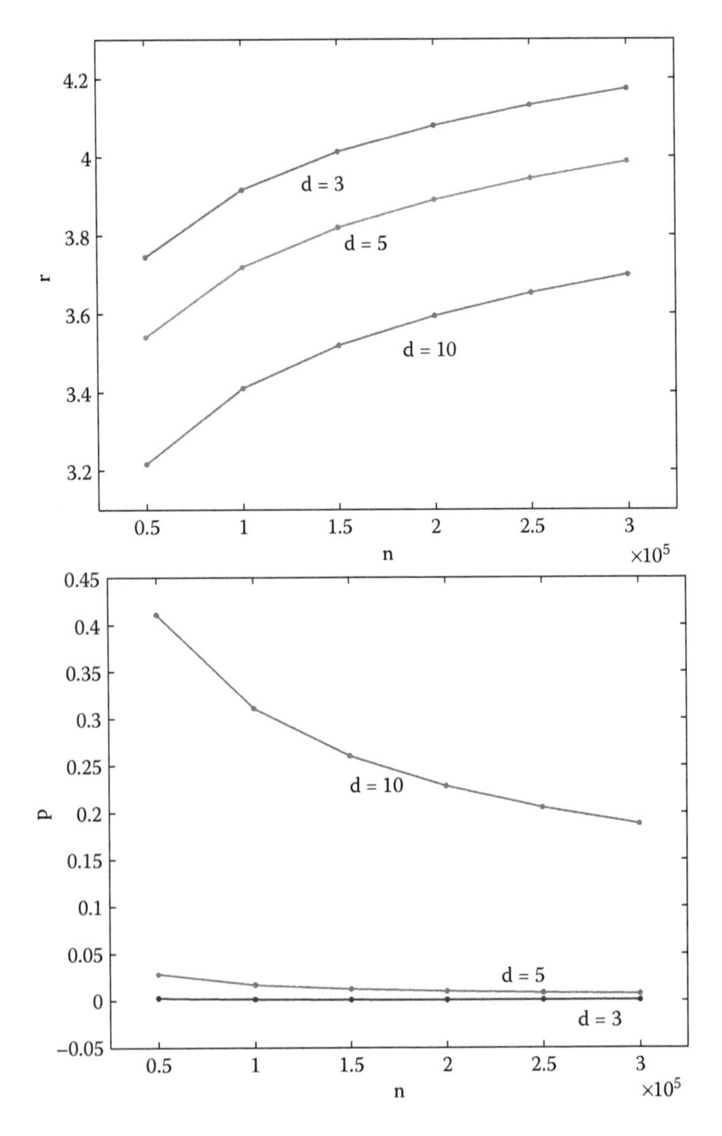

FIGURE 2.5.2 Coarse estimate of the radius of the boundary and the proportion of points that are in the boundary region for a standard normal grid.

half-space drops below $\frac{1}{2}\eta_d$, then the boundary radius is given by

$$r = \Phi^{-1}\left(1 - \frac{1}{2n}\eta_d\right). \tag{2.5.1}$$

In view of the fact that the squared norm of a standard normal variable in d dimensions is a chi-square random variable with d degrees of freedom, the corresponding expected number of boundary points is

$$\mathbb{E}N_b = n(1 - \Psi(r^2, d)) \tag{2.5.2}$$

where $\Psi(\cdot, d)$ is the chi-square cumulative distribution function with d degrees of freedom. Plots of the radius and expected number of boundary points according to this reasoning are presented in Figure 2.5.2 for $d = 3, 5, 10$ and n up to 300,000. For instance, for $d = 10$ and $n = 10^5$, the estimated proportion of boundary points is about 30%. Experimentally we find that equation (2.5.1) underestimates the implied radius for lower dimensions and overestimates it for higher dimensions. The numerical results are discussed in subsection 2.6.7.

2.6 EXPERIMENTS

A major hurdle in testing algorithms for pricing high-dimensional American options is the difficulty of verifying results. One method is using out-of-sample paths to estimate the value of the exercise and hedging strategies implied by the model. Another, which we use here, is to use benchmark results from a special case that can be solved accurately. In the following we introduce benchmark results and then test the proposed method against those results.

2.6.1 Geometric Average Options

We choose to focus on geometric average options in a multivariate Black–Scholes framework, since the pricing problem for these options can be reduced to a one-dimensional problem. The one-dimensional problem can be solved to a high degree of accuracy, thus providing benchmark results for the algorithm.

A geometric average put option written on d assets has payoff function

$$\psi(S) = \left(K - \left(\prod S_i\right)^{1/d}\right)^+ \tag{2.6.1}$$

where S is the asset value and K is the strike price of the option. Assuming a complete and arbitrage-free market with log asset prices following a multivariate Brownian motion with constant covariance Σ, we have a constant risk-neutral drift

$$\mu = r\,\mathbb{1} - \frac{1}{2}\operatorname{diag}\Sigma. \tag{2.6.2}$$

2.6.2 Benchmarks

Defining $Y = \frac{1}{d}\sum_{i=1}^{d} X_i$, where the state variables X_i represent log asset prices, we find that Y satisfies

$$dY(s) = \tilde{\mu}\, ds + \tilde{\sigma}\, dW(s) \tag{2.6.3}$$

where the parameters of the diffusion are given by

$$\tilde{\mu} = r - \frac{1}{2d}\sum_{i=1}^{d}\sigma_i^2 \tag{2.6.4}$$

$$\tilde{\sigma}^2 = \frac{1}{d^2}\sum_{i=1}^{d}\left(\sum_{j=1}^{d}\rho_{ij}\right)^2. \tag{2.6.5}$$

The option is thus equivalent to a standard put option on an asset with starting value $\exp\{\overline{X}_0\}$, strike price K, risk-free rate r, and continuous dividend stream

$$\delta = \frac{1}{2}\left(\frac{1}{d}\sum\sigma_i^2 - \tilde{\sigma}^2\right). \tag{2.6.6}$$

In Table 2.6.1 we provide benchmark results for geometric put options written on up to ten assets, with starting asset values $S_i = 40$ for all i, and strike price 40. The risk-free rate is taken as 0.06, the volatilities $\sigma_i = 0.2$ for all i, and correlations $\rho_{ij} = 0.25$ for $i \neq j$.

TABLE 2.6.1 Benchmark Results for Geometric Average Options in Dimensions 1–10. Also Displayed are the Variance $\tilde{\sigma}^2$ and Continuous Dividend δ for the Equivalent One-Dimensional Problem

d	$\tilde{\sigma}^2 \times 10^2$	$\delta \times 10^2$	European	Bermudan	American
1	4.000	0.000	2.0664	2.2930	2.3196
2	2.500	0.750	1.5553	1.7557	1.7787
3	2.000	1.000	1.3468	1.5380	1.5597
4	1.750	1.125	1.2318	1.4193	1.4392
5	1.600	1.200	1.1585	1.3421	1.3625
6	1.500	1.250	1.1077	1.2893	1.3094
7	1.429	1.286	1.0703	1.2504	1.2703
8	1.375	1.313	1.0416	1.2207	1.2404
9	1.333	1.333	1.0189	1.1971	1.2167
10	1.300	1.350	1.0004	1.1779	1.1974

2.6.3 Experimental Details

Using the methodology proposed in section 2.3, we conducted experiments to find the value of the geometric average put options given above.

We used six different grid sizes ranging from 50,000 to 300,000, and two types of grids consisting of normal Sobol points and normal low distortion points. The grid covariance was taken to correspond to 1.5 times the process covariance at expiry, following the discussion in section 2.3. The transition matrices were generated using distributed computing software in a Matlab environment. A maximum of $20\eta_d$ nearest neighbors were considered when trying to satisfy the local consistency conditions, where η_d is as defined in equation (2.3.5). Implied boundary points were treated as absorbing states in the approximating Markov chain.

We consider the pricing problem for European options, for Bermudan options with ten exercise opportunities, and for true American options that can be exercised at any time up to expiry. For the European and Bermudan problems we used the Crank–Nicolson method with 100 time steps. For solving the linear systems, we used the conjugate gradient squared (CGS) and generalized minimum residual (GMRES) methods, the latter being slower but more robust. For the American problems we used projected successive overrelaxation (PSOR) to solve the linear complementarity problems, with 1000 time steps. While it is not necessary to use such a large number of time steps in practice, we wanted to focus on the error with respect to the space discretization. Having a small enough δt causes the error resulting from time discretization to be negligible in comparison, and thus allows a more accurate assessment of the error resulting from space discretization.

To achieve an improved level of precision, we also appeal to a simple bias control technique. As the bias control variable, we use the European option price because it is in itself a much easier problem to solve, and its solution using the irregular grid technique is highly correlated to the corresponding American option price, where the correlation is seen in respect to our grid choice. The method simply involves using the irregular-grid technique to solve both the American and European problems, and then adding the difference (i.e., the estimated early exercise premium) to the accurately calculated European price.

2.6.4 Experimental Results

We present results in Tables 2.6.2–2.6.4 for prices obtained using normal Sobol grids for the Bermudan, American, and European cases, respectively. The corresponding results for low-distortion grids are presented in Tables 2.6.7–2.6.9.

TABLE 2.6.2 Results for Bermudan Geometric Average Put Options in Dimensions 3–10 Using Normal Sobol Grids

d	5×10^4	10×10^4	15×10^4	20×10^4	25×10^4	30×10^4
3	1.5370	1.5375	1.5376	1.5376	1.5377	1.5377
4	1.4135	1.4147	1.4155	1.4161	1.4163	1.4166
5	1.3300	1.3329	1.3345	1.3360	1.3365	1.3371
6	1.2532	1.2630	1.2667	1.2757	1.2766	1.2780
7	1.1981	1.2133	1.2137	1.2305	1.2311	1.2313
8	1.1489	1.1664	1.1672	1.1891	1.1938	1.1807
9	1.1116	1.1255	1.1351	1.1530	1.1514	1.1612
10	1.0901	1.1080	1.1078	1.1129	1.1242	1.1218

TABLE 2.6.3 Results for American Geometric Average Put Options in Dimensions 3–10 on Normal Sobol Grids

d	5×10^4	10×10^4	15×10^4	20×10^4	25×10^4	30×10^4
3	1.5584	1.5588	1.5590	1.5591	1.5592	1.5592
4	1.4332	1.4347	1.4357	1.4362	1.4365	1.4369
5	1.3489	1.3522	1.3537	1.3551	1.3557	1.3563
6	1.2721	1.2818	1.2858	1.2940	1.2951	1.2965
7	1.2182	1.2325	1.2331	1.2482	1.2491	1.2492
8	1.1693	1.1864	1.1870	1.2071	1.2114	1.1993
9	1.1316	1.1460	1.1549	1.1715	1.1700	1.1802
10	1.1102	1.1281	1.1267	1.1324	1.1433	1.1414

TABLE 2.6.4 Results for European Geometric Average Put Options in Dimensions 3–10 on Normal Sobol Grids

d	5×10^4	10×10^4	15×10^4	20×10^4	25×10^4	30×10^4
3	1.3461	1.3463	1.3465	1.3465	1.3465	1.3465
4	1.2274	1.2286	1.2293	1.2302	1.2304	1.2304
5	1.1482	1.1505	1.1520	1.1541	1.1545	1.1549
6	1.0716	1.0813	1.0849	1.0977	1.0984	1.0993
7	1.0156	1.0275	1.0318	1.0527	1.0541	1.0545
8	0.9624	0.9792	0.9848	1.0123	1.0151	0.9943
9	0.9231	0.9406	0.9507	0.9735	0.9755	0.9802
10	0.8966	0.9203	0.9277	0.9340	0.9418	0.9424

Tables 2.6.5 and 2.6.6 show the results on normal Sobol grids for Bermudan and American options when the European is used as the bias control variable. Tables 2.6.10 and 2.6.11 show the same for low-distortion grids.

Figures 2.6.1 and 2.6.2 present the results graphically for normal Sobol grids. The results for low-distortion grids are shown in Figures 2.6.3 and 2.6.4. We see that the error increases with dimension to about 5–10% for

TABLE 2.6.5 Results for Bermudan Geometric Average Put Options in Dimensions 3–10 on Normal Sobol Grids, Using the European Price as a Bias Control

d	5×10^4	10×10^4	15×10^4	20×10^4	25×10^4	30×10^4
3	1.5382	1.5382	1.5382	1.5379	1.5379	1.5379
4	1.4179	1.4178	1.4180	1.4177	1.4177	1.4179
5	1.3403	1.3409	1.3410	1.3404	1.3405	1.3407
6	1.2892	1.2893	1.2894	1.2857	1.2858	1.2863
7	1.2527	1.2560	1.2521	1.2481	1.2473	1.2470
8	1.2281	1.2288	1.2240	1.2184	1.2203	1.2279
9	1.2074	1.2038	1.2033	1.1984	1.1947	1.1999
10	1.1940	1.1881	1.1805	1.1793	1.1829	1.1799

TABLE 2.6.6 Results for American Geometric Average Put Options in Dimensions 3–10 on Normal Sobol Grids, Using the European Price as a Bias Control

d	5×10^4	10×10^4	15×10^4	20×10^4	25×10^4	30×10^4
3	1.5595	1.5596	1.5596	1.5594	1.5594	1.5594
4	1.4376	1.4378	1.4382	1.4378	1.4379	1.4382
5	1.3592	1.3602	1.3603	1.3595	1.3597	1.3599
6	1.3082	1.3082	1.3085	1.3041	1.3044	1.3048
7	1.2728	1.2752	1.2716	1.2658	1.2653	1.2649
8	1.2484	1.2487	1.2437	1.2364	1.2379	1.2465
9	1.2274	1.2242	1.2231	1.2169	1.2133	1.2189
10	1.2141	1.2082	1.1994	1.1988	1.2020	1.1994

TABLE 2.6.7 Results for Bermudan Geometric Average Put Options in Dimensions 3–10 Using Low-Distortion Grids

d	5×10^4	10×10^4	15×10^4	20×10^4	25×10^4	30×10^4
3	1.5372	1.5375	1.5376	1.5377	1.5377	1.5378
4	1.4141	1.4155	1.4160	1.4163	1.4165	1.4166
5	1.3309	1.3338	1.3360	1.3364	1.3370	1.3371
6	1.2695	1.2729	1.2751	1.2777	1.2779	1.2796
7	1.2139	1.2249	1.2255	1.2292	1.2319	1.2321
8	1.1628	1.1773	1.1850	1.1898	1.1899	1.1863
9	1.1234	1.1397	1.1428	1.1548	1.1514	1.1588
10	1.1177	1.1008	1.1131	1.1103	1.1170	1.1242

$d = 10$. The bias control is very effective, reducing the error for $d = 10$ to less than 1%.

When using the European option to reduce the bias, we see that the results are biased upward, whereas the raw results are biased downward.

TABLE 2.6.8 Results for American Geometric Average Put Options in Dimensions 3–10 on Low-Distortion Grids

d	5×10^4	10×10^4	15×10^4	20×10^4	25×10^4	30×10^4
3	1.5583	1.5587	1.5589	1.5590	1.5590	1.5591
4	1.4341	1.4355	1.4361	1.4364	1.4367	1.4369
5	1.3500	1.3528	1.3550	1.3554	1.3561	1.3564
6	1.2875	1.2912	1.2935	1.2961	1.2965	1.2981
7	1.2319	1.2432	1.2433	1.2474	1.2496	1.2502
8	1.1813	1.1952	1.2032	1.2082	1.2080	1.2042
9	1.1412	1.1580	1.1615	1.1730	1.1689	1.1774
10	1.1390	1.1206	1.1315	1.1288	1.1365	1.1434

TABLE 2.6.9 Results for European Geometric Average Put Options in Dimensions 3–10 Using Low-Distortion Grids

d	5×10^4	10×10^4	15×10^4	20×10^4	25×10^4	30×10^4
3	1.3460	1.3463	1.3464	1.3465	1.3465	1.3466
4	1.2287	1.2295	1.2299	1.2301	1.2304	1.2305
5	1.1501	1.1520	1.1535	1.1540	1.1544	1.1546
6	1.0904	1.0947	1.0965	1.0982	1.0987	1.0994
7	1.0394	1.0474	1.0497	1.0523	1.0545	1.0553
8	0.9877	1.0015	1.0078	1.0122	1.0137	1.0131
9	0.9405	0.9605	0.9654	0.9726	0.9729	0.9779
10	0.9080	0.9100	0.9247	0.9291	0.9322	0.9393

TABLE 2.6.10 Results for Bermudan Geometric Average Put Options in Dimensions 3–10 Using Low-Distortion Grids, Using the European Price as a Bias Control

d	5×10^4	10×10^4	15×10^4	20×10^4	25×10^4	30×10^4
3	1.5379	1.5379	1.5379	1.5379	1.5380	1.5380
4	1.4172	1.4178	1.4179	1.4179	1.4179	1.4179
5	1.3393	1.3403	1.3410	1.3409	1.3410	1.3410
6	1.2867	1.2859	1.2863	1.2871	1.2868	1.2878
7	1.2448	1.2478	1.2461	1.2472	1.2477	1.2471
8	1.2167	1.2174	1.2187	1.2191	1.2178	1.2147
9	1.2017	1.1980	1.1964	1.2011	1.1974	1.1998
10	1.2101	1.1913	1.1888	1.1817	1.1853	1.1853

This is probably due to the upward bias introduced by the convexity of the max operator that appears in the Bermudan and American problems, but not in the European problem.

In one and two dimensions, the generator matrix became numerically unstable for the grid sizes we consider; we have thus not presented results

TABLE 2.6.11 Results for American Geometric Average Put Options in Dimensions 3–10 on Low-Distortion Grids, Using the European Price as a Bias Control

d	5×10^4	10×10^4	15×10^4	20×10^4	25×10^4	30×10^4
3	1.5590	1.5592	1.5592	1.5592	1.5593	1.5593
4	1.4372	1.4378	1.4380	1.4380	1.4381	1.4381
5	1.3584	1.3593	1.3601	1.3599	1.3602	1.3603
6	1.3048	1.3042	1.3047	1.3056	1.3054	1.3064
7	1.2628	1.2661	1.2639	1.2654	1.2654	1.2652
8	1.2352	1.2353	1.2369	1.2376	1.2359	1.2327
9	1.2196	1.2164	1.2150	1.2193	1.2149	1.2184
10	1.2315	1.2111	1.2072	1.2002	1.2048	1.2045

for these low dimensions here. This lack of convergence is due to the finite precision arithmetic, and not to instability in the sense that the generator matrix has unstable eigenvalues (i.e., eigenvalues having a positive real part). The method has been found to work very well in one and two dimensions, but for smaller grid sizes.

2.6.5 Error Behavior

Drawing a parallel with regular grid methods, we expect the error to be related to δx, the distance between grid points with positive weights in A. In a regular grid with the same number of points N in each dimension, we have $n = N^d$ points in total, and the distance to the nearest point is simply $n^{-1/d}$. The error when using a standard finite difference method is on the order of δx^2, or $n^{-2/d}$.

We thus propose modeling irregular grid errors as in the regular grid case, allowing for a constant term in the exponent as well as a multiplicative factor:

$$\log|\varepsilon| = c_1 + c_2 \frac{\log n}{d}. \tag{2.6.7}$$

In Figures 2.6.5 and 2.6.6 we present plots of the log absolute error versus $(\log n)/d$, and Tables 2.6.12 and 2.6.13 show the regression results. Referring to our assumption of error behavior in equation (2.6.7), we find that the complexity is accurately modeled by the given relationship in all three cases (for suitable c_1, c_2). The possibility of modeling error accurately may suggest the application of Richardson extrapolation in cases where monotonic behavior of errors for successive grid sizes is observed.

The linear relationships seen in the log-transformed data strongly suggest that the proposed algorithm has exponential complexity. There is little

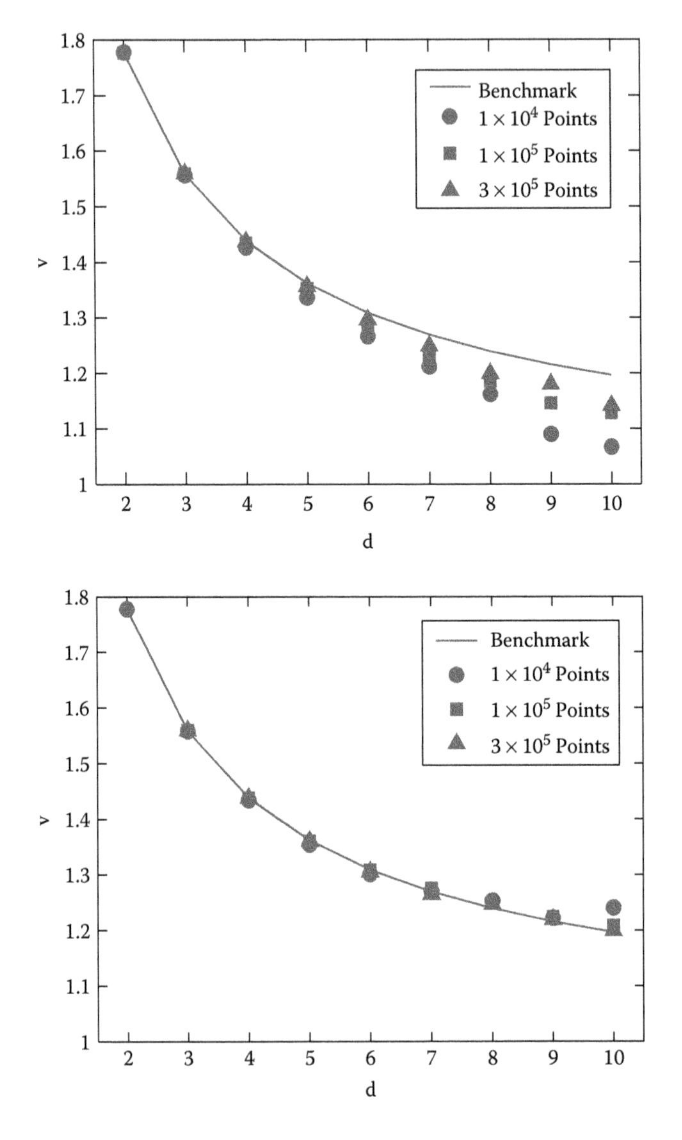

FIGURE 2.6.1 Bermudan pricing results for normal Sobol grids presented raw (top) and using European price as bias control (bottom).

difference in error behavior between the Sobol and low distortion cases. The European and Bermudan prices show about the same asymptotic relationship, while American errors exhibit a slightly faster rate in the Sobol case, although this is barely significant.

The convergence rate for finite-difference methods used to solve partial differential equation (PDE) problems on regular grids is $1/\delta x^2$ or $n^{-2/d}$, which

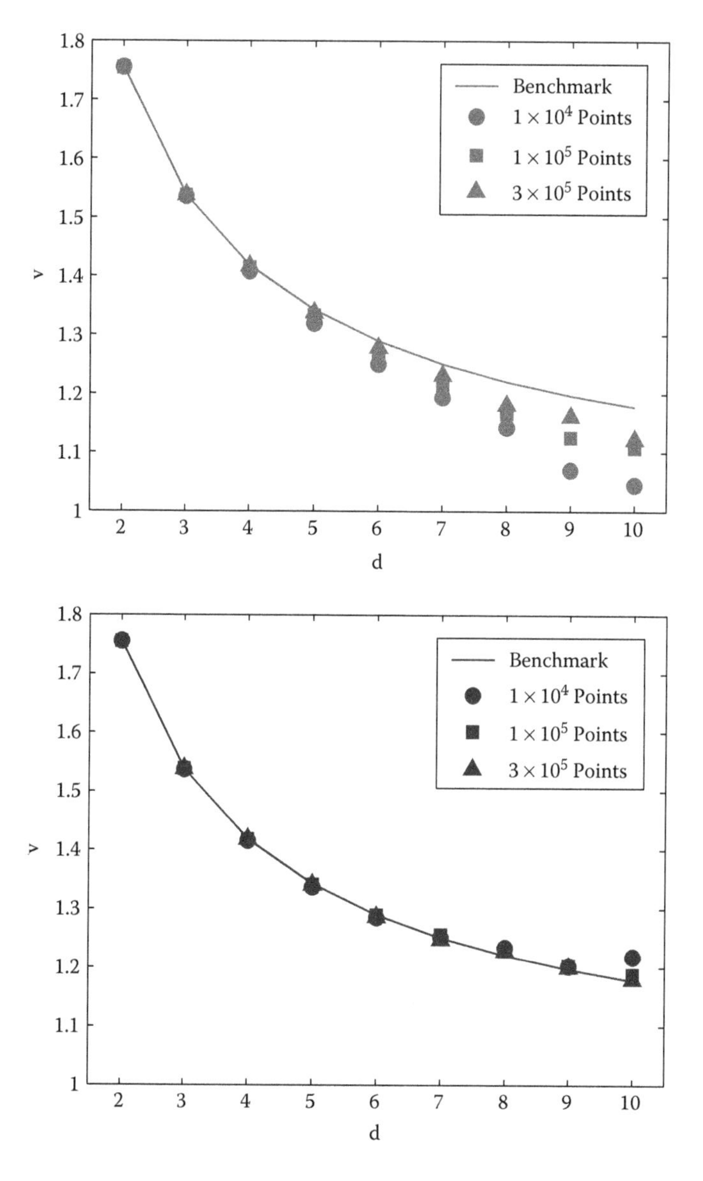

FIGURE 2.6.2 American pricing results for normal Sobol grids presented raw (top) and using European price as bias control (bottom).

here translates to $c_2 = -2$. From this point of view, our method seems to be slightly slower in convergence than the regular grid method. It may be taken into account here that the average δx is larger as a function of the grid size in the irregular grid case.

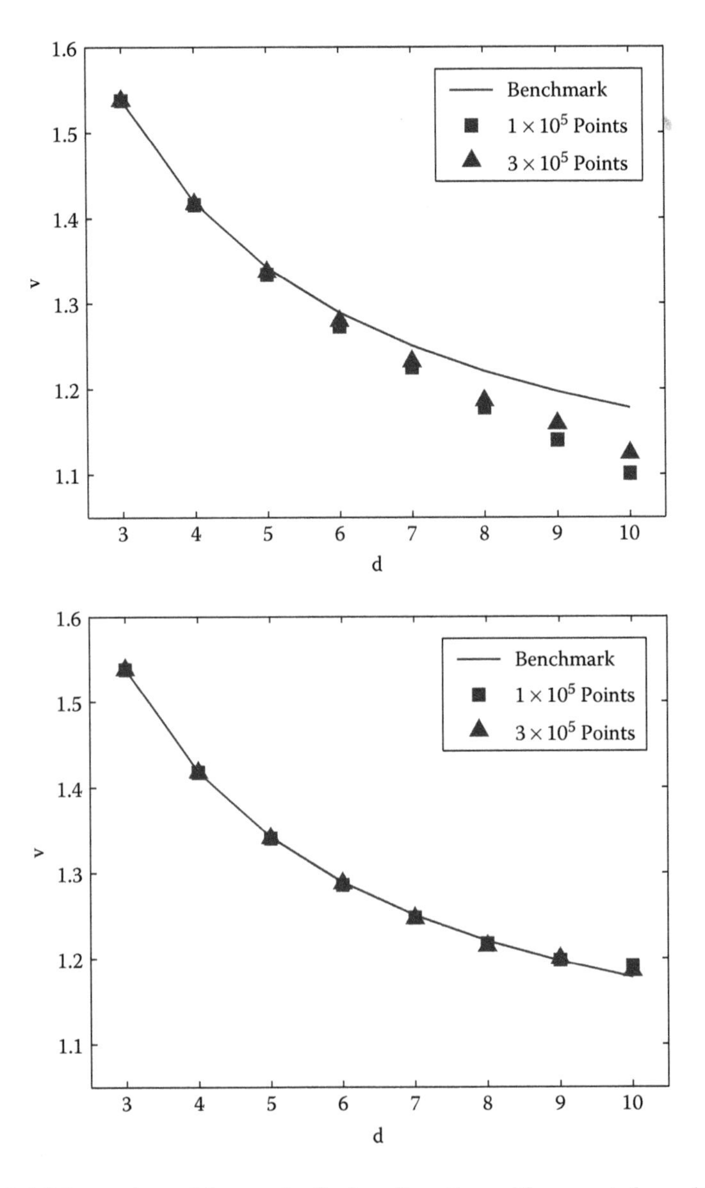

FIGURE 2.6.3 Bermudan pricing results for low-distortion grids presented raw (top) and using European price as bias control (bottom).

The given model for errors implies that the amount of work required to obtain solutions to a certain accuracy increases exponentially with dimension. The algorithm therefore does not break the curse of dimensionality, and in fact the standard finite-difference method appears to do better asymptotically as the number of grid points tends to infinity. However, in high dimensions,

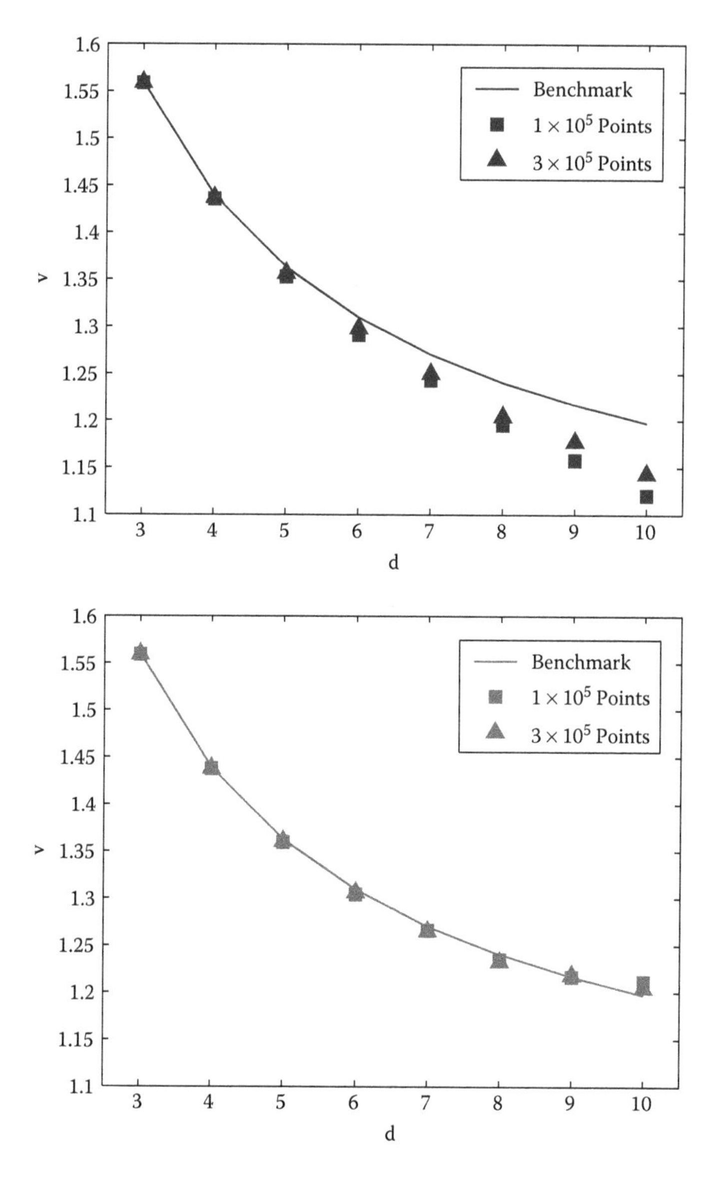

FIGURE 2.6.4 American pricing results for low-distortion grids presented raw (top) and using European price as bias control (bottom).

the advantages of the standard method appear only for impractically large numbers of grid points. The irregular method provides flexibility and freedom in choosing grid sizes. Also, irregular grids do much better than regular grids in terms of the proportion of boundary points for moderate grid sizes; this is discussed in Subsection 2.6.7 below.

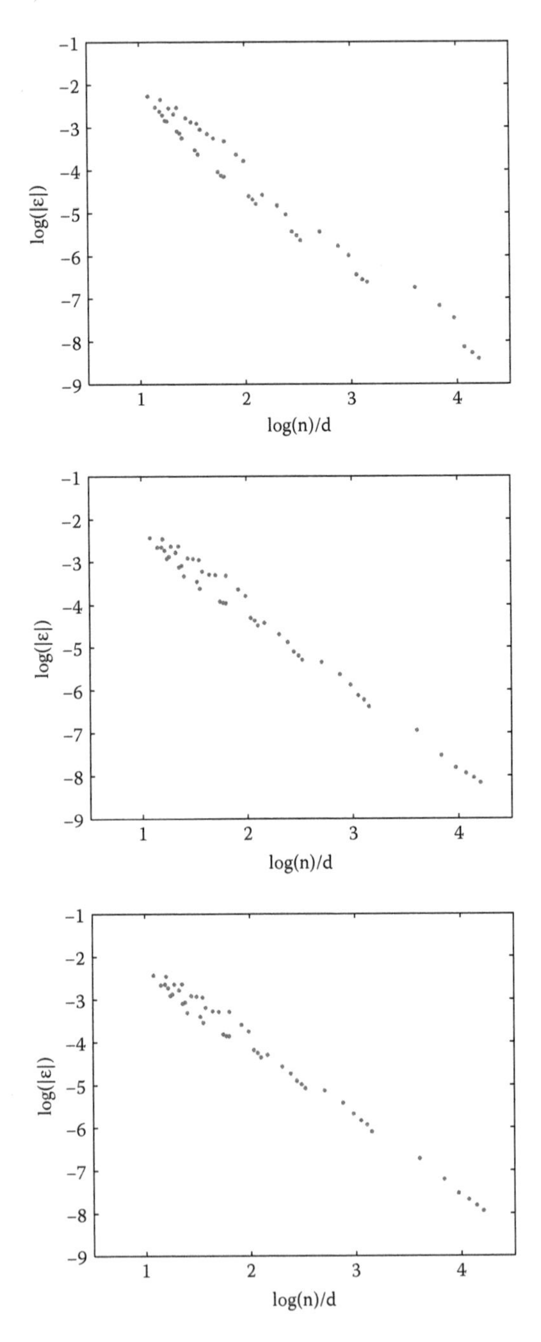

FIGURE 2.6.5 Log of absolute errors for European, Bermudan, and American geometric average options plotted against $(\log n)/d$ for $d = 3, \ldots, 10$ for normal Sobol grids. The points lie in a nearly straight line in all three cases, giving a clear indication of complexity. See Table 2.6.12 for regression results.

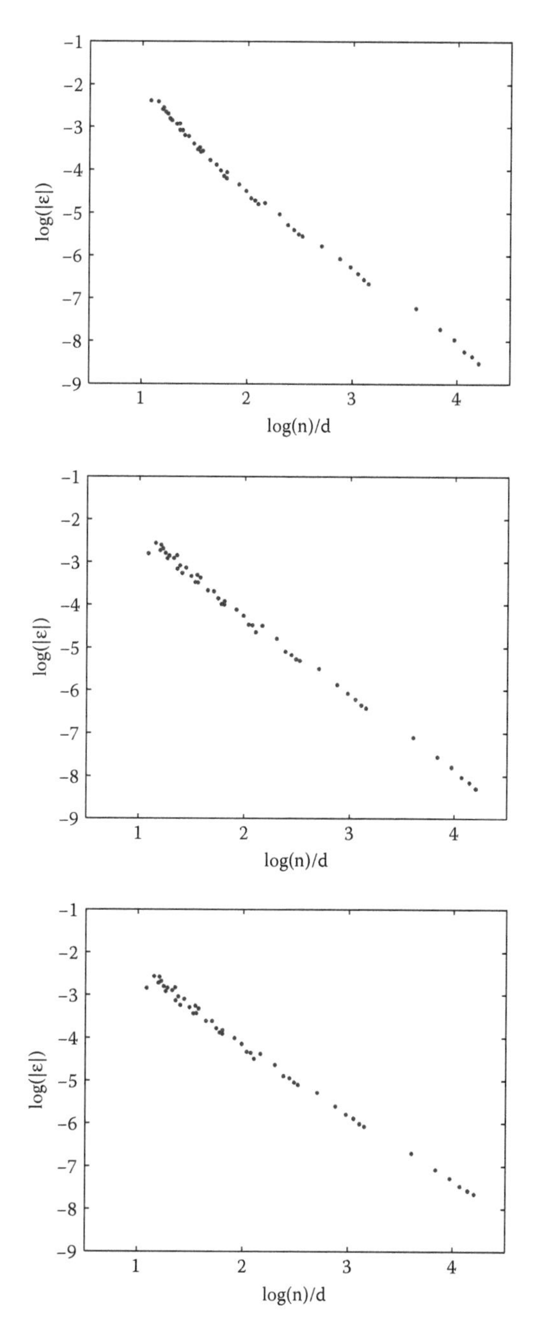

FIGURE 2.6.6 Log of absolute errors for European, Bermudan, and American geometric average options plotted against $(\log n)/d$ for $d = 3, \ldots, 10$ for low-distortion grids. The points lie in a nearly straight line in all three cases, giving a clear indication of complexity. See Table 2.6.13 for regression results.

TABLE 2.6.12 Regression Coefficients for the Error Behavior on Normal Sobol Grids (95% CI in Parentheses)

Option Type	c_1	c_2	R^2
European	−0.35(±0.23)	−1.91(±0.10)	0.971
Bermudan	−0.42(±0.14)	−1.85(±0.06)	0.988
American	−0.55(±0.13)	−1.74(±0.05)	0.989

2.6.6 Timings

The irregular grid method presented in this chapter can be divided into two computationally intensive stages: obtaining the generator matrix and performing the time stepping. Computing transition matrices is not considered here, but it is very similar to computing generator matrices.

Here we provide indications of the timings involved; as usual this depends heavily on the hardware and software used. The software aspect is emphasized here, as there is a huge difference in the performance of different algorithms for solving the linear programming problem and for solving linear systems of equations. The experiments were carried out in Matlab on an 866-MHz Pentium III under Windows 2000.

2.6.6.1 Generator Matrix

In dimension d we are interested in solving a large number of linear programming problems with $\eta_d = d(d+3)/2$ equality constraints and where all variables are nonnegative. The number of variables that should be made potentially active is not known a priori, but it has been found that $5\eta_d$ is sufficient for points close to the center of the grid, and an increased number of $20\eta_d$ is needed closer to the boundary. The strategy is thus to order the points according to their norm and try $5\eta_d$ neighbors until a certain failure rate is reached, then to switch to $20\eta_d$ neighbors on the remaining points.

TABLE 2.6.13 Regression Coefficients for the Error Behavior on Low Distortion Grids (95% CI in Parentheses)

Option Type	c_1	c_2	R^2
European	−0.49(±0.12)	−1.94(±0.05)	0.992
Bermudan	−0.59(±0.08)	−1.84(±0.03)	0.997
American	−0.83(±0.08)	−1.65(±0.03)	0.995

In two dimensions, a single problem takes about 0.06 s and is not sensitive to the number of variables changing from $5\eta_d$ to $20\eta_d$. This is probably due to the relatively large overhead involved in the Matlab routines. In five dimensions we see an increase from 0.07 s for $5\eta_d$ neighbors to 0.10 s for $20\eta_d$. In ten dimensions we see a corresponding increase from 0.31 s to 1.90 s per problem. Given that the number of problems to be solved is in principle equal to the number of grid points,[3] it is clear that the computation of the generator matrix is a time-consuming task. However, the linear programming problems associated with different grid points are independent of each other, so that parallelization can be trivially achieved. The computation of the generator matrix is essentially a model transformation and is not tied to a specific payoff function; different options within the same model can therefore be priced using the same generator matrix. To some extent, it is even possible to reuse generator matrices for different models; several options are discussed by Berridge [3].

2.6.6.2 *Time Stepping*

In dimension d and with n grid points, we use a generator matrix with n rows, each with $\eta_d + 1$ nonzero entries. The complexity of implicit time stepping should thus be quadratic with dimension and linear with grid size.

For 300,000 points in five dimensions, explicit time steps take about 1.5 s and implicit about 29 s with CGS. For ten dimensions, explicit time steps take about 3.0 s and implicit about 21 s with CGS. The fact that implicit solutions can be faster in a higher dimension is due to the conditioning of the matrix, making it more amenable to solution even though it is more dense.

One can thus perform about 10–20 times more explicit than implicit time steps for the same running time. However there is a tradeoff, as the latter generally give much better precision.

2.6.7 Boundary Points

We now compare the observed boundaries presented in Figures 2.6.7 and 2.6.8 with the coarse estimates in section 2.5 and Figure 2.5.2. The proportion of boundary points goes up quickly with dimension, as predicted in section 2.5. A simple calculation reveals that the proportion of boundary points for a regular grid with $n^{1/d}$ steps per dimension is $1 - (1 - 2n^{-1/d})^d$. For example,

---

[3] Some savings may be achieved by assuming that all points beyond a certain radius are boundary points.

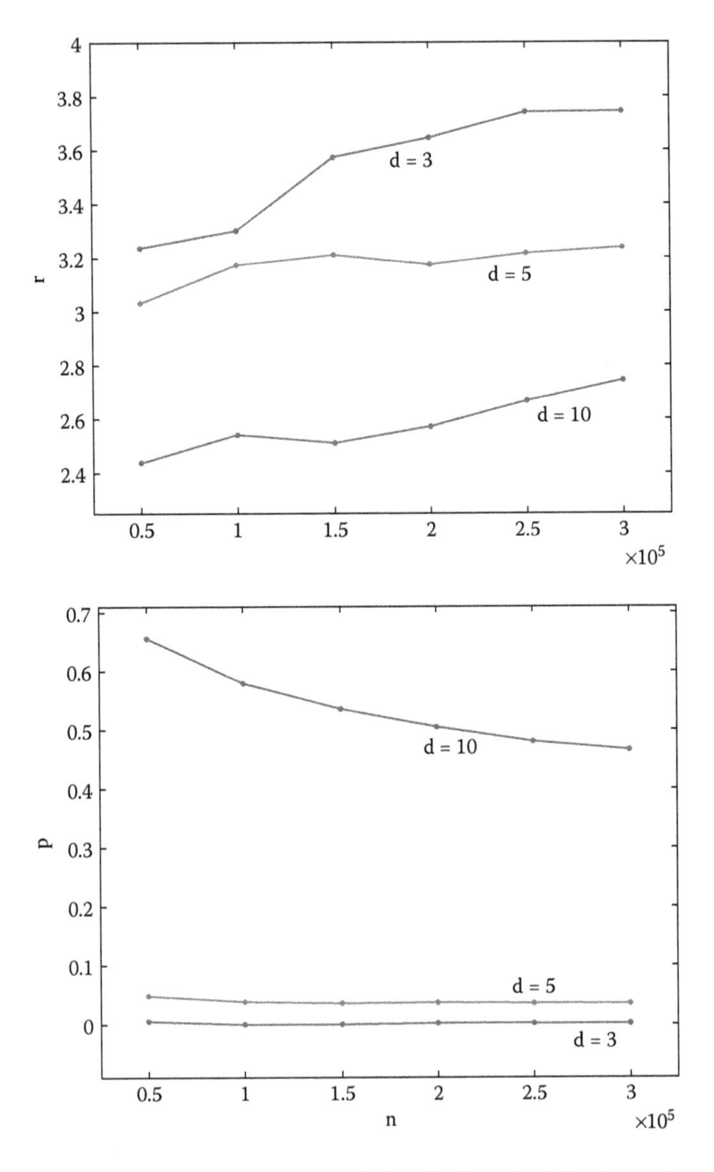

FIGURE 2.6.7 Smallest norms in normal Sobol grids for which local consistency could not be satisfied and proportion of points in the boundary region with $20\eta_d$ nearest neighbors (cf. Figure 2.5.2).

for $d = 10$, one requires a grid size of about 5×10^{14} to bring the proportion of boundary points down to 50%. Using the irregular-grid method, one needs about 3×10^5, as seen in Figures 2.6.7 and 2.6.8.

The reasoning of section 2.5 turns out to deliver usable but rough estimates of the actual proportions of boundary points. A factor that plays a role here

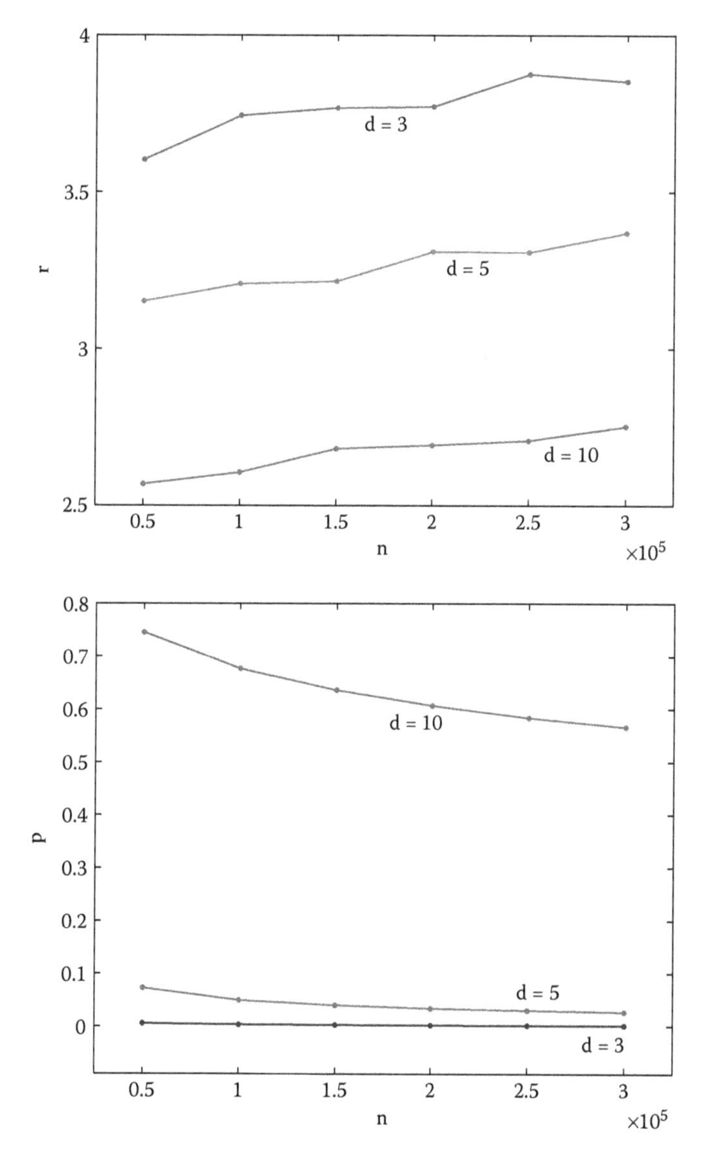

FIGURE 2.6.8 Norms of the smallest points in low-distortion grids for which local consistency could not be satisfied and proportion of points in the boundary region with $20\eta_d$ nearest neighbors (cf. Figure 2.5.2).

is that in the algorithm we used a maximum of $20\eta_d$ neighbors when trying to satisfy local consistency, rather than all grid points. The estimate of the boundary radius on which the reasoning in section 2.5 is based could be more refined.

2.7 CONCLUSIONS

We proposed a method for pricing American and Bermudan options with several underlying assets and an arbitrary payoff structure. The method was tested for geometric average options, which can be easily benchmarked, in dimensions three to ten with very accurate results.

An analysis of the error indicates that the method has complexity that is exponential in the dimension variable. Thus we do not break the curse of dimensionality; rather we extend the range of problems for which computations are still feasible. It was shown that the combination of irregular grids with a bias control technique is effective. We expect that extrapolation can contribute to the accuracy of the method as well, although this was not tested in the present work. In practical terms, the computation of generator matrices is expensive, but this can be easily parallelized. Further computations are cheap for typical grid sizes on the order of 10^4–10^5.

The idea of an irregular grid was previously used by Berridge and Schumacher [2] in combination with a root extraction method to determine the generator matrix, but with that method it appears that stability is difficult to guarantee. Here it was shown that the method of generator matrix construction based on local consistency conditions allows the formulation of simple sufficient conditions for stability.

We found evidence suggesting that the accuracy of the irregular-grid method, as a function of the number of grid points, asymptotically lies somewhat below the accuracy of the regular-grid method. The advantage of the irregular method however is that there is much more freedom in choosing the number of grid points, and that even in fairly high dimensions the ratio between boundary points and interior points is reasonable for typical grid sizes. As an example, suppose that one wants to do a computation in ten dimensions and that one wants to limit the number of grid points to about 10^5. Because $2^{10} \approx 10^3$ and $4^{10} \approx 10^6$, the only option using regular grids is to use three points per dimension. In the resulting grid, all points except the one in the middle are boundary points. When an irregular grid is used, the number of grid points can be chosen arbitrarily, and in the case of 10^5 points in ten dimensions, about one-half are found to be interior (see Figure 2.6.7).

Concerning grid point generation, we tested both Sobol and low-distortion grids. Little difference between the performance of the resulting grids was found. This may suggest that Sobol grids are to be preferred, as these can be obtained with less computational effort. An interesting alternative, which has not been tested in the present work, may be the use of lattice rules [21].

ACKNOWLEDGMENT

Research supported by the Netherlands Organisation for Scientific Research (NWO).

REFERENCES

1. Abraham Berman and Robert J. Plemmons. *Nonnegative Matrices in the Mathematical Sciences*. Academic Press, New York, 1979.
2. S.J. Berridge and J.M. Schumacher. An irregular grid method for high-dimensional free-boundary problems in finance. *Future Generation Computer Systems*, 20: 353–362, 2004.
3. Steffan Berridge. Irregular Grid Methods for Pricing High-Dimensional American Options. Doctoral dissertation, Tilburg University, the Netherlands, 2004.
4. Phelim P. Boyle, Adam W. Kolkiewicz, and Ken Seng Tan. Pricing American-Style Options Using Low Discrepancy Mesh Methods. Technical report IIPR 00-07, University of Waterloo, Ontario, 2000.
5. Phelim P. Boyle, Adam W. Kolkiewicz, and Ken Seng Tan. An improved simulation method for pricing high-dimensional American derivatives. *Mathematics and Computers in Simulation*, 62: 315–322, 2003.
6. Mark Broadie and Paul Glasserman. A stochastic mesh method for pricing high-dimensional American options. *Journal of Computational Finance*, 7: 35–72, 2004.
7. Jacques F. Carrière. Valuation of the early-exercise price for options using simulations and nonparametric regression. *Insurance: Mathematics and Economics*, 19: 19–30, 1996.
8. E. Clément, D. Lamberton, and P. Protter. An analysis of a least squares regression method for American option pricing. *Finance and Stochastics*, 6: 449–471, 2002.
9. Richard W. Cottle, Jong-Shi Pang, and Richard E. Stone. *The Linear Complementarity Problem*. Academic Press, Boston, 1992.
10. Colin W. Cryer. The solution of a quadratic programming problem using systematic overrelaxation. *SIAM Journal on Control*, 9: 385–392, 1971.
11. M.A.H. Dempster and J.P. Hutton. Pricing American stock options by linear programming. *Mathematical Finance*, 9: 229–254, 1999.
12. Jérôme Detemple. *American-Style Derivatives: Valuation and Computation*. Chapman & Hall/CRC, London, 2005.
13. M. Evans and T. Swartz. *Approximating Integrals via Monte Carlo and Deterministic Methods*. Oxford University Press, Oxford, 2000.
14. Paul Glasserman. *Monte Carlo Methods in Financial Engineering*. Springer, New York, 2004.
15. R. Glowinski, J.L. Lions, and R. Trémolières. *Numerical Analysis of Variational Inequalities*. North-Holland, Amsterdam, 1981.

16. Martin B. Haugh and Leonid Kogan. Pricing American options: a duality approach. *Operations Research*, 52: 258–270, 2004.
17. Jacqueline Huang and Jong-Shi Pang. Option pricing and linear complementarity. *Journal of Computational Finance*, 2: 31–60, 1998.
18. Willem Hundsdorfer and Jan Verwer. *Numerical Solution of Time-Dependent Advection-Diffusion-Reaction Equations*. Springer, Berlin, 2003.
19. Patrick Jaillet, Damien Lamberton, and Bernard Lapeyre. Variational inequalities and the pricing of American options. *Acta Applicandae Mathematicae*, 21: 263–289, 1990.
20. Anastasia Kolodko and John Schoenmakers. Iterative construction of the optimal Bermudan stopping time. *Finance and Stochastics*, 10: 27–49, 2006.
21. Frances Kuo and Ian Sloan. Lifting the curse of dimensionality. *Notices of the AMS*, 52: 1320–1328, 2005.
22. Harold J. Kushner and Paul G. Dupuis. *Numerical Methods for Stochastic Control Problems in Continuous Time*. Springer, New York, 1992.
23. Francis A. Longstaff and Eduardo S. Schwartz. Valuing American options by simulation: a simple least squares approach. *Review of Financial Studies*, 14: 113–147, 2001.
24. Harald Niederreiter. *Random Number Generation and Quasi-Monte Carlo Methods*. SIAM, Philadelphia, 1992.
25. Gilles Pagès. A space quantization method for numerical integration. *Journal of Computational and Applied Mathematics*, 89: 1–38, 1997.
26. L.C.G. Rogers. Monte Carlo valuation of American options. *Mathematical Finance*, 12: 271–286, 2002.
27. Alexander Schrijver. *Theory of Linear and Integer Programming*. Wiley, Chichester, U.K. 1998.
28. Lars Stentoft. Convergence of the least squares Monte-Carlo approach to American option valuation. *Management Science*, 50: 1193–1203, 2004.
29. George Tauchen. Finite state Markov chain approximations to univariate and vector autoregressions. *Economic Letters*, 20: 177–181, 1986.
30. James A. Tilley. Valuing American options in a path simulation model. *Transactions of the Society of Actuaries*, 45: 499–520, 1993.
31. John N. Tsitsiklis and Benjamin Van Roy. Regression methods for pricing complex American-style options. *IEEE Transactions on Neural Networks*, 12: 694–703, 2000.

CHAPTER 3

Adverse Interrisk Diversification Effects for FX Forwards

Thomas Breuer and Martin Jandačka
PPE Research Centre, FH Vorarlberg, Dornbirn, Austria

Contents

Abstract: We describe the phenomenon of negative interrisk diversification effects between credit and market risk. For portfolios of FX forwards, integrated market and credit risk may be larger than the sum of both by a factor of 200 to 400. This phenomenon occurs for portfolios hedged against market risk. The result implies that measuring market and credit risk in an integrated way spots risks of adverse interaction between credit and market events that are hidden to a simple addition of pure market and credit risk numbers.

3.1 INTRODUCTION

Interactions of market and credit risk can give rise to risks above and beyond the sum of market plus credit risk, as the following case indicates. During the Russian ruble (RUR) crisis of August 1998, some Western banks incurred severe losses when Russian banks defaulted on their FX forward contracts.[1] On August 17, 1998, Russia announced a devaluation of the RUR and a moratorium on servicing official short-term debt. Subsequently, the RUR depreciated more than 70% against the U.S. dollar (USD), the government imposed conditions on most of its foreign and domestic debt, and several Russian financial institutions became insolvent.

Some Western banks had USD/RUR forwards with Russian banks and matching RUR/USD forwards with mostly Western companies and banks hedging their exchange rate risk. These positions were fully hedged against moves in the USD/RUR exchange rate. Furthermore, default risk was irrelevant to the banks as long as the exchange rate did not move; if a counterparty defaulted, it was always possible to get the currency deliverable to the other counterparty on the market at no loss if the exchange rate did not move. And a move of exchange rates was very improbable in those times of the managed RUR exchange rate regime. So from both a pure market risk and a pure credit risk point of view, the risk of the portfolio was close to zero.

For this reason, banks considered the narrow spread between the matching forwards as an almost riskless profit.[2] However, during the crisis, adverse credit events and market moves occured *simultaneously*. The Russian counterparties defaulted and at the same time the value of the RUR dropped dramatically. The USD deliverable to the Western companies and banks had to be purchased on the market, and the RUR they got in return were not much worth. This led to enormous losses for the banks involved.[3] How could it happen that banks suffered such heavy losses on portfolios that were almost perfectly hedged

[1] Bank Austria, for example, suffered a loss of ATS 4.7 billion, which amounts to EUR 341 million, according to its official Annual Report for 1998 [2, p.63].

[2] Alejandro Eduardoff, in charge of the Moscow dependency of Bank Austria, was quoted by the Austrian weekly *Profil* [11, p. 40] on September 28, 1999: "Why call this 'betting,' why call this 'speculation'? These terms do not apply here. After all, one could assume justifiedly even the day before the RUR devaluation of August 17 that the RUR exchange rate would remain in its narrow corridor. And under this assumption our derivative was a quite acceptable product" [our translation].

[3] Losses would have been much higher had the RUR exchange rate not peaked around September 15. On September 17, 1998, *The Economist* [13] wrote: "And the banks have been helped by some extraordinary manipulation of the RUR exchange rate. Having fallen by more than two-thirds in the first three weeks after mid-August, from six to 22 RURs to the USD, it rocketed to 7.5 on September 15 only to crash to 13.5 the next day. The explanation for this lies in $2 billion or so outstanding in non-deliverable forward contracts, taken out by investors in short-term RUR debt in order to hedge their currency risk, which came due on no prizes for guessing September 15. Although no RURs actually change hands (hence the tag non-deliverable), the value of these contracts depends on the exchange rate. Russian banks would have been faced with huge losses at a rate of 22 RURs to the USD. They suffered almost none at the more favorable exchange rate which was so mysteriously but conveniently reached on Tuesday of this week."

against both credit risk and market risk? The work presented in this chapter suggests that one answer to this question might be due to adverse interactions between market and credit risk.

The rest of the chapter is structured as follows. The relevant literature on the size of the interrisk diversification between market and credit risk is reviewed in section 3.2. Sections 3.3 and 3.4 describe the portfolio, the valuation model, and the data. In section 3.5 we give the main results. Section 3.6 concludes the chapter.

3.2 RELATED RESEARCH

The integration of market and credit risk has been a subject of intense research over the past decade. Some of the most important topics are the introduction of market risk factors in credit risk models, in particular the development of credit risk models with stochastic interest rates; the dependence between default frequencies, macroeconomic variables, and recovery rates; and modeling the joint distribution of market and credit risk factors at a common time horizon.

However, there are relatively few papers quantifying the risk effects of simultaneous moves of market and credit risk factors. Duffie and Singleton [6, chap. 13], reporting on Duffie and Pan [5], compared value at risk (VaR) numbers in the absence of credit risk to VaR numbers when default intensities are correlated weakly or strongly to some market event. This comparison was done for a loan portfolio and an option portfolio. For the loan portfolio, they found that VaR numbers in the case of high correlation are roughly five times higher than in the case of low correlation, which in turn are 12 times higher than risk numbers if no credit risk at all is present. For the option portfolio, VaR numbers in the cases of high and low correlation are very similar to the numbers where no credit risk is present. Duffie and Pan [5] compared pure market risk (in the absence of credit risk) with integrated risk and found—for the loan portfolio—that integrated risk is higher. In contrast, this chapter compares integrated risk with the *sum* of pure market risk and pure credit risk.

Dimakos and Aas [4] decomposed the joint distribution of market, credit, and operational risk factors into a set of conditional probabilities and required conditional independence to write total risk as a sum of conditional marginals plus unconditional credit risk. They found that integrated risk is 10–20% smaller than the sum of individual risks, depending on the quantile.

Kuritzkes et al. [8] assumed joint normality of risks and arrived at closed-form solutions. According to their results, the integrated risk is about 15% smaller than the sum of individual risks of typical banks.

Walder [14] used a framework based on the mark-to-future approach [7] of algorithmics to analyze the contribution of market and credit risk to portfolio risk. He found that the integration of market and credit risk makes interrisk diversification benefits possible. Walder says his result is valid for every portfolio type analyzed, but portfolios with an equilibrated exposure to market and credit risk have the highest potential for integration benefits. Consequently, according to Walder [14, p. 33], the determination of capital requirements by adding market and credit risk overestimates true integrated risk.

Rosenberg and Schuermann [12] modeled the joint risk distribution of market, credit, and operational risk factors using the method of copulas. On the basis of regulatory reports, they designed their portfolio to resemble large, internationally active banks. For these banks, they explore the impact of business mix and interrisk correlations on total risk, whether measured by VaR or expected shortfall. For the portfolios and models Rosenberg and Schuermann consider, total integrated risk is 40–60% smaller than the sum of market and credit risk (see section 6 in Rosenberg and Schuermann [12]).

All the studies mentioned above find integrated risk to be *smaller* than the sum of credit and market risk. While these results are highly plausible, they are not universally valid. In particular, all of the above references [4–6, 8, 12, 14] restrict attention to portfolios without short positions. Short positions are essential for the hedged portfolios we consider in this chapter. We point to a portfolio where integrated risk is much *larger* than the sum of credit and market risk. This negative interrisk diversification effect seems to occur only for portfolios well hedged against market risk.

3.3 THE MODEL

In this chapter, we analyze integrated market and credit risk of an FX forward portfolio. We start from a valuation function specifying the value of the portfolio as a function of various market and credit risk factors. As credit risk factors, we take default probabilities (PDs). Alternatively, any specific credit risk model could be used to determine the PDs from other risk factors. To arrive at a parsimonious model, we do not model recovery rates stochastically but assume them to be constant.

The next step is to model the joint distribution of market risk factors and PD by determining the marginals and the copula. The marginal distributions are modeled with an AR (1) term (autoregressive), a GARCH (1,1) term (generalized autoregressive conditional hateroscedastic), and a residual distribution with historic returns as 80% body and Pareto-fitted tails, as in

McNeil et al. [10]. The details of the model for the marginals are described by Breuer et al. [3]. The model was chosen on the basis of out-of-sample density forecast tests, which are also described by Breuer et al. [3]. For the copula we chose the student copula.

We measure risk by expected shortfall (ES) rather than by value at risk (VaR). The main reason for this is that VaR is not subadditive. Therefore it might occur that VaR for some portfolio is larger than the sum of VaR numbers of its subportfolios. This can happen if the risk factor changes are not distributed elliptically, or if the portfolio value is not a linear function of the risk factors. To exclude nonsubadditivity of the risk measure as a possible explanation for the negative interrisk diversification effect, we use ES as a risk measure. The literature provides several definitions of ES or of related concepts, such as worst conditional expectation, tail conditional expectation, conditional value at risk (CVaR), and tail mean. We use the definition of Acerbi and Tasche [1, Def. 2.6]:

$$ES_\alpha := -\alpha^{-1}(E[X\mathbf{1}_{\{X<x_\alpha\}}] + x_\alpha(\alpha - P[X \leq x_\alpha])), \tag{3.3.1}$$

where x_α is the α-quantile of the distribution X. Expected shortfall is a coherent risk measure [1, Prop. 3.1].

Our goal is to analyze the difference between summing up separate risk numbers for market and credit risk as opposed to the risk number for integrated market and credit risk. Summing up separate risk numbers for credit and market risks will be done for a bank with independent credit and market risk-management units, each modeling the distributions of their respective risk factors, generating scenarios, and calculating risk measures independently. These results are presented in Table 3.5.1. We calculated the expected shortfall of the portfolio from three profit–loss distributions:

Distribution (0) reflects the portfolio value changes due to pure market risk, assuming default probabilities to be zero. Because the benchmark portfolio is perfectly hedged to market risk, ES is equal to zero.

Distribution (1) reflects the profit and losses due to pure credit risk, with market risk factors assumed constant.

Distribution (2) reflects the profits and losses from joint moves in the market risk factor and in the default frequency.

The results in Table 3.5.1 display the ES numbers for distributions (0), (1), and (2) along with the 95% confidence intervals for the ES numbers. Confidence intervals were calculated using the method of Manistre and Hancock [9].

The last column gives the interrisk diversification effect indicator

$$I = \frac{ES(\text{integrated})}{ES(\text{market}) + ES(\text{credit})}.$$

If I is greater than 1, integrated risk is higher than the sum of pure market and pure credit risk, which amounts to the negative interrisk diversification-effect. If it is smaller than 1, integrated risk is smaller than the sum of separate risks.

We use the term "negative interrisk diversification" between credit risk and market risk to denote the fact that integrated risk is greater than the sum of credit risk and market risk. This use of the term "diversification" is nonstandard because the effect is not about adding up risks of different portfolios, but about "adding up" different kinds of risk for the same portfolio.

3.4 PORTFOLIO AND DATA

To understand the effect hitting banks during the Russian crisis, we consider a similar portfolio of FX forwards today. Instead of RUR/USD forwards, we look at EUR/USD forwards. Although the moves in the USD/EUR exchange rate are much less dramatic than the 70% plunge in the USD/RUR rate of August 1998, the effects of the interaction of market and credit risk will still be astonishing.

The portfolio consists of 1000 long and 1000 short USD/EUR forwards, each with a different counterparty. The holder agrees with long counterparties to buy 1 dollar for K euros in 3 months, and the holder agrees with the short counterparties to sell 1 dollar for K euros at the same date. (Actually there is a small spread between the buy and sell prices. It is this spread that makes the portfolio profitable, independent of the future exchange rate, as long as no counterparty defaults. But we do not model this spread here because it is irrelevant for risk-management purposes.) At the time of the agreement, the strike K is set to

$$K = S_0[(100 + r_{\text{EUR}})/(100 + r_{\text{USD}})]^{1/4},$$

where S_0 is the USD/EUR spot rate, r_{USD} is the USD 3-month interest rate, and r_{USD} is the EUR 3-month interest rate, all at the time of the agreement. With this strike K, the value of all the forwards is zero at the time of the agreement.

At maturity, after 3 months, the USD/EUR spot rate will have moved to a new value S, a certain number d_1 of counterparties with long contracts will have defaulted, and a number d_2 of short counterparties will have defaulted.

Then the value of the $1000 - d_1$ long contracts with nondefaulted counterparties will be $(1000 - d_1)(S - K)$. The value of the $1000 - d_2$ short contracts with nondefaulted counterparties will be $-(1000 - d_2)(S - K)$. We assume that the value of long contracts with defaulted counterparties is (a) zero if $S > K$, as the counterparty cannot pay, and (b) $K - S$ if $S < K$, as we have to pay our dues to the counterparty, even if it has defaulted. Thus the contracts with the defaulted d_1 counterparties with long contracts are worth $d_1 \min(S - K, 0)$. Similarly, the value of the contracts with the defaulted d_2 counterparties with the short contracts is $d_2 \min(K - S, 0)$, which equals $-d_2 \max(S - K, 0)$. In total, the value of the portfolio at maturity is

$$(d_2 - d_1)(S - K) + d_1 \min(S - K, 0) - d_2 \max(S - K, 0).$$

We make the calculation as of December 31, 2004. Contracts are due at the end of December 2004. Market data for the exchange rate and the interest rates in the period 1987–2004 were taken from Datastream. The *PD* data were computed from quarterly transition matrices provided by Standard & Poor's CreditPro Corporate Ratings (http://creditpro.sandp.com) starting Q1/1987 and ending Q4/2004.

3.5 RESULTS

Table 3.5.1 shows the ES numbers for market risk, credit risk, and integrated risk of the FX forward portfolio. Column (0) represents the pure market-risk perspective. The ES numbers arise from moves in the market-risk factor only, which in this case is the USD/EUR exchange rate. Default probabilities are assumed to be zero. Because the portfolio is perfectly hedged against exchange rate moves, ES numbers in the pure market-risk perspective are equal to zero. Column (1) shows the pure credit-risk perspective. The exchange rate S is assumed to be constant at the level S_0 that was in force at the time of the agreement of contracts. The default probabilities are stochastic, and the default numbers d_1, d_2 are Bernoulli distributed from 1000 draws, with success rate equal to the default probabilities. Pure credit-risk numbers are small but not zero, although the contracts have value zero at the time of agreement. This is because the strike K is not equal to S_0 due to the interest-rate differential between the USD and the EUR. Therefore the contracts have a small nonzero value if the exchange rate at maturity is S_0.

The main result of Table 3.5.1 is in the last column. Integrated risk is greater than the sum of separate risks by a factor of 209 to 385, depending on

TABLE 3.5.1 ES Numbers for Market Risk, Credit Risk, and Integrated Risk of the FX Forward Portfolio. Integrated Risk Exceeds the Sum of Credit and Market Risk by a Factor of 209 to 385, and the Effect Is Larger for More Extreme Quantiles α

ES	(0)		(1)		(2)		(2)/[(0) + (1)]	
	MR		CR		Integrated		Diversification	
α(%)	No CR		No MR		MR and CR		Effect I	
0.1	0	(0, 0)	18.72	(18.32, 19.12)	7 222.21	(6 939.90, 7 504.52)	385.73	(363.0, 409.6)
0.25	0	(0, 0)	15.20	(14.98, 15.42)	5 286.60	(5 152.21, 5 421.00)	347.71	(334.1, 361.9)
0.5	0	(0, 0)	12.93	(12.79, 13.07)	4 152.38	(4 075.40, 4 229.35)	321.17	(311.8, 330.7)
1	0	(0, 0)	10.97	(10.88, 11.05)	3 233.04	(3 189.05, 3 277.03)	294.82	(288.6, 301.2)
2.50	0	(0, 0)	8.75	(8.71, 8.80)	2 288.00	(2 267.09, 2 308.91)	261.36	(257.6, 265.1)
5	0	(0, 0)	7.34	(7.31, 7.37)	1 730.18	(1 718.27, 1 742.10)	235.71	(233.1, 238.3)
10	0	(0, 0)	6.09	(6.08, 6.11)	1 274.27	(1 267.45, 1 281.08)	209.08	(207.4. 210.7)

Note: Recovery rates are constant at 60%. The 95% confidence intervals are in parentheses.

the quantile α. The dramatic negative interrisk diversification effect of Table 3.5.1 cannot be due to a failure of subadditivity of our risk measure, because ES is subadditive. In this example, the sum of market and credit risk gives no indication at all about the size of integrated risk. It was integrated risk that hit banks in the Russian crisis.

3.6 CONCLUSIONS

The key contribution of this chapter is a description of negative interrisk diversification effects between credit and market risk. Specifically, integrated market and credit risk may be *larger* than the sum of market risk and credit risk. This phenomenon occurs for portfolios hedged almost perfectly against market risk. The result implies that measuring market and credit risk in an integrated way spots risks that are hidden to a simple addition of pure market- and credit-risk numbers.

Our integrated market- and credit-risk model is strongly simplified in at least two ways. First, defaults of different counterparties are assumed to be independent, although probabilities of default vary stochastically, one cause of which may be default correlation dyamics. A better model of dependent defaults is needed. Second, recovery rates are assumed to be constant. This neglects the dependence between recovery rates and default probabilities. It remains to be seen whether the negative interrisk-diversification effect also persists for such integrated models of market and credit risk. It also remains to be seen whether similar negative interrisk diversification effects occur not only for derivatives portfolios, but also for plain bond portfolios.

3.7 ACKNOWLEDGMENT

M. J. is supported by the Internationale Bodenseehochschule. We are grateful to Clemens Thym of Standard and Poor's for providing us with the transition matrices S&P's Credit Pro Corporate Ratings.

REFERENCES

1. Acerbi C., Tasche D.: On the coherence of expected shortfall. *Journal of Banking and Finance*, 26(7): 1487–1503, 2002.
2. Bank Austria: Annual Report for 1998. Bank Austria Aktiengesellschaft, Vienna, 1999.
3. Breuer T., Krenn G., Jandačka M.: Towards an integrated measurement of credit and market risk. Arbeitsberichte Prozess- und Produkt-Engineering:

Methoden 6, Fachhochschule Vorarlberg, 2005. Online at http://www.bis.org/bcbs/events/rtf05breuer.pdf.

4. Dimakos X. K., Aas K.: Integrated risk modelling. *Statistical Modelling*, 4: 265–277, 2004.
5. Duffie D., Pan J.: Analytical value-at-risk with jumps and credit risk. *Finance and Stochastics*, 5: 155–180, 2001.
6. Duffie D., Singleton K. J.: *Credit Risk: Pricing, Measurement, and Management*. Princeton University Press, Princeton, NJ, 2003.
7. Iscoe I., Kreinin A., Rosen D.: An integrated market and credit risk portfolio model. *Algo Research Quarterly*, 2(3): 21–38, 1999.
8. Kuritzkes A., Schuermann T., Weiner S. M.: Risk measurement, risk management, and capital adequacy of financial conglomerates, in R. Herring, R. E. Litan (eds.): *Brookings-Wharton Papers on Financial Services 2003*. Brookings Institution Press, Washington, DC.
9. Manistre B. J., Hancock G. H.: Variance of the CTE estimator. *North American Actuarial Journal*, 9(2): 129–156, 2005.
10. McNeil A. J., Frey R.: Estimation of tail related risk measures for heteroscedastic financial time series: an extreme value approach. *Journal of Empirical Finance*, 7: 271–300, 2000.
11. Profil: Ein durchaus vertretbares Produkt. *Profil*, September 28, 40–41, 1998.
12. Rosenberg J. V., Schuermann T.: A general approach to integrated risk management with skewed, fat-tailed risks. *Journal of Financial Economics*, 79(3): 569–614, 2006.
13. The Economist: The undead. *The Economist*, September 17, 1998.
14. Walder R.: Integrated Market and Credit Risk Management of Fixed Income Portfolios. Technical report 62, FAME research paper, 2002. Online at www.fame.ch/library/EN/RP62.pdf.

CHAPTER 4

Counterparty Risk Pricing under Correlation between Default and Interest Rates

Damiano Brigo and Andrea Pallavicini
Banca IMI, Milan, Italy

Contents

Abstract: We consider counterparty risk for interest rate payoffs in the presence of correlation between the default event and interest rates. The previous analysis of Brigo and Masetti (2006), assuming independence, is further extended to interest-rate payoffs different from simple swap portfolios. A stochastic intensity model is adopted for the default event. We find that correlation between interest rates and default has a relevant impact on the positive adjustment to be subtracted from the default-free price to take into account counterparty risk. We analyze the pattern of such impacts as product characteristics and tenor structures change through some relevant numerical examples. We find the counterparty risk adjustment to decrease with the correlation for receiver payoffs, while the analogous adjustment for payer payoffs increases. The impact of correlation decreases when the default probability increases.

Keywords: counterparty risk, interest-rate default correlation, risk neutral valuation, default risk, interest rate models, default intensity models

4.1 INTRODUCTION

In this chapter we consider counterparty risk for interest-rate payoffs in the presence of correlation between the default event and interest rates. In particular we analyze in detail counterparty-risk (or default-risk) interest-rate swaps (IRS), continuing the work of Sorensen and Bollier (1994) and of Brigo and Masetti (2006), where no correlation is taken into account. We also analyze option payoffs under counterparty risk. In general, the reason to introduce counterparty risk when evaluating a contract is linked to the fact that many financial contracts are traded over the counter, so that the credit quality of the counterparty can be relevant. This is particularly appropriate when thinking of the different defaults experienced by some important companies in recent years. Regulatory issues related to the IAS 39 framework also encourage the inclusion of counterparty risk into valuation.

We are looking at the problem from the viewpoint of a safe (default-free) counterparty entering a financial contract with another counterparty having a positive probability of defaulting before the final maturity. We formalize the general and reasonable fact that the value of a generic claim subject to counterparty risk is always smaller than the value of a similar claim having a null default probability, expressing the discrepancy in precise quantitative terms.

When evaluating default-risky assets, one has to introduce the default probabilities in the pricing models. We consider credit default swaps (CDS)

as liquid sources of market default probabilities. Different models can be used to calibrate CDS data and obtain default probabilities. In Brigo and Morini (2006), for example, *firm value models* (or structural models) are used, whereas in Brigo and Alfonsi (2005), a stochastic intensity model is used. In this work we adopt the second framework, as this lends itself more naturally to interact with interest rate modeling and allows for a very natural way to correlate the default event to interest rates.

In this chapter we find that counterparty risk has a relevant impact on the prices of products and that, in turn, correlation between interest rates and default has a relevant impact on the adjustment due to counterparty risk on an otherwise default-free interest-rate payout. We analyze the pattern of such impacts as product characteristics and tenor structures change through some relevant numerical examples and find stable and financially reasonable patterns.

In particular, we find the (positive) counterparty risk adjustment to be subtracted from the default free price to decrease with correlation for receiver payoffs. The analogous adjustment for payer payoffs increases with correlation. We analyze products such as standard swaps, swap portfolios, and European and Bermudan swaptions, mostly of the receiver type. We also consider CMS spread options and ratchets, which being based on interest rate spreads are out of our "payer/receiver" classification.

In general, our results confirm the counterparty risk adjustment to be relevant and the impact of correlation on counterparty risk to be relevant in turn. We comment on our findings in more detail in the conclusion in section 4.6.

The chapter is structured as follows: in section 4.2 we introduce the general counterparty risk formula result, both exact and with discrete default-monitoring approximations. In section 4.3 we introduce the stochastic interest rate and default (intensity) models, explaining in detail the calibration of the default model to counterparty CDS data and the way to induce correlation between interest rates and default intensity. In section 4.4 we hint at the Monte Carlo and discretization techniques used in the chapter, and in section 4.5 we introduce the list of products we are considering as counterparty risky interest-rate payoffs, explaining in which tables of the appendix we present the related outputs. In section 4.6 we comment on the results, explaining the main observed risk and correlation patterns across products and why they are financially significant. Finally, the appendix contains the detailed outputs. An earlier and reduced version of this chapter is in Brigo and Pallavicini (2006).

4.2 GENERAL VALUATION OF COUNTERPARTY RISK

We denote by τ the default time of the counterparty, and we assume the investor who is considering a transaction with the counterparty to be default free. We place ourselves in a probability space $(\Omega, \mathcal{G}, \mathcal{G}_t, \mathbb{Q})$. The filtration $(\mathcal{G}_t)_t$ models the flow of information of the whole market, including credit. $\mathbb{Q}$ is the risk-neutral measure. This space is endowed also with a right-continuous and complete subfiltration $\mathcal{F}_t$ representing all the observable market quantities but the default event. (Hence $\mathcal{F}_t \subseteq \mathcal{G}_t := \mathcal{F}_t \vee \mathcal{H}_t$ where $\mathcal{H}_t = \sigma(\{\tau \leq u\} : u \leq t)$ is the right-continuous filtration generated by the default event.) We set $\mathbb{E}_t(\cdot) := \mathbb{E}(\cdot|\mathcal{G}_t)$, the risk-neutral expectation leading to prices.

Let us call T the final maturity of the payoff we need to evaluate. If $\tau > T$ there is no default of the counterparty during the life of the product, and the counterparty has no problems in repaying the investors. On the contrary, if $\tau \leq T$ the counterparty cannot fulfill its obligations and the following happens. At τ the net present value (NPV) of the residual payoff until maturity is computed. If this NPV is negative (respectively positive) for the investor (defaulted counterparty), it is completely paid (received) by the investor (counterparty) itself. If the NPV is positive (negative) for the investor (counterparty), only a recovery fraction Rec of the NPV is exchanged.

Let us call $\Pi^D(t, T)$ (sometimes abbreviated as $\Pi^D(t)$) the discounted payoff of a generic defaultable claim at t, and let $\text{CashFlows}(u, s)$ be the net cash flows of the claim without default between time u and time s, discounted back at u, with all payoffs seen from the point of view of the "investor" (i.e., the company facing counterparty risk). Then we have $\text{NPV}(\tau) = \mathbb{E}_\tau\{\text{CashFlows}(\tau, T)\}$ and

$$\Pi^D(t) = \mathbf{1}_{\{\tau > T\}}\text{CashFlows}(t, T) + \mathbf{1}_{\{t < \tau \leq T\}}\left[\text{CashFlows}(t, \tau) + D(t, \tau)(\text{Rec}(\text{NPV}(\tau))^+ - (-\text{NPV}(\tau))^+)\right] \tag{4.2.1}$$

with $D(u, v)$ being the stochastic discount factor at time u for maturity v. This last expression is the general price of the payoff under counterparty risk. Indeed, if there is no early default, this expression reduces to risk neutral valuation of the payoff (first term on the right-hand side); in case of early default, the payments due before default occurs are received (second term), and then, if the residual net present value is positive, only a recovery of it is received (third term), whereas if it is negative it is paid in full (fourth term).

Calling $\Pi(t)$ the discounted payoff for an equivalent claim with a default-free counterparty, i.e., $\Pi(t) = \text{CashFlows}(t, T)$, it is possible to prove the following

PROPOSITION 4.2.1 (GENERAL COUNTERPARTY RISK PRICING FORMULA) *At valuation time t, and on $\{\tau > t\}$, the price of our payoff under counterparty risk is*

$$\mathbb{E}_t\{\Pi^D(t)\} = \mathbb{E}_t\{\Pi(t)\} - \underbrace{\mathrm{L_{GD}}\, \mathbb{E}_t\{\mathbf{1}_{\{t<\tau\leq T\}} D(t,\tau)\,(NPV(\tau))^+\}}_{\text{Positive counterparty-risk adjustment}} \tag{4.2.2}$$

where $\mathrm{L_{GD}} = 1 - \mathrm{R_{EC}}$ is the loss given default *and the recovery fraction $\mathrm{R_{EC}}$ is assumed to be deterministic.*

For a proof, see for example Brigo and Masetti (2006). It is clear that the value of a defaultable claim is the value of the corresponding default-free claim minus an option part, specifically a call option (with zero strike) on the residual NPV giving nonzero contribution only in scenarios where $\tau \leq T$. Counterparty risk adds an optionality level to the original payoff.

Notice finally that the previous formula can be approximated as follows. Take $t = 0$ for simplicity and write, on a discretization time grid, $T_0, T_1, \ldots, T_b = T$,

$$\begin{aligned}\mathbb{E}[\Pi^D(0,T_b)] &= \mathbb{E}[\Pi(0,T_b)] - \mathrm{L_{GD}} \textstyle\sum_{j=1}^{b} \mathbb{E}[\mathbf{1}_{\{T_{j-1}<\tau\leq T_j\}} D(0,\tau)(\mathbb{E}_\tau \Pi(\tau,T_b))^+] \\ &\approx \mathbb{E}[\Pi(0,T_b)] - \underbrace{\mathrm{L_{GD}} \sum_{j=1}^{b} \mathbb{E}[\mathbf{1}_{\{T_{j-1}<\tau\leq T_j\}} D(0,T_j)(\mathbb{E}_{T_j} \Pi(T_j,T_b))^+]}_{\text{approximated (positive) adjustment}}\end{aligned} \tag{4.2.3}$$

where the approximation consists in postponing the default time to the first T_i following τ. From this last expression, under independence between Π and τ, one can factor the outer expectation inside the summation in products of default probabilities times the option prices. This way we would not need a default model but only survival probabilities and an option model for the underlying market of Π. This is only possible, in our case, if the default/interest-rate correlation is zero. This is what led to earlier results on swaps with counterparty risk in Brigo and Masetti (2006). In this chapter we do not assume zero correlation, so that in general we need to compute the counterparty risk without factoring the expectations. To do so we need a default model, to be correlated with the basic interest-rate market.

4.3 MODELING ASSUMPTIONS

In this section we consider a model that is stochastic both in the interest rates (underlying market) and in the default intensity (counterparty). Joint stochasticity is needed to introduce correlation. The interest-rate sector is modeled according to a short-rate Gaussian shifted two-factor process (hereinafter G2++), and the default-intensity sector is modeled according to a square-root process (hereinafter CIR++). Details for both models can be found, for example, in Brigo and Mercurio (2001, 2006). The two models are coupled by correlating their Brownian shocks.

4.3.1 G2++ Interest Rate Model

We assume that the dynamics of the instantaneous-short-rate process under the risk-neutral measure is given by

$$r(t) = x(t) + z(t) + \varphi(t;\alpha), \quad r(0) = r_0, \tag{4.3.1}$$

where α is a set of parameters and the processes x and z are $\mathcal{F}_t$ adapted and satisfy

$$\begin{aligned} dx(t) &= -ax(t)dt + \sigma dZ_1(t), \quad x(0) = 0, \\ dz(t) &= -bz(t)dt + \eta dZ_2(t), \quad z(0) = 0, \end{aligned} \tag{4.3.2}$$

where (Z_1, Z_2) is a two-dimensional Brownian motion with instantaneous correlation $\rho_{1,2}$ described by

$$d\langle Z_1, Z_2\rangle_t = \rho_{1,2}dt,$$

where r_0, a, b, σ, η are positive constants, and where $-1 \leq \rho_{1,2} \leq 1$. These are the parameters entering φ, in that $\alpha = [r_0, a, b, \sigma, \eta, \rho_{1,2}]$. The function $\varphi(\cdot;\alpha)$ is deterministic and well defined in the time interval $[0, T^*]$, with T^* being a given time horizon, typically 10, 30, or 50 (years). In particular, $\varphi(0;\alpha) = r_0$. This function can be set to a value automatically calibrating the initial zero-coupon curve observed in the market. In our numerical tests we use the market inputs listed in Tables 4.6.1 and 4.6.2 corresponding to parameters α given by

$$a = 0.0558, \; b = 0.5493, \; \sigma = 0.0093, \; \eta = 0.0138, \; \rho_{1,2} = -0.7$$

4.3.2 CIR++ Stochastic Intensity Model

For the stochastic intensity model we set

$$\lambda_t = y_t + \psi(t;\beta), \quad t \geq 0, \tag{4.3.3}$$

where ψ is a deterministic function, depending on the parameter vector β (which includes y_0), that is integrable on closed intervals. The initial condition y_0 is one more parameter at our disposal; we are free to select its value as long as

$$\psi(0;\beta) = \lambda_0 - y_0.$$

We take y to be a Cox-Ingersoll-Ross (CIR) process (see for example Brigo and Mercurio [2001, 2006]):

$$dy_t = \kappa(\mu - y_t)dt + \nu\sqrt{y_t}dZ_3(t),$$

where the parameter vector is $\beta = (\kappa, \mu, \nu, y_0)$, with κ, μ, ν, y_0 being positive deterministic constants. As usual, Z is a standard Brownian motion process under the risk neutral measure, representing the stochastic shock in our dynamics. We assume the origin to be inaccessible, i.e.,

$$2\kappa\mu > \nu^2.$$

We will often use the integrated quantities

$$\Lambda(t) = \int_0^t \lambda_s ds, \quad Y(t) = \int_0^t y_s ds, \quad \text{and} \quad \Psi(t,\beta) = \int_0^t \psi(s,\beta)ds.$$

4.3.3 CIR++ Model: CDS Calibration

Assume that the intensity λ, and the cumulated intensity Λ, are independent of the short rate, r, and of interest rates in general. Because in our Cox process we set $\tau = \Lambda^{-1}(\xi)$, with ξ being exponential and independent of interest rates, in this zero-correlation case the default time τ and interest rate quantities $r, D(s,t), \ldots$ are independent. It follows that (approximated no-accrual receiver) CDS valuation becomes model independent and is given by the formula

$$\begin{aligned} \mathrm{CDS}_{a,b}(0,R) = R \sum_{i=a+1}^{b} P(0,T_i)\alpha_i \mathbb{Q}(\tau \geq T_i) \\ -\mathrm{L}_{\mathrm{GD}} \sum_{i=a+1}^{b} P(0,T_i)\mathbb{Q}(\tau \in [T_{i-1}, T_i]) \end{aligned} \qquad (4.3.4)$$

(see for example the "credit" chapters in Brigo and Mercurio [2006] for the details). Here R is the periodic premium rate (or "spread") received by the protection seller from the premium leg, until final maturity or until the first T_i following default, whereas $L_{GD} = 1 - R_{EC}$ is the loss given default protection payment to be paid to the protection buyer in the default (or protection) leg in case of early default, at the first T_i following default.

This formula implies that if we strip survival probabilities from CDS in a model-independent way, to calibrate the market CDS quotes we just need to make sure that the survival probabilities we strip from CDS are correctly reproduced by the CIR++ model. Because the survival probabilities in the CIR++ model are given by

$$\mathbb{Q}(\tau > t)_{model} = \mathbb{E}(e^{-\Lambda(t)}) = \mathbb{E}\exp(-\Psi(t,\beta) - Y(t)), \tag{4.3.5}$$

we just need to make sure that

$$\mathbb{E}\exp\left(-\Psi(t,\beta) - Y(t)\right) = \mathbb{Q}(\tau > t)_{market},$$

from which

$$\Psi(t,\beta) = \ln\left(\frac{\mathbb{E}(e^{-Y(t)})}{\mathbb{Q}(\tau > t)_{market}}\right) = \ln\left(\frac{P^{\text{CIR}}(0,t,y_0;\beta)}{\mathbb{Q}(\tau > t)_{market}}\right). \tag{4.3.6}$$

We choose the parameters β in order to have a positive function ψ (i.e., an increasing Ψ), and P^{CIR} is the closed form expression for bond prices in the time-homogeneous CIR model with initial condition y_0 and parameters β (see for example Brigo and Mercurio [2001, 2006]). Thus, if ψ is selected according to this last formula, as we will assume from now on, the model is easily and automatically calibrated to the market survival probabilities (possibly stripped from CDS data).

This CDS calibration procedure assumes zero correlation between default and interest rates, so in principle when taking nonzero correlation we cannot adopt it. However, we have seen in Brigo and Alfonsi (2005) and further in Brigo and Mercurio (2006) that the impact of interest-rate/default correlation is typically negligible on CDSs, so that we can retain this calibration procedure even under nonzero correlation, and we will do so in this chapter.

Once we have done this and calibrated CDS data through $\psi(\cdot,\beta)$, we are left with the parameters β, which can be used to calibrate further products. However, this will be interesting when single name option data on the credit derivatives market become more liquid. Currently the bid-ask spreads for single-name CDS options are large, suggesting the need to consider these quotes with caution. At the moment, we content ourselves to calibrate only

CDSs for the credit part. To help in specifying β without further data, we set some values of the parameters by implying possibly reasonable values for the implied volatility of hypothetical CDS options on the counterparty.

In our tests we take stylized flat CDS curves for the counterparty, assuming they imply initial survival probabilities at time zero consistent with the following hazard function formulation,

$$\mathbb{Q}(\tau > t)_{\text{market}} = \exp(-\gamma t),$$

for a constant deterministic value of γ. This is to be interpreted as a quoting mechanism for survival probabilities and not as a model. Assuming our counterparty CDSs at time zero for different maturities to imply a given value of γ, we will value counterparty risk under different values of γ. This assumption on CDS spreads is stylized, but our aim is checking impacts rather than having an extremely precise valuation.

In our numerical examples, the intensity volatility parameters are assigned the following values:

$$y_0 = 0.0165, \quad \kappa = 0.4, \quad \mu = 0.026, \quad \nu = 0.14.$$

Paired with stylized CDS data consistent with survivals $\mathbb{Q}(\tau > t)_{\text{market}} = \exp(-\gamma t)$ for several possible values of γ, these parameters imply the CDS volatilities[1] listed in Table 4.6.3.

4.3.4 Interest-Rate/Credit-Spread Correlation

We take the short interest-rate factors x and z and the intensity process y to be correlated, by assuming the driving Brownian motions Z_1, Z_2, and Z_3 to be instantaneously correlated according to

$$d\langle Z_i, Z_3\rangle_t = \rho_{i,3}dt, \quad i \in \{1, 2\}.$$

Notice that the instantaneous correlation between the resulting short rate and the intensity, i.e., the instantaneous interest-rate/credit-spread correlation is

$$\bar{\rho} = \text{Corr}(dr_t, d\lambda_t) = \frac{d\langle r, \lambda\rangle_t}{\sqrt{d\langle r, r\rangle_t \, d\langle \lambda, \lambda\rangle_t}} = \frac{\sigma\rho_{1,3} + \eta\rho_{2,3}}{\sqrt{\sigma^2 + \eta^2 + 2\sigma\eta\rho_{1,2}}}.$$

We find the limit values of -1, 0, and 1 according to Table 4.6.4.

[1] See Brigo (2005, 2006) for a precise notion of CDS implied volatility.

4.4 NUMERICAL METHODS

A Monte Carlo simulation is used to value all the payoffs considered in the present chapter. We adopt the following prescriptions to effectively implement the algorithm. The standard error of each Monte Carlo run is, at most, on the last digit of the numbers reported in the tables.

4.4.1 Discretization Scheme

Payoff present values can be calculated with the joint interest-rate and credit model by means of a Monte Carlo simulation of the three underlying variables x, z, and y, whose joint transition density is needed. The transition density for the G2++ model is known in closed form, but the CIR++ model requires a discretization scheme, leading to a three-dimensional Gaussian local discretization. For CIR++ we adopt a discretization with a weekly step and we find similar convergence results both with the *full truncation scheme* introduced by Lord, Koekkoek, and Van Dijk (2006) and with the *implied scheme* by Brigo and Alfonsi (2005). In the following, we adopt the former scheme.

4.4.2 Forward Expectations

The simulation algorithm allows the counterparty to default on the contract payment dates, unless the time interval between two payment dates is longer than 2 months. In such a case, additional checks on counterparty default are added to ensure that the gap between allowed default times is at most 2 months. The calculation of the forward expectation, required by counterparty risk evaluation, as given in equation (4.2.3) (inner expectation $\mathbb{E}_{T_j}$) is taken by approximating the expectation at the effective default time T_j with a polynomial series in the interest-rate model underlyings, x and z, valued at the first-allowed default time after τ, i.e., at T_j. The coefficients of the series expansion are calculated by means of a least-square regression, as is usually done to price Bermudan options, by means of the algorithm by Longstaff and Schwarz (2001).

4.4.3 Callable Payoffs

Counterparty risk for callable payoffs is calculated in two steps. First, given a riskless version of the payoff, the payoff exercise boundary is calculated by a Monte Carlo simulation with the Longstaff and Schwarz algorithm. Because the default time is unpredictable from the point of view of the interest-rate sector of the model, the same exercise boundary, as a function of the underlying processes at exercise date, is assumed to hold also for the default-risky payoff. Then the risky payoff along with the exercise boundary is treated as a standard European default-risky option, given that the continuation value at any relevant time is now a function of the underlying processes.

4.5 RESULTS AND DISCUSSION

We consider the pricing of different payoffs in the presence of counterparty risk for three different default probability scenarios (as expressed by hazard rates $\gamma = 3\%, 5\%$, and 7%) and for three different correlation scenarios ($\bar{\rho} = -1$, 0, and 1). For a detailed description of the payoffs, the reader is referred to Brigo and Mercurio (2006).

4.5.1 Single Interest Rate Swaps (IRS)

In the following, we consider payoffs depending on at-the-money fix-receiver forward interest-rate-swap (IRS) paying on the EUR market. These contracts reset a given number of years from trade date and start accruing 2 business days later. The IRS's fixed legs pay annually a 30E/360 strike rate, and the floating legs pay LIBOR twice per year. The first products we analyze are simple IRS of this kind. We list in Table 4.6.5 the counterparty risk adjustment for the 10-year IRS and the impact of correlation for different levels of default probabilities.

4.5.2 Netted Portfolios of IRS

After single IRS, we consider portfolios of at-the-money IRS either with different starting dates or with different maturities. In particular we focus on the following two portfolios:

1. ($\Pi 1$) given a set of annually spaced dates $\{T_i : i = 0, \ldots, N\}$, with T_0 at 2 business days from trade date, consider the portfolio of swaps maturing at each T_i, with $i > 0$, and all starting at T_0. The netting of the portfolio is equal to an amortizing swap with decreasing outstanding.
2. ($\Pi 2$) given the same set of annually spaced dates, consider the portfolio of swaps starting at each T_i, with $i < N$, and all maturing at T_N. The netting of the portfolio is equal to an amortizing swap with increasing outstanding.

We list in Table 4.6.5 the counterparty risk adjustment for both portfolios.

4.5.3 European Swaptions

We consider contracts giving the opportunity to enter a receiver IRS at an IRS's reset date. The strike rate in the swap to be entered is fixed at the at-the-money forward swap level observed at option inception, i.e., at trade date. We list in Table 4.6.6 the price of both the riskless and the risky contract. In Table 4.6.7, the same data are cast in terms of Black implied swaption volatility, i.e., we compute the Black swaption volatility that would match the counterparty risk-adjusted swaption price when put in a default-free Black formula for swaptions. In Table 4.6.8 we show an example with payer swaptions instead.

4.5.4 Bermudan Swaptions

We consider contracts giving the opportunity to enter a portfolio of IRS, as defined in section 4.5.2, every 2 business days before the starting of each accruing period of the swap's fix leg. We list in Table 4.6.9 the price of entering each portfolio, risky and riskless, along with the price of entering, at the same exercise dates, the contained IRS of longest tenor.

4.5.5 CMS Spread Options

We consider a contract[2] on the EUR market starting within 2 business days that pays, quarterly on an ACT/360 basis and up to maturity t_M, the following exotic index:

$$(L(S_a(t_i) - S_b(t_i)) - K)^+$$

where L and K are positive constants, and $S_k(t_i)$, with $k \in \{a, b\}$ and $i = 0, \ldots, M$, is the constant maturity swap rate (hereinafter CMS) fixing 2 business days before each accruing period starting date t_i, i.e., the at-the-money rate for an IRS with tenor of k years fixing at t_i. We list in Table 4.6.10 the option prices, both default risky and riskless.

4.5.6 CMS Spread Ratchets

We consider a contract on the EUR market starting within 2 business days that pays, quarterly on an ACT/360 basis and up to maturity t_M, the following exotic index:

$$L(I(t_i) - K(t_i))^+$$

[2] See also Mercurio and Pallavicini (2005) for a detailed discussion of CMS spread option pricing.

where

$$I(t_i) := S_a(t_i) - S_b(t_i) - C$$

$$K(t_{i+1}) := \min(K(t_i) + \Delta, \max(K(t_i) - \Delta, I(t_i)))$$

where L, Δ, and C are positive constants; the initial strike value is $K(t_0) = 0$; and $S_k(t_i)$, with $k \in \{a, b\}$ and $i = 0, \ldots, M$, is the CMS rate fixing 2 business days before each accruing period starting date t_i. We list in Table 4.6.11 the option prices, both default risky and riskless.

4.6 RESULTS INTERPRETATION AND CONCLUSIONS

In this chapter we have found that counterparty risk has a relevant impact on interest-rate payoff prices and that, in turn, correlation between interest rates and default (intensity) has a relevant impact on the adjustment due to counterparty risk. We have analyzed the pattern of such impacts as changes in product characteristics and tenor structures through some fundamental numerical examples, and we have found stable and reasonable behaviors. In particular, the (positive) counterparty risk adjustment to be subtracted from the default-free price decreases with correlation for receiver payoffs (IRS, IRS portfolios, European and Bermudan swaptions). This is to be expected. If default intensities increase with high positive correlation, their correlated interest rates will increase more than with low correlation. When interest rates increase, a receiver swaption value decreases; thus we see that, ceteris paribus, a higher correlation implies a lower value for the swaptions impacting the adjustment, so that with higher correlation the adjustment absolute value decreases. In contrast, the analogous adjustment for payer payoffs increases with correlation, as is to be expected.

Our results, including the CMS spread options and ratchets, generally confirm the relevance of counterparty risk adjustment as well as the impact of correlation on counterparty risk. We have found the following further stylized facts, which hold throughout all payoffs. As the default probability implied by the counterparty CDS increases, the size of the adjustment due to counterparty risk increases as well, but the impact of correlation on it decreases. This is financially reasonable: given large default probabilities for the counterparty, fine details on the dynamics, such as the correlation with interest rates, become less relevant, as everything is being wiped out by massive defaults anyway. On the contrary, with smaller default probabilities, the fine structure of the dynamics, and of correlation in particular, becomes more important.

The conclusion is that we should take into account interest-rate/default correlation in valuing counterparty-risky interest-rate payoffs, especially when the default probabilities are not extremely high.

APPENDIX: DETAILED OUTPUT TABLES

TABLE 4.6.1 EUR Zero-Coupon Continuously Compounded Spot Rates (ACT/360) Observed on June 23, 2006

Date	Rate	Date	Rate	Date	Rate	Date	Rate
26-Jun-06	2.83%	20-Sep-07	3.46%	27-Jun-16	4.19%	27-Jun-28	4.51%
27-Jun-06	2.83%	19-Dec-07	3.52%	27-Jun-17	4.23%	27-Jun-29	4.51%
28-Jun-06	2.83%	19-Mar-08	3.57%	27-Jun-18	4.27%	27-Jun-30	4.52%
04-Jul-06	2.87%	19-Jun-08	3.61%	27-Jun-19	4.31%	27-Jun-31	4.52%
11-Jul-06	2.87%	18-Sep-08	3.65%	29-Jun-20	4.35%	28-Jun-32	4.52%
18-Jul-06	2.87%	29-Jun-09	3.75%	28-Jun-21	4.38%	27-Jun-33	4.52%
27-Jul-06	2.88%	28-Jun-10	3.84%	27-Jun-22	4.41%	27-Jun-34	4.52%
28-Aug-06	2.92%	27-Jun-11	3.91%	27-Jun-23	4.43%	27-Jun-35	4.52%
20-Sep-06	2.96%	27-Jun-12	3.98%	27-Jun-24	4.45%	27-Jun-36	4.52%
20-Dec-06	3.14%	27-Jun-13	4.03%	27-Jun-25	4.47%	27-Jun-46	4.49%
20-Mar-07	3.27%	27-Jun-14	4.09%	29-Jun-26	4.48%	27-Jun-56	4.46%
21-Jun-07	3.38%	29-Jun-15	4.14%	28-Jun-27	4.50%		

TABLE 4.6.2 Market at-the-Money Swaption Volatilities Observed on June 23, 2006

	Tenor						
Expiry	1y	2y	5y	7y	10y	15y	20y
1y	17.51%	15.86%	14.63%	14.20%	13.41%	12.14%	11.16%
2y	16.05%	15.26%	14.55%	14.09%	13.29%	12.03%	11.09%
3y	15.58%	15.06%	14.43%	13.92%	13.10%	11.87%	10.96%
4y	15.29%	14.90%	14.20%	13.67%	12.85%	11.66%	10.79%
5y	15.05%	14.67%	13.90%	13.36%	12.55%	11.42%	10.60%
7y	14.39%	14.00%	13.22%	12.70%	11.96%	10.95%	10.20%
10y	13.25%	12.94%	12.23%	11.79%	11.17%	10.31%	9.65%
15y	11.87%	11.64%	11.11%	10.76%	10.26%	9.52%	8.89%
20y	11.09%	10.92%	10.45%	10.14%	9.67%	8.91%	8.27%

TABLE 4.6.3 Black Volatilities for CDS Options Implied by CIR++ Model (with Parameters $y_0 = 0.0165$, $\kappa = 0.4$, $\mu = 0.026$, $\nu = 0.14$) for Different Choices of the Default-Probability Parameter γ. Interest Rates Are Modeled According to Section 4.3.1 and $\bar{\rho} = 0$

	σ_{impl}			
γ	1×1	1×4	4×1	1×9
3%	42%	25%	26%	15%
5%	25%	15%	15%	9%
7%	18%	11%	11%	7%

TABLE 4.6.4 Values of Model Instantaneous Correlations $\rho_{1,3}$ and $\rho_{2,3}$ Ensuring Special Interest-Rate/Credit-Spread Instantaneous Correlations for the Chosen Interest-Rate and Intensity Dynamics Parameters

$\bar{\rho}$	$\rho_{1,3}$	$\rho_{2,3}$
−1	4.05%	−74.19%
0	0	0
1	−4.05%	74.19%

TABLE 4.6.5 Counterparty Risk Price for Receiver IRS Portfolio Defined in Section 4.5.2 for a Maturity of 10 Years, along with the Counterparty Risk Price for a 10 Year Swap. Every IRS, Constituting the Portfolios, Has Unitary Notional. Prices Are in Basis Points

γ	$\bar{\rho}$	Π1	Π2	IRS
3%	−1	−140	−294	−36
	0	−84	−190	−22
	1	−47	−115	−13
5%	−1	−181	−377	−46
	0	−132	−290	−34
	1	−99	−227	−26
7%	−1	−218	−447	−54
	0	−173	−369	−44
	1	−143	−316	−37

TABLE 4.6.6 Counterparty Risk Price for European Receiver Swaptions Defined in Section 4.5.3 for Different Expiries and Tenors. Riskless Prices Are Listed Too. Contracts Have Unitary Notional. Prices Are in Basis Points

γ	$\bar{\rho}$	1 × 5	5 × 5	10 × 5	20 × 5
3%	−1	−14	−37	−53	−56
	0	−9	−27	−42	−48
	1	−6	−19	−34	−41
5%	−1	−19	−50	−71	−70
	0	−14	−41	−61	−65
	1	−11	−35	−55	−61
7%	−1	−23	−61	−84	−79
	0	−19	−53	−77	−75
	1	−16	−47	−72	−73
riskless		106	205	215	157
γ	$\bar{\rho}$	1 × 10	5 × 10	10 × 10	20 × 10
3%	−1	−38	−78	−98	−98
	0	−25	−56	−78	−83
	1	−16	−43	−64	−72
riskless		184	342	353	256
γ	$\bar{\rho}$	1 × 20	5 × 20	10 × 20	20 × 20
3%	−1	−87	−140	−160	−150
	0	−61	−107	−129	−131
	1	−45	−83	−107	−114
riskless		261	474	486	354

TABLE 4.6.7 Counterparty Risk Implied Volatilities for European Receiver Swaptions Defined in Section 4.5.3 for Different Expiries and Tenors. Riskless Implied Volatilities Are Listed Too. Contracts Have a Unitary Notional

γ	$\bar{\rho}$	1 × 5	5 × 5	10 × 5	20 × 5
3%	−1	−1.96%	−2.52%	−3.06%	−3.74%
	0	−1.26%	−1.82%	−2.38%	−3.20%
	1	−0.77%	−1.32%	−1.93%	−2.78%
5%	−1	−2.60%	−3.40%	−4.06%	−4.71%
	0	−1.96%	−2.78%	−3.51%	−4.37%
	1	−1.54%	−2.35%	−3.16%	−4.09%
7%	−1	−3.19%	−4.14%	−4.81%	−5.32%
	0	−2.62%	−3.60%	−4.39%	−5.06%
	1	−2.22%	−3.23%	−4.11%	−4.89%
riskless		14.63%	13.90%	12.23%	10.45%
γ	$\bar{\rho}$	1 × 10	5 × 10	10 × 10	20 × 10
3%	−1	−2.74%	−2.86%	−3.14%	−3.72%
	0	−1.84%	−2.08%	−2.50%	−3.17%
	1	−1.19%	−1.59%	−2.03%	−2.75%
riskless		13.41%	12.55%	11.17%	9.67%
γ	$\bar{\rho}$	1 × 20	5 × 20	10 × 20	20 × 20
3%	−1	−3.71%	−3.14%	−3.19%	−3.53%
	0	−2.63%	−2.40%	−2.57%	−3.09%
	1	−1.95%	−1.87%	−2.14%	−2.68%
riskless		11.16%	10.60%	9.65%	8.27%

TABLE 4.6.8 Counterparty Risk Price for European Payer Swaptions Defined in Section 4.5.3 for Different Expiries and Tenors. Riskless Prices Are Listed Too. Contracts Have Unitary Notional. Prices Are in Basis Points

γ	$\bar{\rho}$	1 × 5	5 × 5	10 × 5	20 × 5
3%	−1	−6	−20	−33	−40
	0	−10	−28	−44	−50
	1	−16	−39	−56	−58
riskless		106	205	215	157

TABLE 4.6.9 Counterparty Risk Price for Callable Receiver IRS Portfolio Defined in Section 4.5.4 for a Maturity of 10 years, along with the Counterparty Risk Price for a Spot-Starting 10 Year Bermuda Swaption. Riskless Prices Are Listed Too. Every IRS, Constituting the Portfolios, Has Unitary Notional. Prices Are in Basis Points

γ	$\bar{\rho}$	Π1	Π2	IRS
3%	−1	−197	−387	−47
	0	−140	−289	−34
	1	−101	−219	−25
5%	−1	−272	−528	−65
	0	−223	−446	−54
	1	−188	−387	−46
7%	−1	−340	−652	−80
	0	−295	−578	−70
	1	−266	−529	−63
	riskless	1083	1917	240

TABLE 4.6.10 Counterparty Risk Price for CMS Spread Options Defined in Section 4.5.5 with $L = 15$, $K = 15\%$, $a = 10y$, $b = 2y$, and Three Different Maturities $t_M \in \{5y, 10y, 15y\}$. Riskless Prices Are Listed Too. Prices Are in Basis Points

γ	$\bar{\rho}$	5y	10y	20y
3%	−1	−5	−16	−34
	0	−4	−11	−24
	1	−2	−8	−18
5%	−1	−7	−22	−44
	0	−6	−17	−37
	1	−5	−15	−31
7%	−1	−9	−26	−52
	0	−7	−23	−46
	1	−6	−20	−42
	riskless	58	122	182

TABLE 4.6.11 Counterparty Risk Price for CMS Spread Ratchets Defined in Section 4.5.6 with $L = 15$, $\Delta = 1\%$, $C = 1\%$, $a = 10y$, $b = 2y$, and Three Different Maturities $t_M \in \{5y, 10y, 15y\}$. Riskless Prices Are Listed Too. Prices Are in Basis Points

γ	$\bar{\rho}$	5y	10y	20y
3%	−1	−27	−86	−232
	0	−27	−88	−239
	1	−27	−90	−246
	riskless	555	1049	1748

REFERENCES

Brigo, D. (2005). Market Models for CDS Options and Callable Floaters. *Risk*, January. Also in: *Derivatives Trading and Option Pricing*, ed. Dunbar, N. Risk Books, London, 89–94.

Brigo, D. (2006). Constant Maturity Credit Default Swap Valuation with Market Models. *Risk*, June, 78–83.

Brigo, D., and Alfonsi, A. (2005). Credit Default Swaps Calibration and Derivatives Pricing with the SSRD Stochastic Intensity Model. *Finance and Stochastic*, 9 (1).

Brigo, D., and Masetti, M. (2006). Risk Neutral Pricing of Counterparty Risk. In *Counterparty Credit Risk Modeling: Risk Management, Pricing and Regulation*, ed. Pykhtin, M. Risk Books, London.

Brigo, D., and Mercurio, F. (2001). *Interest Rate Models: Theory and Practice*. Springer-Verlag, Heidelberg.

Brigo, D., and Mercurio, F. (2006). *Interest Rate Models: Theory and Practice — with Smile, Inflation and Credit*, 2nd ed. Springer-Verlag, Heidelberg.

Brigo, D., and Morini, M. (2006). Structural Credit Calibration. *Risk*, April, 78–83.

Brigo, D., and Pallavicini, A. (2006). Counterparty Risk Valuation under Correlation between Interest Rates and Default. Working paper, available at SSRN.com.

Cherubini, U. (2005). Counterparty Risk in Derivatives and Collateral Policies: The Replicating Portfolio Approach. In: *Proceedings of the Counterparty Credit Risk 2005 C.R.E.D.I.T. Conference*. Venice, Sept. 22–23, Vol. 1.

Longstaff, F.A., and Schwarz, E.S. (2001). Valuing American Options by Simulation: A Simple Least-Squares Approach. *Review of Financial Studies*, 14, 113–147.

Lord, R., Koekkoek, R., and Van Dijk, D.J.C. (2006). A Comparison of Biased Simulation Schemes for Stochastic Volatility Models. Working paper. Located at http://ssrn.com/abstract=903116.

Mercurio, F., and Pallavicini, A. (2005). Mixing Gaussian Models to Price CMS Derivatives. Working paper, available at: http://ssrn.com/abstract=872708.

Sorensen, E.H., and Bollier, T.F. (1994). Pricing Swap Default Risk. *Financial Analysts Journal*, 50, 23–33.

CHAPTER 5

Optimal Dynamic Asset Allocation for Defined Contribution Pension Plans

Andrew J.G. Cairns
Maxwell Institute, Edinburgh, and Heriot-Watt University, Edinburgh, United Kingdom

David Blake
City University, London, United Kingdom

Kevin Dowd
Nottingham University Business School, Nottingham, United Kingdom

Contents

5.1 SUMMARY OF CAIRNS, BLAKE, AND DOWD

In this short chapter we summarize some of the results of Cairns, Blake, and Dowd (2006). The chapter considers asset-allocation strategies that might be adopted by members of defined-contribution pension plans. The underlying model incorporates asset, salary, and interest-rate risk. We assume that the member measures utility in terms of the replacement ratio at the time of retirement: the ratio of pension to final salary. The plan member's objective is to maximize their expected terminal utility.

In the general model we have the state variables:

$Y(t)$ = salary or labor income
$W(t)$ = accumulated pension wealth
$r(t)$ = risk-free interest rate (one-factor model)

For the assets and salary risk we have $N+1$ sources of risk: $Z_0(t),\ Z_1(t), \ldots,$ $Z_N(t)$. Within this we have a cash account, $R_0(t)$:

$$dR_0(t) = r(t)R_0(t)dt$$
$$\text{where } dr(t) = \mu_r(r(t))dt + \sum_{j=1}^{N} \sigma_{rj}(r(t))dZ_j(t).$$

For the risky assets, $R_1(t), \ldots, R_N(t)$ we have:

$$dR_i(t) = R_i(t)\left[\left(r(t) + \sum_{j=1}^{N} \sigma_{ij}\xi_j\right) dt + \sum_{j=1}^{N} \sigma_{ij} dZ_j(t)\right]$$

where $C = (\sigma_{ij})$ = nonsingular volatility matrix $(N \times N), \xi = (\xi_j)$ = market prices of risk $(N \times 1)$.

For the salary model we have:

$$dY(t) = Y(t)\left[(r(t) + \mu_Y(t))\,dt + \sum_{j=1}^{N} \sigma_{Yj} dZ_j(t) + \sigma_{Y0} dZ_0(t)\right]$$

where $\mu_Y(t)$ is deterministic. Finally, for the plan member's pension wealth, $W(t)$, we have:

$$p(t) = (p_1(t), \ldots, p_N(t)) = \text{proportion of wealth in risky assets}$$

$$dW(t) = W(t)[(r(t) + p(t)'C\xi)dt + p(t)'CdZ(t)] + \pi Y(t)dt.$$

The form of the terminal utility is $U(W(T), Y(T), r(T)) = f(W(T)/[a(r(T))\, Y(T)])$ where $a(r(T))$ is the price at T for a level annuity payable from T, which depends on interest rates at T. For example, $f(\cdot)$ might be a power function. This form means that we are able to replace the state variables $W(t)$ and $Y(t)$ with $X(t) = W(t)/Y(t)$.

If $\sigma_{Y0} \neq 0$, then we have nonhedgeable salary risk, and its presence means that the plan member's wealth cannot be allowed to become negative: that is, she cannot borrow against future contributions. From the computational

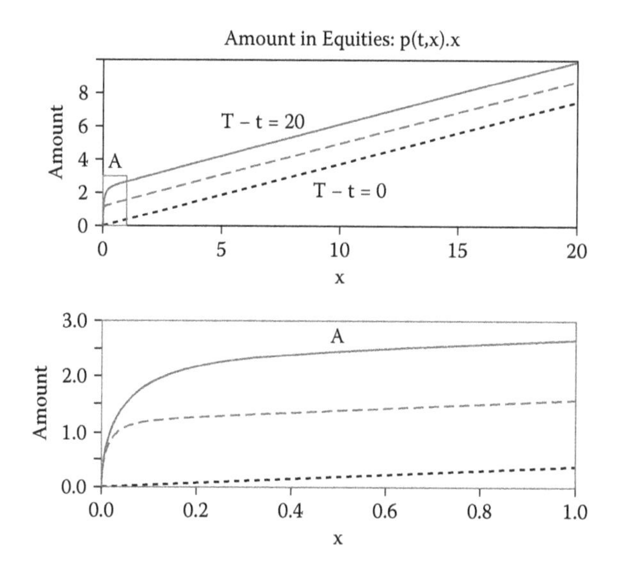

FIGURE 5.1.1 Optimal amount in equities for different times to retirement.

point of view, this results in a singularity on the optimal value function, $V(t,x,r)$, at $x = 0$. We develop a numerical solution to this by transforming x to $\log x$. By way of illustration, the optimal strategy in the case of one risky asset and $r(t) = r$ is presented in Figure 5.1.1.

In this figure we can see how the amount in equities, for a given value of x, increases the further she is from retirement. For smaller x, less is invested in equities, and the amount in equities tends to zero as x tends to zero. This final observation is a necessary condition for the wealth process to remain positive. Closer investigation of the numerical solutions suggests that the amount invested in equities is either $O(\sqrt{x})$ or $o(\sqrt{x})$. The latter would guarantee that the process never hits 0, whereas the former would need additional conditions on the parameter values. Further work remains to determine which rate of convergence we have.

By completing the market with a fictitious asset, we can construct an upper bound for the optimal value function. In general terms, this upper bound is close to our numerical solution for $V(t,x)$. However, it fails to enlighten us on the nature of the solution as $x \to 0$.

REFERENCES

Cairns, A.J.G., Blake, D., and Dowd, K. (2006) "Stochastic Lifestyling: Optimal Dynamic Asset Allocation for Defined Contribution Pension Plans," *Journal of Economic Dynamics and Control*, 30: 843–877.

CHAPTER 6

On High-Performance Software Development for the Numerical Simulation of Life Insurance Policies

S. Corsaro, P.L. De Angelis, Z. Marino, and F. Perla
University of Naples Parthenope, Naples, Italy

Contents

Abstract: In this work we focus on the numerical issues involved in evaluating an important class of financial derivatives: participating life insurance contracts. We investigate the impact of different numerical methods on accuracy and efficiency in the solution of main computational kernels generally arising from mathematical models describing the financial problem. The main kernels involved in the evaluation of these financial derivatives are multidimensional integrals and stochastic differential equations. For this reason we consider different Monte Carlo simulations and various stochastic-differential-equation discretization schemes. We have established that a combination of the Monte Carlo method with the antithetic variates (AV) variance-reduction technique and the fully implicit Euler scheme developed by Brigo and Alfonsi (2005) provides high efficiency and good accuracy.

Keywords: life insurance policies, multidimensional integrals, stochastic differential equations

Mathematics Subject: 91B30, 65C20, 65C05 **JEL**: C15, C63, C88

6.1 INTRODUCTION

This work was carried out within a project focused on the numerical implementation of efficient and accurate methods for financial evaluation of participating life insurance policies. Some preliminary results concerning previous work by authors on this subject are shown in Corsaro et al. (2005). Our aim in this framework is the analysis of numerical methods and algorithms for the solution of the main computational kernels in the mathematical models describing the problem, i.e., the evaluation of multidimensional integrals and the solution of stochastic differential equations (SDEs). The former can represent expected values; the latter often models diffusion processes describing the time evolution of interest-rate risk à la Cox, Ingersoll, and Ross (1985) (CIR) as well as stock index à la Black and Scholes (1973). We discuss the development of algorithms and software based on different numerical schemes, and investigate their impact on accuracy and efficiency in the solution.

High-dimensional integrals are usually solved via the Monte Carlo (MC) method. It becomes particularly attractive with respect to deterministic integration methods when the dimension is large, as its convergence rate does not decrease dramatically as dimension increases. Evaluating financial derivatives in many cases reduces to computing expectations that can be written as integrals of large dimension. In this setting, Monte Carlo proves very promising.

On the other hand, the MC method suffers from the disadvantage that the rate of convergence is quite low: this motivated the search for methods with faster convergence. The expected error of the classical MC method depends on the variance of the integrand; therefore, it decreases if the variance is reduced. Different variance reduction techniques, to be used in combination with the MC algorithm, are well known in the literature (see, for example, Boyle, Broadie, and Glasserman [1998] and Glasserman [2004]), such as, for instance, *importance sampling, control variates, and antithetic variates*. One of the simplest and most widely used techniques is the antithetic variates (AV), and it is the first technique we consider to improve the MC method.

In an attempt to avoid the deficiencies of the MC algorithm, many deterministic methods have been proposed as well. One class of such deterministic algorithms, the quasi–Monte Carlo methods (QMC), is based on *low-discrepancy sequences*, i.e., deterministic sequences chosen to be more evenly dispersed through the domain of integration than random ones. In this work we test the sequences proposed by Faure (1992), using the implementation reported in Glasserman (2004). Concerning the other main computational kernel, that is, the solution of stochastic differential equations, the first method we consider is the well-known explicit stochastic Euler scheme. One drawback of the method is that stability requirements could impose stringent restrictions on the step size, so we take into account the implicit Euler scheme. Further, to improve the accuracy of the discrete solution, we consider a third, higher order method that is a slightly different form of the method proposed by Milstein (1978). Finally, we test a fully implicit positivity-preserving Euler scheme, proposed recently by Brigo and Alfonsi (2005), that preserves the monotonicity of the continuous CIR process.

The rest of this work is organized as follows: in section 6.2 we introduce the financial problem with the aim of describing the main computational kernels involved in the solution. In section 6.3 we present and analyze different numerical methods to solve the kernels with the aim of finding the most promising methods from the point of view of accuracy and efficiency. In section 6.4 we present the mathematical model proposed in Pacati (2000) that we use as a benchmark model for testing the efficiency and accuracy of the considered numerical methods. In section 6.5 we report the numerical experiments we performed using the considered numerical methods. The obtained results lead to the selection of a combination of two methods that show the best behavior in the solution of the model.

6.2 COMPUTATIONAL KERNELS IN PARTICIPATING LIFE INSURANCE POLICIES

In this chapter we consider an important class of financial derivatives, participating life insurance contracts. In particular, we analyze portfolios of level-premium mixed life participating policies with benefits indexed to the annual return of a specified investment portfolio. A level-premium mixed life participating policy is a typical example of a profit-sharing policy that has been widely sold in past years by Italian companies, and it is still widespread nowadays.

The basic idea of the participating rule is the following: the insurance company invests the mathematical reserve of the policy in a fund, the segregated fund, whose yearly return is shared between the company and the insured. A readjustment rate is contractually defined; it depends on the return and is applied to raise, for the same year, the insured capital according to a rule that depends on the policy type. Due to the participating rule, the benefits of the policy are random variables with regard to both actuarial and financial uncertainties. The former is connected with all the events influencing the duration of the policies. Typical risk drivers are mortality/longevity risk and surrender risk. For the latter, the most relevant types of risk are the interest-rate risk and the stock-price risk. The policy is a derivative contract, in which the return of the segregated fund is the underlying asset. Therefore, the simulation of this financial instrument is highly complex and computationally intensive, mainly due to the huge number of involved variables and conditions to take into account for accurate forecasts. The literature on this topic is rich, and we recall De Felice and Moriconi (2001, 2005), Bacinello (2003), Andreatta and Corradin (2003), Ballotta and Haberman (2006), Grosen and Jørgensen (2000), and Jensen, Jørgensen, and Grosen (2001). Much effort has been spent on the development of mathematical models and related algorithms to perform mark-to-market evaluation of this kind of contract that, in contrast to the traditional valuation framework, would allow consideration of the financial uncertainty affecting the benefits (see, for example, De Felice and Moriconi [2001, 2005] and Pacati [2000]). The mark-to-market evaluation requires a mathematical description model for the bond and stock markets. The valuation of the financial components is performed using a stochastic pricing model based on the no-arbitrage principle. The model is calibrated on market data to capture the current interest-rate levels, the interest-rate volatilities, the stock-price volatilities, and correlations.

We consider a market model with two sources of uncertainty: interest-rate risk and stock-market risk. We model interest-rate risk through the one-factor CIR model (see Cox, Ingersoll, and Ross [1985]): if $r(t)$ is the market spot rate at time t, we assume that it follows a square-root mean-reverting diffusion process

$$dr(t) = \alpha[\gamma - r(t)]\,dt + \rho\sqrt{r(t)}\,dZ_r(t) \tag{6.2.1}$$

where Z_r is a standard Brownian motion, and α, γ, and ρ are positive constant parameters, with the condition $2\alpha\gamma > \rho^2$, which ensures that the process r remains positive. Furthermore, we assume that the market price of interest-rate risk is of the form

$$q(r(t),t) = \pi \frac{\sqrt{r(t)}}{\rho} \tag{6.2.2}$$

with $\pi \in \mathbb{R}$ a constant parameter.

Stock market uncertainty is considered by modeling the stock index $S(t)$ as a Black–Scholes log-normal process with constant drift and volatility parameters $\mu > 0$ and $\sigma > 0$

$$dS(t) = \mu S(t)dt + \sigma S(t)\, dZ_S(t) \tag{6.2.3}$$

where Z_S is a standard Brownian motion.

It is well known that this model is complete and arbitrage-free, and the risk-neutral dynamics of the state variables are

$$dr(t) = \tilde{\alpha}[\tilde{\gamma} - r(t)]\, dt + \rho\sqrt{r(t)}\, d\tilde{Z}_r(t) \tag{6.2.4}$$

$$dS(t) = r(t)S(t)dt + \sigma S(t)\, d\tilde{Z}_S(t) \tag{6.2.5}$$

where $\tilde{Z}_r$ and $\tilde{Z}_S$ are the risk-neutral Girsanov transformations of the two Brownian motions Z_r and Z_S, $\tilde{\alpha} = \alpha - \pi$, and $\tilde{\gamma} = \alpha\gamma/\tilde{\alpha}$. We finally assume that the two Brownian motions driving $r(t)$ and $S(t)$ are correlated by a constant correlation factor.

In this framework, a standard no-arbitrage argument shows that the market price at time t of a random payment $X(r, S; v)$ at time v with $t < v$, subject only to financial uncertainty, is given by:

$$V(t, X) = \tilde{\mathrm{E}}_t\left[X(r, S; v)\; \mathrm{e}^{-\int_t^v r(u)\, du}\right] \tag{6.2.6}$$

where $\tilde{\mathrm{E}}_t$ is the risk-neutral expectation implied by the risk-neutral version of the model, conditional to the market information at time t.

6.3 NUMERICAL METHODS FOR THE COMPUTATIONAL KERNELS

In this section we focus on numerical issues in the solution of the financial problem. As pointed out in the previous section, the main computational kernels are involved in the evaluation of multidimensional integrals representing expectation values according to equation (6.2.6) and the solution of stochastic differential equations.

6.3.1 Numerical Methods for High-Dimensional Integration

High-dimensional integrals are usually solved via the MC method. The main idea underlying the Monte Carlo algorithm for multivariate integration is to replace a continuous average with a discrete one over randomly selected points. It is well known that the expected error in the Monte Carlo method is proportional to the ratio $\sigma_f/\sqrt{N}$, where σ_f^2 is the variance of the integrand function and N is the number of computed trajectories. In this formula, the value σ_f depends on the integrand function and thus on the dimension of the integral, but the factor $1/\sqrt{N}$ does not. In particular, the $O(1/\sqrt{N})$ convergence rate holds for every dimension. This shows why MC becomes increasingly attractive as the dimension of the integral increases in comparison with deterministic methods for numerical integration, which are conversely characterized by a rate of convergence strongly decreasing with respect to the dimension. On the other hand, the MC method presents two deficiencies: the rate of convergence is only proportional to $N^{-1/2}$, and special care has to be taken in generating independent random points because we actually deal with pseudo-random numbers. Since, as already pointed out, the expected error of the MC method depends on the variance of the integrand, convergence can be speeded up by decreasing the variance. One of the simplest and most widely used variance-reduction techniques is the *antithetic variates* (AV), which we use in our experiments in conjunction with the MC method. For brevity, in the following, we refer to the AV reduction technique combined with the MC method as the *antithetic-variates method.* This method attempts to reduce variance by introducing negative dependence between pairs of replications; in particular, in a simulation driven by independent standard normal variables Z_i, as in our case, this technique can be implemented by pairing the sequence Z_i with the sequence $-Z_i$. If the Z_i are used to simulate the increments of a Brownian path, then the $-Z_i$ simulate the increments of the reflection of the Brownian path about the origin; this suggests that it can result in a smaller variance. We look to Glasserman (2004) for a deeper insight into the matter. The use of the antithetic-variates method can approximately double the computational complexity with respect to a classical Monte Carlo simulation because, for each trajectory, two realizations of the Brownian path have to be simulated. Therefore, its application is effective if we obtain an estimator with a variance smaller than the one corresponding to a classical Monte Carlo simulation performed with a double number of trajectories.

Another way to improve the convergence rate of the MC method is to use deterministic sequences, called *low-discrepancy sequences*, that are chosen to be more evenly dispersed through the region of integration than random sequences. Low-discrepancy sequences are sometimes referred to as *quasi-random* sequences. Numerical integration methods based on them are named *low-discrepancy methods* or quasi–Monte Carlo (QMC) methods. For a complete description of these methods, we refer the reader to Niederreiter (1992). QMC methods provide deterministic error bounds proportional to $(\log N)^d/N$ for suitably chosen deterministic sequences, where d is the dimension of the integral. Different low-discrepancy sequences are well known in literature; here we confine ourselves to Faure sequences. Studies using these low-discrepancy sequences in finance applications have found that the errors produced are substantially smaller than the corresponding errors generated by crude Monte Carlo (see, for example, Joy, Boyle, and Tan [1998], Papageorgiou and Traub [1997], Paskov and Traub [1995], Perla [2003]). However, as pointed out in Sloan and Wozniakowsky (1998), the existing theory of the worst-case error bounds of QMC algorithms does not explain this phenomenon. All of these considerations serve as a useful caution against assuming that QMC methods will outperform MC methods in all situations.

6.3.2 Numerical Solution of Stochastic Differential Equations

In this section we discuss numerical schemes for the solution of the two linear stochastic differential equations (6.2.4) and (6.2.5). In particular, we just focus on equation (6.2.4), as all our considerations extend to equation (6.2.5) in a natural way.

In the numerical solution of SDEs, the convergence and numerical stability properties of the schemes play a fundamental role as well as in a deterministic framework. Regarding the stability of a numerical method for SDEs, an important role is played by its region of *absolute stability*, as discussed in Kloeden and Platen (1992), because it defines possible restrictions on the maximum allowed step size, ensuring that errors will not propagate in successive iterations. Generally, implicit methods reveal larger stability regions than explicit ones, as the bounds imposed on the values of step size are less stringent than for explicit methods.

We consider four numerical schemes for the solution of SDEs. Let $[0, T]$ be a time interval; we consider, for simplicity, a time grid $0 = t_0 < t_1 < \cdots < t_N = T$ with fixed time step $h > 0$, that is, $t_i = ih$, $i = 0, \ldots, N$. Let us furthermore denote by $\overline{r}$ a time-discrete approximation of function r in equation (6.2.4) on the aforementioned time grid. In the following, we will sometimes use the notation

$$\overline{r}(t_i) = \overline{r}(ih) = \overline{r}_i. \tag{6.3.1}$$

The first method we consider is the well-known explicit Euler stochastic scheme. The explicit Euler approximation of equation (6.2.4) is defined by

$$\bar{r}(t_{i+1}) = \bar{r}(t_i) + \tilde{\alpha}[\tilde{\gamma} - \bar{r}(t_i)](t_{i+1} - t_i) + \rho\sqrt{(t_{i+1} - t_i)\bar{r}(t_i)}\tilde{Z}_{i+1} \quad (6.3.2)$$

with $\bar{r}(0) = r(0)$, and $\tilde{Z}_1, \tilde{Z}_2, \ldots$ being independent, standard normal random variables. Using equation (6.3.1), we can write

$$\bar{r}_{i+1} = \bar{r}_i + \tilde{\alpha}[\tilde{\gamma} - \bar{r}_i]\, h + \rho\sqrt{h\bar{r}_i}\ \tilde{Z}_{i+1}. \quad (6.3.3)$$

The explicit Euler scheme achieves order-one weak convergence if appropriate hypotheses on the coefficients of the equation, reported in Kloeden and Platen (1992) and Glasserman (2004), are satisfied. Stability theory usually takes into account a class of test functions for which the conditions to ensure stability are stated. For the test equation corresponding to the parameters in our model, the region of absolute stability is defined by the condition $\tilde{\alpha}h < 2$.

As already pointed out, because implicit schemes can reveal better stability properties, we have taken into account the implicit Euler scheme as reported in Kloeden and Platen (1992, p. 396). This implicit scheme is obtained by making implicit only the purely deterministic term of the equation, while at each time step, the coefficients of the random part of the equation are retained from the previous step. Using the same notations as in equation (6.3.3), we have, at each time step,

$$\bar{r}_{i+1} = \bar{r}_i + \tilde{\alpha}[\tilde{\gamma} - \bar{r}_{i+1}]\, h + \rho\sqrt{h\bar{r}_i}\ \tilde{Z}_{i+1}. \quad (6.3.4)$$

The implicit Euler scheme has the same weak order of convergence as the corresponding explicit Euler scheme, but the step size can be chosen arbitrarily large. From a computational point of view, it is not more expensive than equation (6.3.3), but our numerical experiments revealed that it can provide better accuracy, as shown in the next section.

Moreover, we consider a higher order method proposed by Milstein (1978), having order-two weak convergence. More precisely, we consider a simplified version of the scheme for practical implementation, as shown in Glasserman (2004, p. 351) and in Kloeden and Platen (1992, p. 465), approximating the diffusion process shown in equation (6.2.4) by the following expansion

$$\bar{r}_{i+1} = \bar{r}_i + ah + b\sqrt{h}\tilde{Z}_{i+1} + \frac{1}{2}\left(a'b + ab' + \frac{1}{2}b^2b''\right)h\sqrt{h}\tilde{Z}_{i+1}$$
$$+ \frac{1}{2}bb'h[\tilde{Z}^2_{i+1} - 1] + \left(aa' + \frac{1}{2}b^2a''\right)\frac{1}{2}h^2 \quad (6.3.5)$$

where $a = \alpha[\gamma - r(t)]$ and $b = \rho\sqrt{r(t)}$, and a, b and their derivatives are all evaluated at time t_i. This scheme is more accurate than the Euler method, but it is computationally more expensive.

It is well known that the Euler scheme for the CIR process can lead to negative values, as the Gaussian increment is not bounded from below. Then, finally, we test a fully implicit, positivity-preserving Euler scheme introduced by Brigo and Alfonsi (2005). According to this scheme, the discrete values of r are obtained by means of the following recursion

$$\overline{r}_{i+1} = \left(\frac{\rho\sqrt{h}\tilde{Z}_{i+1} + \sqrt{\rho^2 h \tilde{Z}_{i+1}^2 + 4(\overline{r}_i + \delta h)(1 + \alpha h)}}{2(1 + \alpha h)} \right)^2 \tag{6.3.6}$$

with $\delta = \alpha\gamma - \rho^2/2$.

6.4 A BENCHMARK MATHEMATICAL MODEL

In this section we introduce the mathematical model to describe participating life insurance policies studied in Pacati (2000), in which the revaluation of the insured capital is proportional to the number of paid premiums. We use this as a benchmark model to evaluate the impact of the various considered numerical methods on efficiency and accuracy in the solution of the problem.

Let us consider, at time t, a level-premium mixed life participating policy, with term n years for an insured of age x at the inception of the contract. Let P be the net constant annual premium paid by the policyholder at the beginning of the year, C_0 be the sum that is initially insured, and $i \geq 0$ the technical interest rate. We denote with a the number of years between the inception date of the contract and our valuation date t. We assume that a is an integer, and therefore the policy starts exactly a years before the valuation date t. The time to maturity is $m = n - a$. We consider the interest-crediting mechanism proposed in Pacati (2000), supposing that benefit payments occur at integer payments dates $a + 1, a + 2, \dots, a + m$. Let C_a be the sum insured at the inception date $t - a$; then the evaluation of the insured capital C_{a+k} at time $a + k,\ k = 1, \dots, m$, is given by

$$C_{a+k} = C_a \Phi(t, k) - \frac{C_0}{n} \Psi(t, m, k) \tag{6.4.1}$$

where

$$\Phi(t,k) = \prod_{l=1}^{k}(1+\rho_{t+l}) \tag{6.4.2}$$

$$\Psi(t,m,k) = \sum_{l=0}^{k-1}(m-k+l)\rho_{t+k-l}\prod_{j=k-l+1}^{k}(1+\rho_{t+j}) \tag{6.4.3}$$

and the product is equal to 1 if the index set is empty. The readjustment rate at the end of the year just terminated, ρ_t, is defined as:

$$\rho_t = \max\left(\frac{\beta R_t - i}{1+i}, s_{\min}\right). \tag{6.4.4}$$

R_t is the return of the segregated fund in the same year, $\beta \in (0,1]$ is the so-called *participation coefficient*, and $s_{\min} \geq 0$ is the yearly minimum guaranteed. The technical rate i, the coefficient $s_{\min}$, and the participation coefficient β are contractually specified; thus their values are fixed at time zero. The quantity βR_t in equation (6.4.4) represents the portion of the fund return that is credited to the policyholder (by increasing the insured sum); the remaining portion, $(1-\beta)R_t$, is retained by the insurer and determines its investment gain.

An analysis of equation (6.4.1) shows that the insured capital appreciated at year $a+k$ is computed by subtracting from the full revaluation, $C_a\Phi(t,k)$, a quantity that is proportional to C_0/n and depends on future interest rates. Obviously, for $k > 1$, $\Phi(t,k)$ and $\Psi(t,m,k)$ are random at time t due to the randomness of the future interest rates $\rho_{t+1}, \rho_{t+2}, \ldots, \rho_{t+m}$. The statutory technical reserve of the policy, that is, the level of funding that the company has to maintain by law, is the *net premium mathematical reserve* and is defined in a traditional actuarial setting. In Pacati (2000), two possible decompositions of the insured capital in the case of survival C_{a+m} are described, namely the *put* and *call decompositions*; in the same way, two analogous decompositions of C_{a+k} are shown: the insured capital in the case of death at time $t+k$ corresponding to the time to expire $a+k$.

In contrast to the traditional framework, the mark-to-market approach is able to consider the financial uncertainty affecting the benefits. This uncertainty comes from the market where the fund's manager invests the policy reserve. Then we consider the *mark-to-market reserve* of the policy, also called the *stochastic reserve*, to emphasize the fact that the evaluation is done in a mark-to-market setting while considering a stochastic evolution of interest rates, in contrast to "traditional" constant-rate evaluation.

The stochastic reserve $R(t)$ is defined as the difference between the mark-to-market value of the future obligations of the company $R_Y(t)$ and the mark-to-market value of the future obligations of the insured $R_X(t)$. For the considered policies, we obtain:

$$R_X(t) = P \sum_{k=1}^{m-1} {}_k p_{x+a} v(t, t+k) \tag{6.4.5}$$

where ${}_k p_{x+a}$ is the expectation of life of an insured of age $x + a$ years after k years. On the other hand, taking into account that, at the term, the insured is alive or not, we obtain:

$$R_Y(t) = \sum_{k=1}^{m} V(t, C_{a+k})\, {}_{k-1|1}q_{x+a} + V(t, C_n)\, {}_m p_{x+a} \tag{6.4.6}$$

where ${}_{k-1|1}q_{x+a}$ is the probability that the insured of age $x + a$ dies between the year $k - 1$ and the year k, and V denotes the market price, as defined in equation (6.2.6). Using the relation in equation (6.4.1) and the linear property of the price for $k = 1, 2, \ldots, m$ we obtain:

$$V(t, C_{a+k}) = C_a V(t, \Phi(t,k)) - \frac{C_0}{n} V(t, \Psi(t,m,k)) \tag{6.4.7}$$

and then, defining

$$\phi(t,k) = V(t, \Phi(t,k)) \tag{6.4.8a}$$

$$\psi(t,m,k) = V(t, \Psi(t,m,k)) \tag{6.4.8b}$$

the relation in equation (6.4.6) can be rewritten as follows:

$$R_Y(t) = \sum_{k=1}^{m} {}_{k-1|1}q_{x+a} \left[C_a \phi(t,k) - \frac{C_0}{n} \psi(t,m,k) \right]$$

$$+ {}_m p_{x+a} \left[C_a \phi(t,m) - \frac{C_0}{n} \psi(t,m,m) \right]. \tag{6.4.9}$$

Because functions ϕ and ψ depend neither on x nor on a, from equation (6.4.9) we note that we can evaluate once the factors ϕ and ψ for any value m

in the portfolio and for any value of $k = 1, \ldots, m$, and successively substitute them into equation (6.4.9). All our considerations can be extended to the put and call decompositions of $R_Y(t)$, obtained according to the put and call decompositions of C_{a+k}. If, for $k = 1, 2, \ldots, m$, we denote with

$$\phi^{\text{base}}(t,k) = V\left(t, \frac{1}{(1+i)^k}\prod_{l=1}^{k}(1+\beta R_{t+l})\right) \tag{6.4.10a}$$

$$\psi^{\text{base}}(t,m,k) = V\left(t, \sum_{l=0}^{k-1}(m-k+l)\frac{\beta R_{t+k-l}-i}{(1+i)^{l+1}}\prod_{j=k-l+1}^{k}(1+\beta R_{t+k})\right) \tag{6.4.10b}$$

$$\phi^{\text{put}}(t,k) = \phi(t,k) - \phi^{\text{base}}(t,k) \tag{6.4.10c}$$

$$\psi^{\text{put}}(t,m,k) = \psi(t,m,k) - \psi^{\text{base}}(t,m,k) \tag{6.4.10d}$$

$$\phi^{\text{guar}}(t,k) = V(t,(1+s_{\min})^k) \tag{6.4.10e}$$

$$\psi^{\text{guar}}(t,m,k) = V\left(t, s_{\min}\sum_{l=0}^{k-1}(m-k+l)(1+s_{\min})^l\right) \tag{6.4.10f}$$

$$\phi^{\text{call}}(t,k) = \phi(t,k) - \phi^{\text{guar}}(t,k) \tag{6.4.10g}$$

$$\psi^{\text{call}}(t,m,k) = \psi(t,m,k) - \psi^{\text{guar}}(t,m,k) \tag{6.4.10h}$$

the basis, put, guaranteed, and call components of ϕ and ψ, by combining them with equation (6.4.9), we can evaluate the basis, put, guaranteed, and call components of $R_Y(t)$, as described in Pacati (2000).

An analysis of equation (6.4.10) shows that, having computed the factors in equations (6.4.8a) and (6.4.8b), we only have to evaluate equations (6.4.10a), (6.4.10b), (6.4.10e), and (6.4.10f), as all the others can be obtained by difference.

From the computational point of view, the most important kernel for the evaluation of $R_Y(t)$, and then of $R(t)$, is constituted by the evaluation of the factors defined in equations (6.4.8a), (6.4.8b), (6.4.10a), and (6.4.10b), as they depend on the future returns of the fund and are then affected by the market uncertainty.

6.5 NUMERICAL EXPERIMENTS

In this section we show some of the numerical experiments we performed. The test case we use refers to a specific date of evaluation of the bond market: January 4, 1999. The bond market data have been estimated following Pacati (1999), and the parameters used for the CIR model are reported in Table 6.5.1. All the experiments refer to an $n = 20$ years term policy for a 30-year-old insured. The residual maturity is $a = 10$; the technical interest rate is set to 4%; and the yearly minimum guaranteed is supposed to be $s_{\min} = 0\%$ and $\beta = 0.8$. The initial capital is set to $C_0 = 100$. The values of the expectation of life have been computed by the life tables SIM81. Finally, the correlation factor between $d\tilde{Z}_S(t), d\tilde{Z}_r(t)$ in equation (6.2.4) and (6.2.5) is set to -0.1.

As already pointed out in the previous section, our main computational kernels are multidimensional integrals and stochastic differential equations. We focus on them separately. In the discussion about techniques for the numerical computation of multidimensional integrals representing the mean values to be estimated, we use the Euler scheme as a test method for the solution of the involved SDEs.

We tested and compared performances of MC, antithetic variates, and QMC methods. In our simulations, the routine snorm of the package ranlib, written by Brown, Lovato, and Russell, available through the Netlib repository, has been used to generate standard normally distributed values; these values have afterwards been properly modified to obtain the fixed level of correlation. In the QMC method, the values of the Faure sequences have been mapped to values from standard normal random variables via the routine dinvnr of the package dcdflib, written by Brown, Lovato, and Russell (1994), available through the Netlib repository as well. The routine approximates the inverse normal cumulative function via Newton's method, as described in Glasserman (2004).

First we show our results in the estimation of the obligations of the insurance company R_Y given by equation (6.4.9), of functions $\phi, \phi^{\text{base}}, \psi$ and ψ^{base}

TABLE 6.5.1 Parameters for the CIR Model

	$t = 01/04/1999$
$r(t)$	0.0261356909
$\tilde{\alpha}$	0.0488239077
$\tilde{\gamma}$	0.1204548842
ρ	0.1056548588

TABLE 6.5.2 Obligations of the Company, R_Y in Equation (6.4.9). N Is the Number of Simulated Trajectories; in the Second Column the Expected Value, That Is, the Sample Mean Computed Via AV with $N = 20 \times 10^6$, Is Reported

N	Expected Value	MC	AV	QMC
1,000	85.530725	85.330736	85.538849	103.593865
2,500	85.530725	85.446784	85.513984	94.824499
5,000	85.530725	85.490321	85.526349	94.824499
10,000	85.530725	85.515832	85.529525	89.144140
20,000	85.530725	85.556925	85.532012	87.049162
50,000	85.530725	85.531398	85.532053	85.537668
100,000	85.530725	85.535615	85.532456	85.317408

expressed by equation (6.4.10), and of the spot interest rates r satisfying equation (6.2.4). A monthly discretization step has been considered in the numerical solution of the involved SDEs. The integral of function r in equation (6.2.6) has been evaluated by means of the trapezoidal rule.

In Table 6.5.2 we report the values of the obligations of the company estimated via the three integration methods for different values of the number N of simulated trajectories.

To estimate the error, an "almost true" value is needed: we assumed as true expected value the sample mean computed via the antithetic variates method with a number of replications equal to 20×10^6. We observe that, with the classical MC method, we obtain three significant digits for $N \geq 10^4$; applying the antithetic variates method, the same accuracy is reached for just $N = 10^3$. Moreover, to obtain four significant digits with antithetic variates, we need $N = 2 \times 10^4$ simulations, but with MC we need $N = 5 \times 10^4$ simulations. Because the application of the antithetic variates technique at most doubles the computational cost, we deduce that efficiency is strongly improved in these cases. All the experiments we performed confirmed this.

On the other hand, the QMC method, with the Faure sequences, does not perform well in this framework; to deliver two significant digits, QMC needs, at least $N = 5 \times 10^4$.

It is well known that special care has to be taken in generating pseudo-random points. To show the sensitivity of the MC and antithetic variates methods to the initial seed, in Table 6.5.3 we report the minimum and the maximum values of the obligations of the company estimated via the Monte Carlo method and antithetic variates, repeating each MC and antithetic variates run 20 times. We observe that MC exhibits a sensitivity greater than the one shown

TABLE 6.5.3 To Evaluate the Sensitivity of MC and AV to Initial Seed, We Repeat Each Run 20 Times. The Second and Fourth Columns Report the Minimum Resulting Values for R_y; the Third and Fifth Columns Report the Maximum Resulting Values That R_y Assumes

	MC		AV	
N	R_y Min	R_y Max	R_y Min	R_y Max
1,000	84.448579	86.314370	85.098095	85.880454
2,500	84.962291	86.099855	85.387740	85.704434
10,000	85.097654	86.042048	85.431023	85.605205

by antithetic variates to the seed of the pseudo-random number generator and that both minimum and maximum estimations computed by antithetic variates deliver two significant digits for all considered values of N.

In Table 6.5.4 we show the CPU time spent by the Monte Carlo simulation and antithetic variates methods combined with a Euler scheme. To evaluate the overhead of antithetic variates with respect to the MC method, we also reported the ratio between the two values. We observe that the execution time of antithetic variates is never the double of execution time of the Monte Carlo method, even though it requires the generation of a number of simulations that is double with respect to MC. The ratio between the two execution times is always about 1.2.

In Table 6.5.5 we deal with functions ϕ, ϕ^{base}, ψ, and ψ^{base}. A total of $N = 10^4$ MC repetitions have been simulated, and the results have been compared with the ones obtained with the antithetic variates method for $N = 1000$, $N = 2500$, and $N = 5000$. Two different step sizes h in the integration of the SDEs have now been considered, a monthly discretization and a weekly one. In the first column the amplitudes of 95% confidence

TABLE 6.5.4 MC and AV CPU Times in Seconds for Different Values of the Number N of Trajectories

	N						
	1,000	2,500	5,000	10,000	20,000	50,000	100,000
MC	5.84	17.75	32.90	66.24	130.75	334.19	668.16
AV	7.87	19.73	39.53	79.01	157.13	393.59	786.06
$\frac{AV}{MC}$	1.34	1.11	1.2	1.19	1.2	1.18	1.18

TABLE 6.5.5 Comparison between MC and AV in the Evaluation of ϕ, ϕ^{base}, ψ, and ψ^{base} for Different Values of Simulated Trajectories. For Each Method and Function, the First Column Reports the Amplitude of the 95% Confidence Interval, and the Second Column Reports the Sample Variance. Two Different Step Sizes in the Integration of the SDEs Are Considered: a Monthly Discretization and a Weekly Discretization

	MC with $N = 10,000$		AV with $N = 5,000$		AV with $N = 2,500$		AV with $N = 1000$	
	95% Conf. Int.	σ_N^2	95% Conf. Int.	σ_N^2	95% Conf. Int.	σ_N^2	95% Conf. Int.	σ_N^2
				Discretization Step 30 days				
ϕ	1.743123e-02	1.977351e-02	2.814142e-03	2.576854e-03	3.957107e-03	2.547555e-03	6.327318e-03	2.605357e-03
ϕ^{base}	4.146804e-02	1.119064e-01	1.548429e-02	7.801544e-02	2.189052e-02	7.796150e-02	3.460240e-02	7.791845e-02
ψ	3.219449e-02	6.745141e-02	1.044438e-02	3.549465e-02	1.490610e-02	3.614897e-02	2.331941e-02	3.538857e-02
ψ^{base}	1.737034e-02	1.963562e-02	1.699363e-03	9.396592e-04	2.396800e-03	9.346120e-04	3.830252e-03	9.547343e-04
				Discretization Step 7 days				
ϕ	5.588971e-03	2.032786e-02	2.683103e-03	2.342462e-03	3.784740e-03	2.330451e-03	6.036319e-03	2.371222e-03
ϕ^{base}	1.323779e-02	1.140405e-01	1.555592e-02	7.873892e-02	2.199212e-02	7.868680e-02	3.476027e-02	7.863107e-02
ψ	9.734121e-03	6.166253e-02	9.952142e-03	3.222783e-02	1.409371e-02	3.231608e-02	2.228629e-02	3.232238e-02
ψ^{base}	5.303716e-03	1.830578e-02	1.649245e-03	8.850505e-04	2.336331e-03	8.880482e-04	3.751378e-03	9.158186e-04

intervals are given; in the second one, the values of sample variances, denoted with σ_N^2, of computed sample means are reported. In most of the cases, the size order of the amplitude of 95% confidence intervals obtained by the MC method for $N = 10^4$ is nearly the same as that estimated by the antithetic variates method, even for $N = 10^3$; sometimes the antithetic variates method gives an even more accurate result: for instance, it happens almost always in the estimation of function ϕ. An analogous behavior is reflected in the estimation of sample variances. It is worth noting the regular reduction by a factor of about two for the amplitudes of the confidence intervals obtained with the antithetic variates when the number of replications halves. Further, we note that the values of the sample variances are nearly the same for all the considered values of N, even for $N = 10^3$, that is, the method reveals better stability properties.

All the experiments done until now show that the use of the antithetic variates method provides the same accuracy as the MC method with a number of replications that is reduced by a factor near to four. An important result is that the use of antithetic variates provides good accuracy with a smaller number of replications, resulting in a large advantage in terms of execution time.

We now turn to the numerical solution of the SDEs in equations (6.2.4) and (6.2.5). In particular, all the experiments described below refer to the evaluation of equation (6.2.4). We tested the four SDE discretization schemes—Euler, implicit Euler, Milstein, and Brigo–Alfonsi—described above. To estimate the error, we refer to the *deterministic solution* obtained neglecting the stochastic term in equation (6.2.4). Note that the Euler and implicit Euler schemes can lead to negative values for the CIR process; when this happens, we set the computed negative value to zero.

To confirm our previous statements on the better performance of the antithetic variates method over MC, we represent, in Figures 6.5.1 and 6.5.2, the values of the absolute errors of the interest rates computed at each time with the four different SDE methods and with both the MC and antithetics variates methods, for $N = 5 \times 10^3$ and for $N = 5 \times 10^4$. A monthly discretization step size is used. We observe that, as we expected, from an accuracy point of view, the antithetic variates method also outperforms the MC method in the estimation of spot interest rates. For example, for $N = 5 \times 10^3$ (Figure 6.5.1), the obtained errors using antithetic variates are never greater than 10^{-3}; while MC crosses this value.

In Figure 6.5.3, we plot the same variables obtained with $N = 5 \times 10^3$ replications of MC and with $N = 10^3$ replications of the antithetic variates

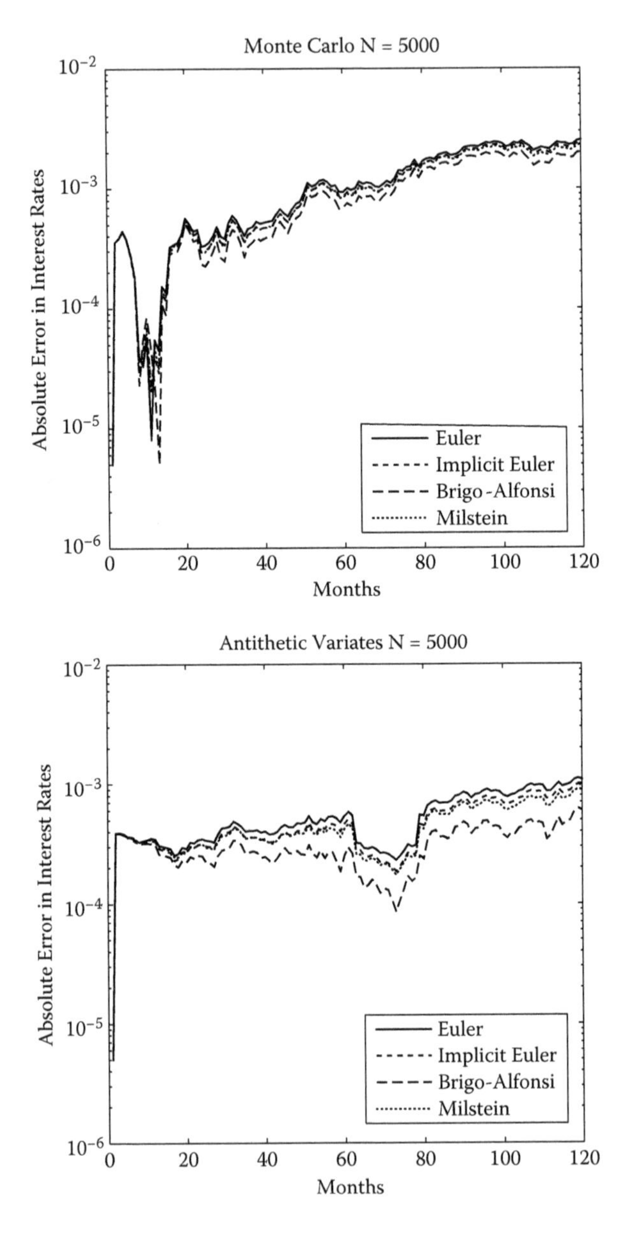

FIGURE 6.5.1 Absolute errors in the estimation of interest rates computed with MC and antithetic variates vs. time. $N = 5 \times 10^3$ trajectories have been simulated; the discretization step size is monthly. At each time t the average over trajectories is represented.

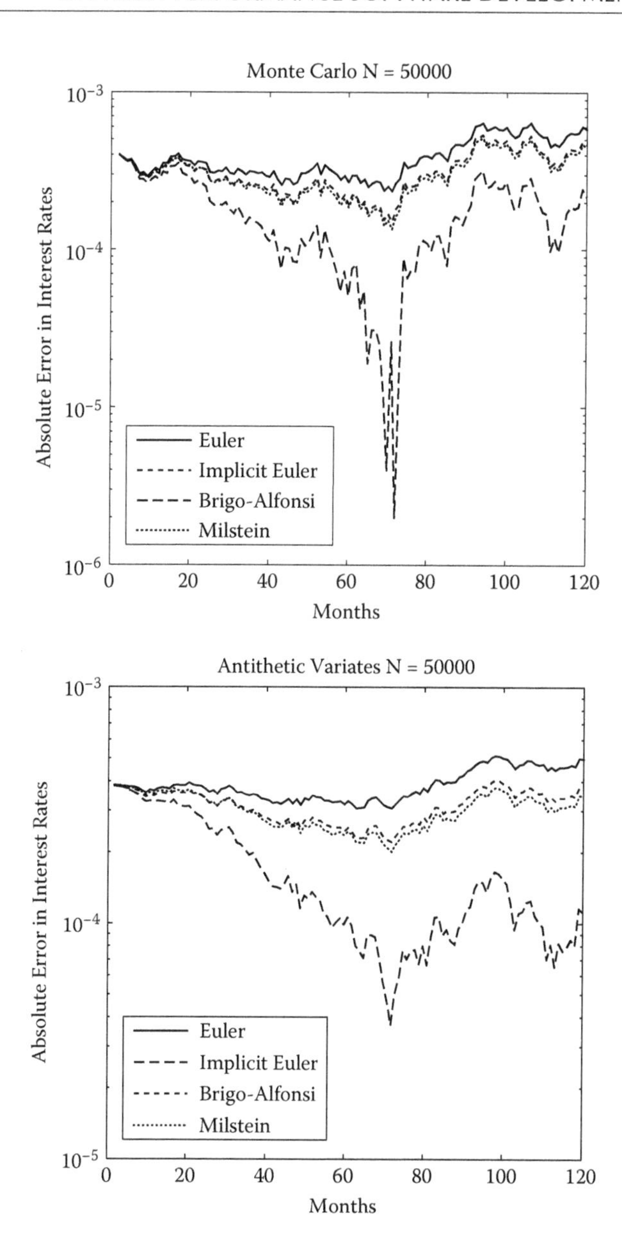

FIGURE 6.5.2 Absolute errors in the estimation of interest rates computed with MC and antithetic variates vs. time. $N = 5 \times 10^4$ trajectories have been simulated; the discretization step size is monthly. At each time t the average over trajectories is represented.

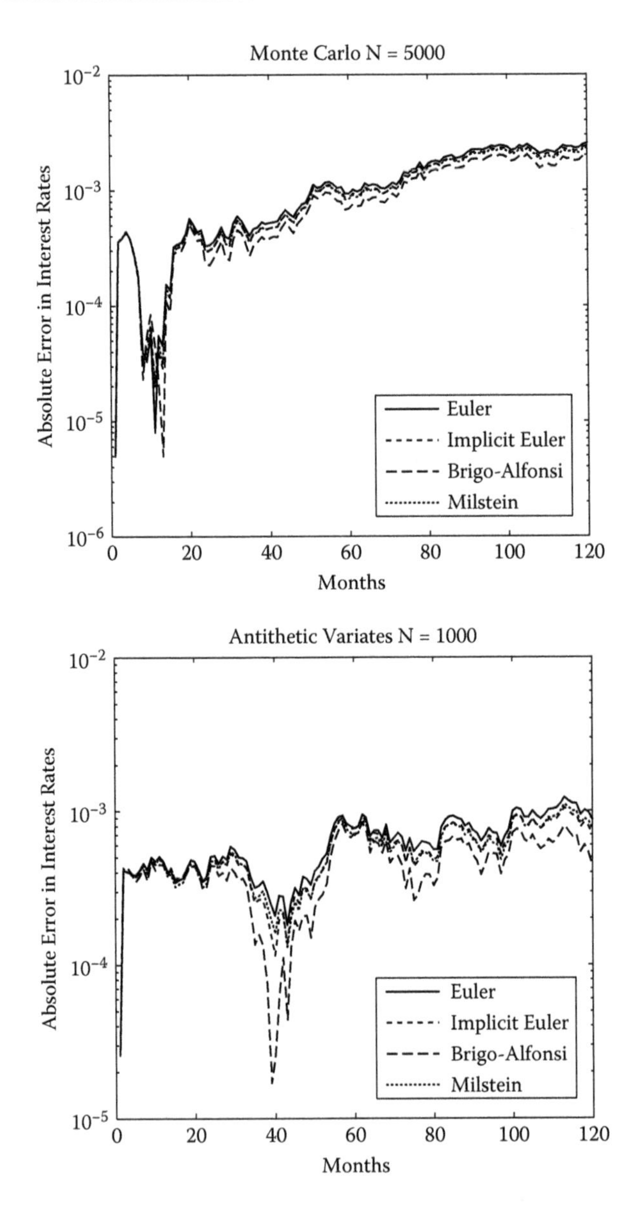

FIGURE 6.5.3 Absolute errors in the estimation of interest rates computed with MC for $N = 5000$ and with antithetic variates for $N = 1000$ vs. time; the discretization step size is monthly. At each time t the average over trajectories is represented.

TABLE 6.5.6 Execution Times (s) for the AV Method with Four Different SDE Schemes for Different Values of N Trajectories the SDE

	N						
SDE Scheme	1,000	2,500	5,000	10,000	20,000	50,000	100,000
Euler	7.87	19.73	39.53	79.01	157.13	393.59	786.06
Implicit Euler	8.30	20.80	41.63	83.69	166.34	414.45	829.81
Brigo–Alfonsi	8.65	21.67	43.30	86.69	173.52	433.25	868.01
Milstein	9.98	25.04	49.89	99.77	199.39	498.56	1000.37

method; we observe that errors estimated via antithetic variates with $N = 10^3$ are almost always lower than those estimated via MC with $N = 5 \times 10^3$. Finally, looking to Figures 6.5.1 and 6.5.3, it is possible to observe that results obtained via antithetic variates with $N = 5 \times 10^3$ are more stable than those with $N = 10^3$. Analyzing the behavior of the four different SDE methods, we note that the absolute error for all the methods lies in the interval $[10^{-6}, 10^{-3}]$. In particular, for values of N equal to 1000 and 5000, the accuracy given by the four methods is almost comparable. As the number of simulations increases (Figure 6.5.2), the Euler method exhibits the worst behavior; implicit Euler is comparable with the Milstein scheme, but the computational complexity of the latter is higher; and the Brigo–Alfonsi method reaches the highest level of accuracy.

In Table 6.5.6 we report the execution times for the antithetic variates method using the four different schemes for the SDEs. The results show that the Milstein scheme is more time-consuming than the others; the Brigo–Alfonsi method is slightly more expensive than the implicit Euler one. Finally, Figure 6.5.4 reports, for both the MC and antithetic variates methods, the values of the root-mean-square (RMS) absolute error defined by

$$RMS = \sqrt{\frac{1}{M}\sum_{i=1}^{M}(\overline{r}_i - r_{det}(t_i))^2}$$

where M is the number of time steps from t_0 to t_N, $\overline{r}_i$ is the computed sample mean at time t_i, and $r_{det}(t_i)$ is value at time t_i of the deterministic solution.

Figure 6.5.4 shows that the RMS errors of all the considered methods are significantly reduced by using the antithetic variates method, and the implicit Euler and Milstein have comparable behavior. The Brigo–Alfonsi method outperforms all the other ones, especially for high values of N.

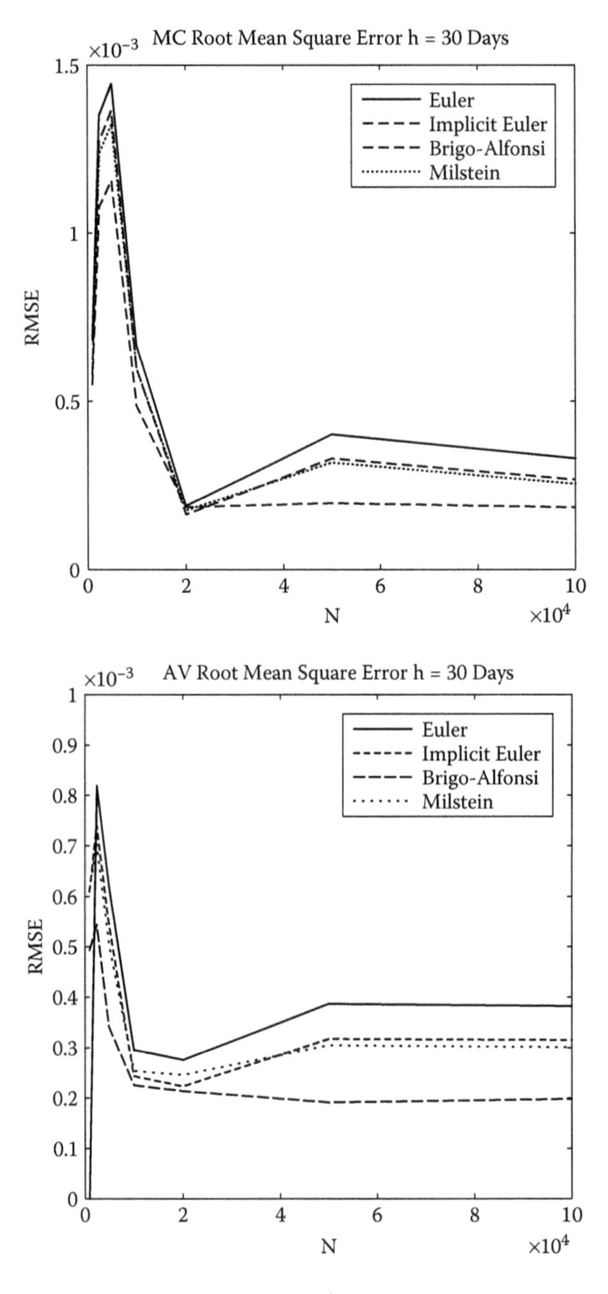

FIGURE 6.5.4 Top: RMS errors in the SDE solutions with MC method vs. N; bottom: RMS errors in the SDE solutions with antithetic variates vs. N. The discretization step size is monthly.

6.6 CONCLUSIONS

In this chapter we focused on the numerical issues involved in evaluating a class of participating life insurance policies. More precisely, we analyzed the impact of different Monte Carlo simulations and various SDE discretization schemes on the solution of our problem.

The main purpose in the evaluation of most financial instruments focuses on the ability to obtain a response in a "useful" time rather than to achieve very high accuracy. Generally, rather low accuracy is sufficient. This motivated our efforts to search for methods that provide a trade-off between accuracy and efficiency.

We have shown that by selecting an appropriate numerical scheme, the number of trajectories, and consequently the execution times, can be reduced while preserving the requested level of accuracy. In particular, the experiments we performed showed that the antithetic variates method allows us to obtain good accuracy values with a small number of replications, thus resulting in a reduction of the execution time compared with the classical MC method.

Among the numerical methods used to solve the SDEs involved in our financial problems, the fully implicit Brigo–Alfonsi method exhibits the best values of accuracy, and it is only slightly more expensive than the implicit Euler method, which is comparable with the Milstein method in terms of accuracy. Thus, we can establish that the best trade-off is realized by a combination of the antithetic variates method and the fully implicit Brigo–Alfonsi scheme.

Future research will deal with deeper analyses of (a) QMC methods to test other deterministic sequences, for example, Sobol sequences (see Glasserman [2004]), randomized sequences (see L'Ecuyer and Lemieux [2002]), and QMC with dimension-reduction techniques (see Imai and Tan [2004]) that could behave in a different way; and (b) other numerical methods for the solution of SDEs that preserve the positivity of the CIR process, such as described in Deelstra and Delbaen (1998) and in Kahl and Schurz (2006).

ACKNOWLEDGMENT

This work was supported by a grant of the National Research Project *Imprese di Assicurazione e Fondi Pensione. Modelli per la Valutazione, per la Gestione e per il Controllo* (MURST PRIN, 2003).

REFERENCES

G. Andreatta, and S. Corradin (2003), Valuing the Surrender Options Embedded in a Portfolio of Italian Life Guaranteed Participating Policies: A Least Squares Monte Carlo Approach, Proceedings of *"Real option theory meets practice", 8th Annual International Conference*, Montreal.

A.R. Bacinello (2003), Fair Valuation of a Guaranteed Life Insurance Participating Contract Embedding a Surrender Option, *Journal of Risk and Insurance*, 70(3), 461–487.

L. Ballotta, and S. Haberman (2006), The Fair Valuation Problem of Guaranteed Annuity Options: the Stochastic Mortality Environment Case, *Insurance: Mathematics and Economics*, 38, 195–214.

F. Black, and M. Scholes (1973), The Pricing of Options and Corporate Liabilities, *Journal of Political Economy*, 81(3), 637–654.

P. Boyle, M. Broadie, and P. Glasserman (1998), Monte Carlo Methods for Security Pricing, in *Monte Carlo Methodologies and Application for Pricing and Risk Management*, Bruno Depire, ed., Risk Books, London, pp. 15–44.

D. Brigo, and A. Alfonsi (2005), Credit Default Swaps Calibration and Option Pricing with the SSRD Stochastic Intensity and Interest-Rate Model, *Finance Stochastics*, 9(1), 29–42.

B.W. Brown, J. Lovato, and K. Russell (1991), http://www.netlib.no/netlib/random/.

B.W. Brown, J. Lovato, and K. Russell (1994), DCDFLIB: a Library of C Routines for Cumulative Distribution Functions, Inverses, and Other Parameters, technical report, Department of Biomathematics, The University of Texas, Houston.

S. Corsaro, P.L. De Angelis, Z. Marino, and F. Perla (2005), Numerical Aspects in Financial Evaluation of Participating Life Insurance Policies, Proceedings of the *VIII Italian-Spanish Meeting on Financial Mathematics* Ver bania.

J.C. Cox, J.E. Ingersoll, and S.A. Ross (1985), A Theory of the Term Structure of Interest Rates, *Econometrica*, 53, 385–407.

G. Deelstra, and F. Delbaen (1998), Convergence of Discretized Stochastic (Interest Rate) Process with Stochastic Drift Term, *Applied Stochastic Models Data Anal.*, 14, 77–84.

M. De Felice, and F. Moriconi (2002), Finanza dell'Assicurazione sulla Vita. Principi per l'Asset-Liability Management e per la Misurazione dell'Embedded Value, Giornale dell' Istituto Italiano degli Attuari, LXI, 13–89.

M. De Felice, and F. Moriconi (2005), Market Based Tools for Managing the Life Insurance Company, *Astin Bulletin*, 35(1), 79–111.

H. Faure (1992), Good Permutations for Extreme Discrepancy, *Journal of Number Theory*, 41, 47–56.

P. Glasserman (2004), *Monte Carlo Methods in Financial Engineering*, Springer-Verlag, New York.

A. Grosen, and P.L. Jørgensen (2000), Fair Valuation of Life Insurance Liabilities: the Impact of Interest Rate Guarantees, Surrender Options and Bonus Policies. *Insurance: Mathematics and Economics*, 26, 37–57.

B. Jensen, P.L. Jørgensen, and A. Grosen (2001), A Finite Difference Approach to the Valuation of Path Dependent Life Insurance Liabilities. *The Geneva Papers on Risk and Insurance Theory*, 26(1), 57–84.

C. Joy, P. Boyle, and K. Tan (1998), Quasi-Monte Carlo Methods in Numerical Finance, in *Monte Carlo Methodologies and Application for Pricing and Risk Management*, Bruno Depire, ed., Risk Books, London, pp. 269–280.

C. Kahl, and H. Schurz (2006), Balanced Milstein Methods for SDEs, *Monte Carlo Methods Appl.*, 12(2), 143–170.

P.E. Kloeden, and E. Platen (1992), *Numerical Solution of Stochastic Differential Equations*, Springer-Verlag, Berlin.

J. Imai, and K.S. Tan (2002), Minimizing Effective Dimension Using Linear Transformation, in *Monte Carlo and Quasi-Monte Carlo Methods 2002*, H. Niederreiter, ed., Springer-Verlag, Berlin, 275–292.

P. L'Ecuyer, and C. Lemieux (2002), Recent Advances in Randomized Quasi-Monte Carlo Methods, in *Modeling Uncertainty: An Examination of Stochastic Theory, Methods, and Applications*, M. Dror, P. L'Ecuyer, and F. Szidarovszki, eds., Kluwer Academic, Boston, 419–474.

G.N. Milstein (1978), A Method of Second-Order Accuracy Integration of Stochastic Differential Equations, *Theory of Probability and Its Applications*, 23, 396–401.

H. Niederreiter (1992), *Random Number Generation and Quasi Monte Carlo Methods*, *SIAM*, Philadelphia.

C. Pacati (1999), Estimating the Euro Term Structure of Interest Rates, technical report 32, Progetto MIUR Modelli per la Finanza Matematica, Siena.

C. Pacati (2000), Valutazione di Portafogli di Polizze Vita con Rivalutazione agli Ennesimi, technical report 38, Progetto MIUR Modelli per la Finanza Matematica, Siena.

A. Papageorgiou, and J.F. Traub (1997), Faster Evaluation of Multidimensional Integrals. *Computers in Physics*, 11(6), 574–578.

S. Paskov, and J.F. Traub (1995), Faster Valuation of Financial Derivatives. *Journal of Portfolio Management*, 22(1), 113–120.

F. Perla (2003), A Parallel Mathematical Software for Asian Option Pricing. *International Journal of Pure and Applied Mathematics*, 5(4), 451–467.

I. Sloan, and H. Wozniakowsky (1998), When Are Quasi Monte Carlo Algorithms Efficient for High Dimensional Integrals? *Journal of Complexity* 14, 1–33.

CHAPTER 7

An Efficient Numerical Method for Pricing Interest Rate Swaptions

Mark Cummins and Bernard Murphy
University of Limerick, Limerick, Ireland

Contents

Abstract: It is not possible to directly apply integral transform and fast Fourier transform (FFT) theory to the problem of interest rate swaption pricing due to the nonaffine model dynamics assumed for the swap rate and underlying factor processes, defined under an equivalent swap measure. However, drawing from a recent result in the fixed-income literature, approximate, affine model dynamics are derived for a family of well-known affine models by replacing identified low-variance martingales with their martingale values. This allows the use of standard integral transform techniques in the pricing of interest-rate swaption contracts. The contribution of this chapter is primarily numerical, the main objective of which is to develop a computationally efficient swaption-pricing technology using fast Fourier transform methods. The pricing algorithms developed will greatly facilitate future empirical research into testing the goodness of fit of underlying term-structure models and in evaluating the dynamic hedging performance of various derivative-pricing models—topics of considerable interest among academics and practitioners alike.

Keywords: fast Fourier transform, interest rate swaption pricing, numerical methods in finance

7.1 INTRODUCTION

The application of the futures option pricing methodology of Black (Black's model) [2] to swaptions is well established. Under Black's model, the forward swap rate is assumed to be given by a lognormal distribution and, hence, a closed-form solution for the price of a swaption is derived. However, closed-form solutions for swaption prices do not exist under more general model dynamics, and so various approximate pricing methods have been developed in the literature. The focus of the literature review here will be on the affine term-structure model literature, but much work has been done within the LIBOR (London interbank offered rate) market model framework, and the interested reader is referred to, for instance, Brace, Gatarek, and Musiela [3] and Andersen and Andreasen [1]. Under the affine model framework, approximate swaption pricing methodologies have been proposed by Wei [23], Munk [20], Collin-Dufresne and Goldstein [7], Singleton and Umantsev [22], and Schrager and Pelsser [21].

Wei [23] and Munk [20], in one-factor and multifactor affine model settings, respectively, show that the price of a European coupon bond option is approximately proportional to the price of a European zero-coupon bond option with a defined maturity equal to the stochastic duration of the coupon bond. Collin-Dufresne and Goldstein [7] propose an approximation methodology that involves the implementation of an Edgeworth expansion of the density of the coupon bond price. Singleton and Umanstev [22] propose a swaption pricing methodology where, by approximating the exercise region by means of line segments, the required probabilities in the swaption pricing formula can be approximated by probabilities associated with events that define an affine relation with the underlying state vector. The fundamental outcome of this approximation is that standard Fourier transform techniques can be used to provide approximate swaption prices.

However, of most importance to this discussion is the alternative approach of Schrager and Pelsser [21], who propose a swaption-pricing methodology where the dynamics of the forward swap rate and underlying factors, defined under the swap measure, are approximated by means of replacing *low-variance martingales* with their martingale values. The key advantage of this approach is that the approximate dynamics derived remain within the affine framework. Hence, the pricing of European-exercise swaptions contracts is possible using the transform-based and extended transform-based pricing methodologies of Duffie, Pan, and Singleton [15]. In fact, a key result likely to be of interest to practitioners is that a "Black-like" option pricing formula can be derived to price European-exercise swaptions.

The approximate swaption pricing methodology of Schrager and Pelsser [21] is detailed in section 7.2. The implementation of the pricing methodology is efficient up to the evaluation of numerical integrations. However, this chapter outlines how the authors independently adapted the technology of Carr and Madan [4] to develop a fast Fourier transform (FFT) implementation that offers a substantially more efficient swaption pricing methodology.[1] Details of this methodology are provided in section 7.3. To illustrate the implementation of this FFT-based swaption pricing methodology, section 7.4

[1] The originality of this extension was independently confirmed by the external examiners at the author's (Dr. Mark Cummins) Ph.D. viva voce in 2006. It subsequently came to the notice of the authors that a similar extension was proposed by Schrager and Pelsser in their August 2005 revised version of the referenced paper.

firstly provides the functional forms of the required transforms and extended transforms under the Schrager and Pelsser [21] affine, approximate swap measure dynamics associated with a range of common affine interest rate models of the Vasicek, generalized Vasicek, and CIR type.[2] It secondly proposes an approximation to the time-dependent drift and innovation components under the Schrager and Pelsser [21] affine approximation. The purpose of the time-dependency approximation is to improve computational efficiency in the evaluation of the associated transforms and extended transforms.

Section 7.4 then finally presents results from the implementation of the FFT-based swaption pricing methodology proposed here. For this a comprehensive analysis of swaption prices and computational run times is given based on the following alternative pricing methodologies: (a) the FFT-based swaption pricing methodology, under which is assumed the Schrager and Pelsser [21] affine approximation only; (b) the FFT-based swaption pricing methodology, under which is assumed both the Schrager and Pelsser [21] affine approximation and time-dependency approximation; (c) the transform-based and extended transform-based swaption pricing methodology proposed by Schrager and Pelsser [21], under which is assumed the Schrager and Pelsser [21] affine approximation and time-dependency approximation; and (d) the Monte Carlo simulation approach to swaption pricing, under which is assumed the exact, nonaffine swap measure dynamics. The computational speeds of the alternative swaption pricing implementations are used to highlight the computational efficiency offered by the FFT.

To conclude, section 7.5 presents an illustrative application of the FFT-based swaption pricing methodology proposed by means of an empirical exercise, which tests a range of the alternative affine interest models considered against an extensive EURIBOR swaptions data set. Specifically, an implied parameter estimation of the candidate models is conducted using a nonlinear ordinary least squares (NL-OLS) framework, where the evaluation of the candidate models is examined on the basis of in-sample pricing performance. By way of completing the analysis, the out-of-sample pricing performance of the estimated affine interest-rate models is also examined. Section 7.6 concludes and discusses various relevant future research directions.

---

[2] Schrager and Pelsser [21] report approximation errors and computational speed comparisons for their swaption pricing methodology relative to the alternative methodologies of Munk [20], Collin-Dufresne and Goldstein [7], and Singleton and Umantsev [22]. Among the models used for this, the authors consider the one-factor Vasicek model and a two-factor CIR model. However, the functional forms and the derivations of the required transforms and extended transforms are not presented.

7.2 PRICING SWAPTIONS USING INTEGRAL TRANSFORMS

In detailing the Schrager and Pelsser [21] swaption pricing methodology, it is first necessary to provide a definition for the date-t value of a payer swap contract, which commences at date T_l and terminates at date T_L, with $T_l < T_L$, and which makes payments at the dates $T_j,\ j = l+1, \dots, L$. It is assumed for convenience that the underlying principle of the swap contract is 1, and that the associated fixed payments are made at the rate K. Let $D(t, T)$ denote the date-t value of a zero-coupon bond with maturity T, and so it follows that the value of a payer swap $V_{l,L}^{pay}(t)$ can be defined as follows:

$$V_{l,L}^{pay}(t) = V_{l,L}^{flo}(t) - V_{l,L}^{fix}(t) = \{D(t,T_l) - D(t,T_L)\} - K \sum_{j=l+1}^{L} \Delta_{j-1} D(t,T_j),$$

where $V_{l,L}^{flo}(t)$ is the value of the floating side of the swap; $V_{l,L}^{fix}(t)$ is the value of the fixed side of the swap; and Δ_{j-1} is the market convention for the day-count fraction for the swap payment at date T_j. Now define the *forward par swap rate* (FPSR) $y_{l,L}(t)$ to be the fixed rate at which the value of the swap contract is zero, i.e.,

$$y_{l,L}(t) = \frac{D(t,T_l) - D(t,T_L)}{\sum_{j=l+1}^{L} \Delta_{j-1} D(t,T_j)} = \frac{D(t,T_l) - D(t,T_L)}{P_{l+1,L}(t)},$$

where the notation $P_{l+1,L}(t)$ is introduced to denote the present value of a basis point (PVBP).

A payer swaption is an option that is written on a payer swap contract. Let PS_{T_l} denote the value of a payer swaption at the maturity date T_l of the contract, and note that by definition PS_{T_l}, is given by

$$\begin{aligned} PS_{T_l} &= \max\left(V_{l,L}^{pay}(T_l), 0\right) \\ &= P_{l+1,L}(T_l)\max(y_{l,L}(T_l) - K, 0). \end{aligned}$$

Assume now a short rate process r_t and define a money market account $M_t = \exp(\int_0^t r_s ds)$. It follows that the value of the payer swaption at the current date t is given by

$$PS_t = M_t E_t^{\mathbb{Q}} \left[\frac{P_{l+1,L}(T_l)}{M_{T_l}} \max(y_{l,L}(T_l) - K, 0) \right],$$

where $\mathcal{Q}$ is an assumed equivalent martingale measure allowing for the risk-neutral valuation of the swaption. Applying the standard change of numeraire technique of Geman, El Karoui, and Rochet [16], the above valuation equation can be rewritten as

$$PS_t = P_{l+1,L}(t) E_t^{\mathcal{Q}_{l+1,L}} [\max(y_{l,L}(T_l) - K, 0)], \tag{7.2.1}$$

where $\mathcal{Q}_{l+1,L}$ is the equivalent martingale measure associated with the choice of $P_{l+1,L}(t)$ as numeraire, herein referred to as the swap measure.

The pricing of such payer swaptions is considered by Schrager and Pelsser [21] under the general affine term structure framework of Duffie and Kan [14], where the short rate r_t is assumed to have the following affine relation with an underlying M-dimensional factor process X_t: $r_t = \omega_0 + \omega_X' X_t$, where ω_0 is scalar and ω_X is an M-dimensional vector. The factor process X_t is assumed to evolve according to the following mean-reverting affine diffusion process under the risk-neutral measure $\mathcal{Q}$:

$$dX_t = \Gamma(\Theta - X_t)\,dt + \Sigma\sqrt{V_t}\,dW_t^{\mathcal{Q}},$$

where Γ and Σ are $M \times M$ matrices; Θ is an M-dimensional vector; $W_t^{\mathcal{Q}}$ is an M-dimensional Brownian motion; and V_t is an $M \times M$ diagonal matrix with diagonal elements

$$(V_t)_{j,j} = \eta_j + \varsigma_j' X_t, \quad j = 1, \ldots, M,$$

where η_j is scalar and ς_j is an M-dimensional vector. Under such an affine term-structure model (ATSM), the price of a zero-coupon bond is given by the following exponential affine function in the factor process X_t: $D(t,T) = \exp\left(A(t,T) - B(t,T) \cdot X_t\right)$, where the functions $A(t,T)$ and $B(t,T)$ solve the system of ordinary differential equations (ODEs) as described in Duffie and Kan [14].

The key result of Schrager and Pelsser [21] involves firstly deriving the diffusion dynamics associated with the swap rate $y_{l,L}(t)$ and the underlying factor process X_t under the swap measure $\mathcal{Q}_{l+1,L}$, and then providing an approximation to this using the technique of replacing *low-variance martingales* with their martingale values. The significance of this approximation is that the derived model dynamics of the swap rate and underlying factor processes remain within the affine term-structure framework, and hence Schrager and Pelsser [21] show that the swaption pricing problem can be addressed using the transform-based and extended transform-based contingent claim pricing methodologies of Duffie, Pan, and Singleton [15].

From the developments of Schrager and Pelsser [21], note first that under the swap measure $\mathcal{Q}_{l+1,L}$ the factor process X_t is governed by the following diffusion process:

$$dX_t = \left[\Gamma\left(\Theta - X_t\right) - \Sigma V_t \Sigma' \left(\sum_{j=l+1}^{L} \Delta_{j-1} B(t, T_j) \frac{D(t, T_j)}{P_{l+1,L}(t)}\right)\right] dt$$

$$+\Sigma \sqrt{V_t} d W_t^{\mathcal{Q}_{l+1,L}}. \tag{7.2.2}$$

For the swap rate it can be shown that $y_{l,L}(t)$ is governed by the following diffusion process:

$$dy_{l,L}(t) = \left(\sum_{j=l}^{L} q_j(t) B(t, T_j)'\right) \Sigma \sqrt{V_t} d W_t^{\mathcal{Q}_{l+1,L}}, \tag{7.2.3}$$

where $q_l(t) = -\frac{D(t,T_l)}{P_{l+1,L}(t)}$, $q_j(t) = \Delta_{j-1} y_{l,L}(t) \frac{D(t,T_j)}{P_{l+1,L}(t)}$, $j = l+1, \ldots, L-1$, and $q_L(t) = (1 + \Delta_{L-1} y_{l,L}(t)) \frac{D(t,T_L)}{P_{l+1,L}(t)}$.

From the context of its use, note that t is understood to represent the current date or the passage of time. However, as in Schrager and Pelsser [21], to assist the exposition here it is necessary to fix the current date $t = 0$, the time at which swaption values are required, and to consider t specifically as running time. Now note that, by definition, the general term $\frac{D(t,T_j)}{P_{l+1,L}(t)}$ is a martingale under the swap measure $\mathcal{Q}_{l+1,L}$ given that it represents the bond price normalized by the associated numeraire $P_{l+1,L}(t)$. From this it follows that the terms $q_j(t)$ are also martingales under the swap measure. It is argued by Schrager and Pelsser [21] that these martingales, which appear in equations (7.2.2) and (7.2.3), represent *low-variance martingales* within the affine framework, in the same manner that such martingales are argued to be low-variance within the LIBOR market model by, for instance, d'Aspremont [12]. Hence, the following approximate dynamics are proposed for the swap rate and factor processes, where the low-variance martingales $\frac{D(t,T_j)}{P_{l+1,L}(t)}$ and $q_j(t)$ are replaced by the martingale values $\frac{D(0,T_j)}{P_{l+1,L}(0)}$ and $q_j(0)$, respectively:[3]

$$dy_{l,L}(t) = \left(\sum_{j=l}^{L} q_j(0)B(t,T_j)'\right)\Sigma\sqrt{V_t}dW_t^{Q_{l+1,L}}$$

$$dX_t = \left[\Gamma(\Theta - X_t) - \Sigma V_t \Sigma'\left(\sum_{j=l+1}^{L}\Delta_{j-1}B(t,T_j)\frac{D(0,T_j)}{P_{l+1,L}(0)}\right)\right]dt$$

$$+\Sigma\sqrt{V_t}dW_t^{Q_{l+1,L}}. \tag{7.2.4}$$

As noted earlier, given the above approximate affine diffusion dynamics for the swap rate and factor processes, Schrager and Pelsser [21] propose the pricing of swaptions according to the transform-based and extended transform-based contingent claim pricing methodologies of Duffie, Pan, and Singleton [15]. The following discussion provides the details of this pricing methodology.

Under the swap measure $Q_{l+1,L}$ consider firstly a general contingent claim with time-T_l payoff function $G_{q_1,q_0}(K) = e^{q_1 \cdot \tilde{X}_{T_l}} I_{\{q_0 \cdot \tilde{X}_{T_l} > K\}}$, where $\tilde{X}_t$ defines an $(M+1)$-dimensional affine factor process; $K \in \mathbb{R}$ is some constant trigger defined under the contingent claim contract details; and $q_0, q_1 \in \mathbb{R}^{M+1}$. It follows that the time-zero value of this contingent claim $V_{G_{q_1,q_0}}(K) = P_{l+1,L}(0)$ $E_0^{Q_{l+1,L}}[e^{q_1 \cdot \tilde{X}_{T_l}} I_{\{q_0 \cdot \tilde{X}_{T_l} > K\}}]$ is given by

$$V_{G_{q_1,q_0}}(K) = P_{l+1,L}(0)\frac{\tilde{\psi}^{Q_{l+1,L}}(q_1,\tilde{X}_t,0,T_l)}{2}$$

$$-\frac{P_{l+1,L}(0)}{\pi}\int_0^{\infty}\frac{imag[\tilde{\psi}^{Q_{l+1,L}}(q_1 - ivq_0,\tilde{X}_t,0,T_l)e^{ivK}]}{v}dv,$$

where $\tilde{\psi}^{Q_{l+1,L}}(\tilde{u},\tilde{X}_t,0,T_l) = E_0^{Q_{l+1,L}}[e^{\tilde{u}\cdot\tilde{X}_{T_l}}]$, for $\tilde{u} \in \mathbb{C}^{M+1}$.

Consider now a general *extended* contingent claim defined with time-T_l payoff function $G_{q_2,q_1,q_0}(K) = (q_2 \cdot \tilde{X}_{T_l})e^{q_1 \cdot \tilde{X}_{T_l}} I_{\{q_0 \cdot \tilde{X}_{T_l} > K\}}$, where $q_0, q_1, q_2 \in \mathbb{R}^{M+1}$, and note that $V_{G_{q_2,q_1,q_0}}(K) = P_{l+1,L}(0)E_0^{Q_{l+1,L}}[(q_2 \cdot \tilde{X}_{T_l})e^{q_1 \cdot \tilde{X}_{T_l}} I_{\{q_0 \cdot \tilde{X}_{T_l} > K\}}]$,

[3] As noted by Schrager and Pelsser [21] this approximation is affine with time-dependent coefficients. The drift change is assumed to be a deterministic function of time, which influences the restrictions on Γ, Θ, Σ, and V_t as detailed by Duffie and Kan [14]. It is emphasized by Schrager and Pelsser [21] that this drift change is small and unlikely to cause problems in practice.

the value at time zero, is given by

$$V_{G_{q_2,q_1,q_0}}(K) = P_{l+1,L}(0)\frac{\tilde{\Psi}^{Q_{l+1,L}}(q_2, q_1, \tilde{X}_t, 0, T_l)}{2}$$

$$-\frac{P_{l+1,L}(0)}{\pi}\int_0^{\infty}\frac{imag[\tilde{\Psi}^{Q_{l+1,L}}(q_2, q_1 - i\upsilon q_0, \tilde{X}_t, 0, T_l)e^{i\upsilon K}]}{\upsilon}d\upsilon,$$

where $\tilde{\Psi}^{Q_{l+1,L}}(\tilde{w}, \tilde{u}, \tilde{X}_t, 0, T_l) = E_0^{Q_{l+1,L}}[(\tilde{w} \cdot \tilde{X}_{T_l})e^{\tilde{u}\cdot\tilde{X}_{T_l}}]$, for $\tilde{w}, \tilde{u} \in \mathbb{C}^{M+1}$.

With these general pricing formulations established, it now possible to consider the specific problem of pricing swaption contracts. From equation (7.2.1) note that the price of a payer swaption at time zero is given by the general pricing equation $PS_0 = P_{l+1,L}(0)E_0^{Q_{l+1,L}}[\max(y_{l,L}(T_l) - K, 0)]$. Defining now the process vector $\tilde{X}_t = [y_{l,L}(t) \;\; X_t]'$, it follows from the above development that

$$PS_0 = V_{G_{\epsilon^{(1)},\mathbf{0},\epsilon^{(1)}}}(K) - K V_{G_{\mathbf{0},\epsilon^{(1)}}}(K), \tag{7.2.5}$$

where $\epsilon^{(1)}$ and $\mathbf{0}$ respectively denote the first basis vector and zero vector in $\mathbb{R}^{M+1}$. Note that $V_{G_{\epsilon^{(1)},\mathbf{0},\epsilon^{(1)}}}(K)$ is the time-zero value of a contingent claim that pays off $y_{l,L}(T_l)$ at time T_l contingent on $y_{l,L}(T_l) > K$, whereas $V_{G_{\mathbf{0},\epsilon^{(1)}}}(K)$ represents the time-zero value of a contingent claim that pays off 1 at time T_l if $y_{l,L}(T_l) > K$.

7.3 PRICING SWAPTIONS USING THE FFT

The transform-based and extended transform-based swaption pricing formulation proposed by Schrager and Pelsser [21] and detailed in equation (7.2.5) provides an efficient methodology for the valuation of a swaptions contract, the implementation of which in practice involves the application of two numerical integrations. However, the FFT offers even greater computational efficiency, allowing for the evaluation of swaption prices over a large range of strike prices.

In order to apply FFT techniques, it is first necessary to enforce square integrability on the two component values in equation (7.2.5), that is, to define $\tilde{V}_{G_{\epsilon^{(1)},\mathbf{0},\epsilon^{(1)}}}(K) \equiv e^{\alpha^* K}V_{G_{\epsilon^{(1)},\mathbf{0},\epsilon^{(1)}}}(K)$ and $\tilde{V}_{G_{\mathbf{0},\epsilon^{(1)}}}(K) \equiv e^{\alpha^* K}V_{G_{\mathbf{0},\epsilon^{(1)}}}(K)$, where $\alpha^* > 0 \in \mathbb{R}$. With these modifications in place it is straightforward to show that the component values can be evaluated by means of the following Fourier inversion: $V_{G_{\epsilon^{(1)},\mathbf{0},\epsilon^{(1)}}}(K) = e^{-\alpha^* K}\int_{-\infty}^{\infty}\phi_{G_{\epsilon^{(1)},\mathbf{0},\epsilon^{(1)}}}(\upsilon)e^{i2\pi\upsilon K}d\upsilon$ and

$V_{G_{0,\epsilon^{(1)}}}(K) = e^{-\alpha^* K} \int_{-\infty}^{\infty} \phi_{G_{0,\epsilon^{(1)}}}(\upsilon) e^{i2\pi \upsilon K} d\upsilon$, where

$$\phi_{G_{\epsilon^{(1)},0,\epsilon^{(1)}}}(\upsilon) = \frac{P_{l+1,L}(0)\tilde{\Psi}^{\mathcal{Q}_{l+1,L}}(\epsilon^{(1)}, i[w(\upsilon)\epsilon^{(1)} - i\alpha^*\epsilon^{(1)}], \tilde{X}_t, 0, T_l)}{\alpha^* + iw(\upsilon)}$$

and

$$\phi_{G_{0,\epsilon^{(1)}}}(\upsilon) = \frac{P_{l+1,L}(0)\tilde{\psi}^{\mathcal{Q}_{l+1,L}}(i[w(\upsilon)\epsilon^{(1)} - i\alpha^*\epsilon^{(1)}], \tilde{X}_t, 0, T_l)}{\alpha^* + iw(\upsilon)},$$

and where $\epsilon^{(1)}$ is the first basis vector in $\mathbb{R}^{M+1}$ and $w(\upsilon) \equiv -2\pi\upsilon$. Approximating each of the component inverse Fourier transforms above by means of an appropriate N-point inverse discrete Fourier transform, and then applying an appropriate inverse FFT routine, gives an N-dimensional vector of swaption prices defined as follows:

$$\mathbf{V}_{\mathbf{G}_{\epsilon^{(1)},0,\epsilon^{(1)}}}(\mathbf{K}) - \mathbf{K} \circ \mathbf{V}_{\mathbf{G}_{0,\epsilon^{(1)}}}(\mathbf{K}),$$

where

$$\mathbf{V}_{\mathbf{G}_{\epsilon^{(1)},0,\epsilon^{(1)}}}(\mathbf{K}) \equiv \left(V_{G_{\epsilon^{(1)},0,\epsilon^{(1)}}}(K_0), V_{G_{\epsilon^{(1)},0,\epsilon^{(1)}}}(K_1), \ldots, V_{G_{\epsilon^{(1)},0,\epsilon^{(1)}}}(K_{N-1})\right)'$$

and

$$\mathbf{V}_{\mathbf{G}_{0,\epsilon^{(1)}}}(\mathbf{K}) \equiv \left(V_{G_{0,\epsilon^{(1)}}}(K_0), V_{G_{0,\epsilon^{(1)}}}(K_1), \ldots, V_{G_{0,\epsilon^{(1)}}}(K_{N-1})\right)',$$

and where $\mathbf{K} \equiv (K_0, K_1, \ldots, K_{N-1})'$ is the strike price vector and the symbol "$\circ$" represents the Hadamard product.

7.4 APPLICATION AND COMPUTATIONAL ANALYSIS

This section considers a number of common affine diffusion interest-rate models studied in the literature and presents the associated transforms and extended transforms under the appropriate specializations of the general approximate, affine dynamics of equation (7.2.4). The interest-rate models considered are of the Vasicek, the generalized Vasicek, and the CIR type. Specifically, the following models are considered: the one- and two-factor Vasicek models, see [24] and [17], respectively; the one-, two-, and three-factor generalized Vasicek models, see [8]; and the one- and two-factor CIR models, see [10] and [5], respectively. The derivations of the transforms and extended transforms are easily derived and follow the general details of Duffie, Pan, and Singleton [15].

With the transforms and extended transforms derived for these various affine interest models, sample swaption prices are generated using the FFT-based swaption pricing methodology proposed in the previous section. The computational speeds of these FFT-based swaption pricing implementations are also presented. For comparative purposes, sample swaption prices and computational speeds are also presented using the swaption pricing methodology of Schrager and Pelsser [21].

7.4.1 Vasicek Models

7.4.1.1 *One-Factor Vasicek Model*

Under the equivalent martingale measure $\mathcal{Q}$, the one-factor Vasicek (1FV) model is given by

$$dr_t = \varkappa_r \left(\bar{r} - r_t\right) dt + \sigma_r dW_{r,t}^{\mathcal{Q}}.$$

Under this specification bond prices are defined by $D(t,T) = \exp(A(t,T) - B(t,T)r_t)$, where $B(t,T) \equiv \frac{1-e^{-\varkappa_r \tau}}{\varkappa_r}$,

$$A(t,T) \equiv \frac{\left(B(t,T) - \tau\right)\left(\varkappa_r^2 \bar{r} - \frac{\sigma_r^2}{2}\right)}{\varkappa_r^2} - \frac{\sigma_r^2 B^2(t,T)}{4\varkappa_r},$$

and, for notational convenience, $\tau \equiv T - t$.

Let $\tilde{X}_t \equiv [y_{l,L}(t) \;\; r_t]'$ and $\tau_l \equiv T_l$, the term-to-maturity of the swaptions contract to be priced. The associated transform of the state vector $\tilde{X}_t$ is given by

$$\tilde{\psi}^{\mathcal{Q}_{l+1,L}}(\tilde{u} \equiv (\tilde{u} \;\; 0)', \tilde{X}_t, 0, T_l) = \exp[\alpha(\tau_l) + \tilde{u} y_{l,L}(0)],$$

where $\alpha(\tau_l)$ solves the ODE

$$\dot{\alpha}(\tau_l) = \frac{1}{2}\left(\tilde{u} g(\tau_l; t)\sigma_r\right)^2,$$

subject to the boundary condition $\alpha(0) = 0$, and $g(\tau_l; t) \equiv \sum_{j=l}^{L} q_j(0) B(t, T_j)$.

The associated extended transform of $\tilde{X}_t$ is given by

$$\tilde{\Psi}^{\mathcal{Q}_{l+1,L}}(\tilde{w} \equiv (\tilde{w} \;\; 0)', \tilde{u}, \tilde{X}_t, 0, T_l) = \tilde{\psi}^{\mathcal{Q}_{l+1,L}}(\tilde{u}, \tilde{X}_t, 0, T_l)(Y(\tau_l) + \tilde{w} y_{l,L}(0)),$$

where $Y(\tau_l)$ solves ODE

$$\dot{Y}(\tau_l)=\tilde{u}\tilde{w}\left(g(\tau_l;t)\sigma_r\right)^2,$$

subject to the boundary condition $Y(0)=0$.

7.4.1.2 Two-Factor Vasicek Model

The two-factor Vasicek (2FV) model is given by

$$d\begin{bmatrix} r_t \\ \bar{r}_t \end{bmatrix} = \begin{bmatrix} \varkappa_r & 0 \\ 0 & \varkappa_{\bar{r}} \end{bmatrix}\left(\begin{bmatrix} \bar{r}_t \\ \bar{\theta} \end{bmatrix} - \begin{bmatrix} r_t \\ \bar{r}_t \end{bmatrix}\right) dt$$

$$+\begin{bmatrix} \sigma_r & 0 \\ \sigma_{\bar{r}}\rho_{r,\bar{r}} & \sigma_{\bar{r}}\sqrt{1-\rho_{r,\bar{r}}^2} \end{bmatrix} d\begin{bmatrix} W_{r,t}^{Q} \\ W_{\bar{r},t}^{Q} \end{bmatrix}.$$

Bond prices are defined by $D(t,T)=\exp\left(A(t,T)-B_1(t,T)r_t-B_2(t,T)\bar{r}_t\right)$, where $B_1(t,T)\equiv\frac{1-e^{-\varkappa_r\tau}}{\varkappa_r}$, $B_2(t,T)\equiv\frac{\varkappa_r}{\varkappa_r-\varkappa_{\bar{r}}}\left(\frac{1-e^{-\varkappa_{\bar{r}}}}{\varkappa_{\bar{r}}}-\frac{1-e^{-\varkappa_r}}{\varkappa_r}\right)$, and

$$A(t,T)\equiv\left(B_1(t,T)-\tau\right)\left(\bar{\theta}-\frac{\sigma_r^2}{2\varkappa_r^2}\right)+B_2(t,T)\bar{\theta}-\frac{\sigma_r^2 B_1^2(t,T)}{4\varkappa_r}$$

$$+\frac{\sigma_{\bar{r}}^2}{2}\left[\frac{\tau}{\varkappa_{\bar{r}}^2}-2\frac{B_2(t,T)+B_1(t,T)}{\varkappa_{\bar{r}}^2}+\frac{1}{(\varkappa_r-\varkappa_{\bar{r}})^2}\frac{1-e^{2\varkappa_{\bar{r}}^2\tau}}{2\varkappa_r}\right.$$

$$\left.-\frac{2\varkappa_r}{\varkappa_{\bar{r}}(\varkappa_r-\varkappa_{\bar{r}})^2}\frac{1-e^{-(\varkappa_r+\varkappa_{\bar{r}})\tau}}{\varkappa_r+\varkappa_{\bar{r}}}+\frac{\varkappa_r^2}{\varkappa_{\bar{r}}^2(\varkappa_r-\varkappa_{\bar{r}})^2}\frac{1-e^{-2\varkappa_{\bar{r}}\tau}}{2\varkappa_{\bar{r}}}\right].$$

Let $\tilde{X}_t\equiv[y_{l,L}(t)\ \ r_t\ \ \bar{r}_t]'$ and note that the associated transform of $\tilde{X}_t$ is given by

$$\tilde{\psi}^{Q_{l+1,L}}(\tilde{u}\equiv(\tilde{u}\ \ 0\ \ 0)',\tilde{X}_t,0,T_l)=\exp[\alpha(\tau_l)+\tilde{u}y_{l,L}(0)],$$

where $\alpha(\tau_l)$ solves the ODE

$$\dot{\alpha}(\tau_l)=\frac{1}{2}\tilde{u}^2\left[\left(g_1(\tau_l;t)\sigma_r+g_2(\tau_l;t)\sigma_{\bar{r}}\rho_{r,\bar{r}}\right)^2+\left(g_2(\tau_l;t)\sigma_{\bar{r}}\sqrt{1-\rho_{r,\bar{r}}^2}\right)^2\right],$$

subject to the boundary condition $\alpha(0)=0$, and $g_i(\tau_l;t)\equiv\sum_{j=l}^{L}q_j(0)$ $B_i(t,T_j), i=1,2$.

The associated extended transform of $\tilde{X}_t$ is given by

$$\tilde{\Psi}^{Q_{l+1,L}}(\tilde{w} \equiv (\tilde{w}\ \ 0)', \tilde{u}, \tilde{X}_t, t, T_l) = \tilde{\psi}^{Q_{l+1,L}}(\tilde{u}, \tilde{X}_t, 0, T_l)$$

$$\times\ (Y(\tau_l) + \tilde{w}\, y_{l,L}(0)),$$

where $Y(\tau_l)$ solves the ODE

$$\dot{Y}(\tau_l) = \tilde{u}\tilde{w}\left[(g_1(\tau_l;t)\sigma_r + g_2(\tau_l;t)\sigma_{\tilde{r}}\rho_{r,\tilde{r}})^2 + \left(g_2(\tau_l;t)\sigma_{\tilde{r}}\sqrt{1-\rho_{r,\tilde{r}}^2}\right)^2\right],$$

subject to $Y(0) = 0$.

7.4.1.3 Time-Dependency Approximation

Time-dependency approximations can be introduced by means of replacing the time-dependent functions $B(t, T_j)$ and $B_i(t, T_j), i = 1, 2$, by the date-zero values $B(0, T_j)$ and $B_i(0, T_j)$ in the approximate, affine dynamics of the swap rate and model variables as defined under the swap measure. The motivation for the introduction of the time-dependency approximation in the case of each model is to improve computational efficiency in the evaluation of the associated transforms and extended transforms. The time-dependency approximation allows for closed-form solutions to the transforms and extended transforms in the case of the Vasicek- and generalized Vasicek-type models, and allows for closed-form solutions to the transforms in the case of the CIR-type models. These computational efficiencies, and the computational error introduced as a result of the time-dependency approximations, are discussed in section 7.4.4.

7.4.2 Generalized Vasicek Models

7.4.2.1 One-Factor Generalized Vasicek Model

The one-factor generalized Vasicek (1FGV) model defines the short rate $r_t = \delta + x_{1,t}$, where $\delta \in \mathbb{R}$ is constant, and

$$dx_{1,t} = -\varkappa_1 x_{1,t}dt + \sigma_1 dW_{1,t}^{Q}.$$

Bond prices are given by $D(t, T) = \exp(A(t, T) + B_{\varkappa_1}(t, T)x_{1,t})$, where, in general, $B_\varkappa(t, T) \equiv \frac{1-e^{-\varkappa\tau}}{\varkappa}$ and

$$A(t, T) \equiv -\delta\tau + \frac{1}{2}\frac{\sigma_1^2}{\varkappa_1^2}\left[\tau - 2B_{\varkappa_1}(t, T) + B_{2\varkappa_1}(t, T)\right].$$

Let $\tilde{X}_t \equiv [y_{l,L}(t)\ x_{1,t}]'$ and note that the associated transform of $\tilde{X}_t$ is given by

$$\tilde{\psi}^{Q_{l+1,L}}(\tilde{u} \equiv (\tilde{u}\ 0)', \tilde{X}_t, 0, T_l) = \exp[\alpha(\tau_l) + \tilde{u} y_{l,L}(0)],$$

where $\alpha(\tau_l)$ solves the ODE

$$\dot{\alpha}(\tau_l) = \frac{1}{2}\left(\tilde{u} g_1(\tau_l; t)\sigma_1\right)^2,$$

subject to the boundary condition $\alpha(0) = 0$, and $g_1(\tau_l; t) \equiv \sum_{j=l}^{L} q_j(0) B_{\varkappa_1}(t, T_j)$.

From this it follows directly that the extended transform of $\tilde{X}_t$ is given by

$$\tilde{\Psi}^{Q_{l+1,L}}(\tilde{w} \equiv (\tilde{w}\ 0)', \tilde{u}, \tilde{X}_t, 0, T_l) = \tilde{\psi}^{Q_{l+1,L}}(\tilde{u}, \tilde{X}_t, 0, T_l)(Y(\tau_l) + \tilde{w} y_{l,L}(0)),$$

where $Y(\tau_l)$ solves the ODE

$$\dot{Y}(\tau_l) = \tilde{u}\tilde{w}\left(g_1(\tau_l; t)\sigma_1\right)^2,$$

subject to the boundary condition $Y(0) = 0$.

7.4.2.2 Two-Factor Generalized Vasicek Model

The two-factor generalized Vasicek (2FGV) model defines the short rate $r_t = \delta + x_{1,t} + x_{2,t}$, where

$$d\begin{bmatrix} x_{1,t} \\ x_{2,t} \end{bmatrix} = \begin{bmatrix} -\varkappa_1 & 0 \\ 0 & -\varkappa_2 \end{bmatrix}\begin{bmatrix} x_{1,t} \\ x_{2,t} \end{bmatrix} dt + \begin{bmatrix} \sigma_1 & 0 \\ \sigma_2\rho_{1,2} & \sigma_2\sqrt{1-\rho_{1,2}^2} \end{bmatrix} d\begin{bmatrix} W_{1,t}^{Q} \\ W_{2,t}^{Q} \end{bmatrix}.$$

Bond prices are given by $D(t, T) = \exp(A(t, T) - \sum_{i=1}^{2} B_{\varkappa_i}(t, T)x_{i,t})$, where, in general, $B_{\varkappa}(t, T) \equiv \frac{1-e^{-\varkappa\tau}}{\varkappa}$ and

$$A(t, T) \equiv -\delta\tau + \frac{1}{2}\sum_{i=1}^{2}\sum_{j=1}^{2}\frac{\sigma_i\sigma_j\rho_{i,j}}{\varkappa_i\varkappa_j} \times \left[\tau - B_{\varkappa_i}(t, T) - B_{\varkappa_j}(t, T) + B_{\varkappa_i+\varkappa_j}(t, T)\right].$$

Let $\tilde{X}_t \equiv [y_{l,L}(t)\ x_{1,t}\ x_{2,t}]'$ and note that the associated transform of $\tilde{X}_t$ is given by

$$\tilde{\psi}^{Q_{l+1,L}}(\tilde{u} \equiv [\tilde{u}\ 0\ 0]', \tilde{X}_t, 0, T_l) = \exp[\alpha(\tau_l) + \tilde{u} y_{l,L}(0)],$$

where $\alpha(\tau_l)$ solves the ODE

$$\dot{\alpha}(\tau_l) = \frac{1}{2}\tilde{u}^2\left[(g_1(\tau_l;t)\sigma_1 + g_2(\tau_l;t)\sigma_2\rho_{1,2})^2 + \left(g_2(\tau_l;t)\sigma_2\sqrt{1-\rho_{1,2}^2}\right)^2\right],$$

subject to the boundary condition $\alpha(0)=0$, and $g_i(\tau_l;t) \equiv \sum_{j=l}^{L} q_j(0) B_{\varkappa_i}(t,T_j), i = 1,2$.

It follows that the extended transform of $\tilde{X}_t$ is given by

$$\tilde{\Psi}^{Q_{l+1,L}}(\tilde{w} \equiv (\tilde{w}\ \ 0\ \ 0)', \tilde{u}, \tilde{X}_t, 0, T_l) = \tilde{\psi}^{Q_{l+1,L}}(\tilde{u}, \tilde{X}_t, 0, T_l) \times (Y(\tau_l) + \tilde{w}\, y_{l,L}(0)),$$

where $Y(0)$ solves the ODE

$$\dot{Y}(\tau_l) = \tilde{u}\tilde{w}\left[(g_1(\tau_l;t)\sigma_1 + g_2(\tau_l;t)\sigma_2\rho_{1,2})^2 + \left(g_2(\tau_l;t)\sigma_2\sqrt{1-\rho_{1,2}^2}\right)^2\right],$$

subject to the boundary condition $Y(0) = 0$.

7.4.2.3 *Three-Factor Generalized Vasicek Model*

The three-factor generalized Vasicek (3FGV) model defines the short rate $r_t = \delta + x_{1,t} + x_{2,t} + x_{3,t}$, where

$$d\begin{bmatrix} x_{1,t} \\ x_{2,t} \\ x_{3,t} \end{bmatrix} = \begin{bmatrix} -\varkappa_1 & 0 & 0 \\ 0 & -\varkappa_2 & 0 \\ 0 & 0 & -\varkappa_3 \end{bmatrix}\begin{bmatrix} x_{1,t} \\ x_{2,t} \\ x_{3,t} \end{bmatrix} dt + \Sigma d\begin{bmatrix} W_{1,t}^{Q} \\ W_{2,t}^{Q} \\ W_{3,t}^{Q} \end{bmatrix}$$

and, for notational convenience,

$$\Sigma \equiv \begin{bmatrix} \sigma_1 & 0 & 0 \\ \sigma_2\rho_{1,2} & \sigma_2\sqrt{1-\rho_{1,2}^2} & 0 \\ \sigma_3\rho_{1,3} & \sigma_3\sqrt{\frac{(\rho_{2,3}-\rho_{1,3}\rho_{1,2})^2}{1-\rho_{1,2}^2}} & \sigma_3\sqrt{\frac{(1-\rho_{1,2}^2)(1-\rho_{1,3}^2)-(\rho_{2,3}-\rho_{1,3}\rho_{1,2})^2}{1-\rho_{1,2}^2}} \end{bmatrix}.$$

Bond prices are given by $D(t,T) = \exp(A(t,T) - \sum_{i=1}^{3} B_{\varkappa_i}(t,T)x_{i,t})$, where, in general, $B_{\varkappa}(t,T) \equiv \frac{1-e^{-\varkappa\tau}}{\varkappa}$ and

$$A(t,T) \equiv -\delta\tau + \frac{1}{2}\sum_{i=1}^{3}\sum_{j=1}^{3}\frac{\sigma_i\sigma_j\rho_{i,j}}{\varkappa_i\varkappa_j}$$

$$\times\left[\tau - B_{\varkappa_i}(t,T) - B_{\varkappa_j}(t,T) + B_{\varkappa_i+\varkappa_j}(t,T)\right].$$

Let $\tilde{X}_t \equiv [y_{l,L}(t)\ x_{1,t}\ x_{2,t}\ x_{3,t}]'$ and note that the associated transform of $\tilde{X}_t$ is given by

$$\tilde{\psi}^{\mathcal{Q}_{l+1,L}}(\tilde{u} \equiv (\tilde{u}\ 0\ 0\ 0)', \tilde{X}_t, 0, T_l) = \exp[\alpha(\tau_l) + \tilde{u}y_{l,L}(0)],$$

where $\alpha(\tau_l)$ solves the ODE

$$\dot{\alpha}(\tau_l) = \frac{1}{2}\tilde{u}^2\{[\Sigma_{11}g_1(\tau_l;t) + \Sigma_{21}g_2(\tau_l;t) + \Sigma_{31}g_3(\tau_l;t)]^2$$

$$+ [\Sigma_{22}g_2(\tau_l;t) + \Sigma_{32}g_3(\tau_l;t)]^2 + [\Sigma_{33}g_3(\tau_l;t)]^2\},$$

subject to the boundary condition $\alpha(0) = 0$, and $g_i(\tau_l;t) \equiv \sum_{j=l}^{L} q_j(0)$ $B_{\varkappa_i}(t,T_j)$, $i = 1,2,3$.

From this it follows that the extended transform of $\tilde{X}_t$ is given by

$$\tilde{\Psi}^{\mathcal{Q}_{l+1,L}}(\tilde{w} \equiv (\tilde{w}\ 0\ 0\ 0)', \tilde{u}, \tilde{X}_t, 0, T_l)$$

$$= \tilde{\psi}^{\mathcal{Q}_{l+1,L}}(\tilde{u}, \tilde{X}_t, 0, T_l)(Y(\tau_l) + \tilde{w}y_{l,L}(0)),$$

where $Y(\tau_l)$ solves the ODE

$$\dot{Y}(\tau_l) = \tilde{u}\tilde{w}\{[\Sigma_{11}g_1(\tau_l;t) + \Sigma_{21}g_2(\tau_l;t) + \Sigma_{31}g_3(\tau_l;t)]^2$$

$$+ [\Sigma_{22}g_2(\tau_l;t) + \Sigma_{32}g_3(\tau_l;t)]^2 + [\Sigma_{33}g_3(\tau_l;t)]^2\},$$

subject to $Y(0) = 0$.

7.4.2.4 *Time-Dependency Approximation*

Similar to that done previously, time-dependency approximations can be introduced by means of replacing the time-dependent functions $B_\kappa(t,T_j)$ by the date-zero values $B_\kappa(0,T_j)$ in the swap measure approximate, affine dynamics of the swap rate and model variables. Computational efficiencies and error are discussed in section 7.4.4.

7.4.3 CIR Models

7.4.3.1 *One-Factor CIR Model*

Under the specification, of the one-factor CIR (1FCIR) model

$$dr_t = \varkappa_r (\bar{r} - r_t)\, dt + \sigma_r \sqrt{r_t} dW^{Q}_{r,t},$$

where bond prices are given by $D(t, T) = \exp(A(t, T) - B(t, T)r_t)$, where $B(t, T) \equiv \frac{2(e^{\delta_r \tau}-1)}{(\delta_r+\varkappa_r)(e^{\delta_r \tau}-1)+2\delta_r}$,

$$A(t, T) \equiv \ln\left[\left(\frac{2\delta_r e^{(\delta_r+\varkappa_r)\tau/2}}{(\delta_r + \varkappa_r)(e^{\delta_r \tau} - 1) + 2\delta_r}\right)^{2\varkappa_r \bar{r}/\sigma_r^2}\right],$$

and $\delta_r \equiv \sqrt{\varkappa_r^2 + 2\sigma_r^2}$.

Let $\tilde{X}_t \equiv [y_{l,L}(t)\;\; r_t]'$ and note that the associated transform of $\tilde{X}_t$ is given by

$$\tilde{\psi}^{Q_{l+1,L}}(\tilde{u} \equiv (\tilde{u}\;\; 0)', \tilde{X}_t, 0, T_l) = \exp[\alpha(\tau_l) + \tilde{u} y_{l,L}(0) + \beta(\tau_l) r_0],$$

where $\beta(\tau_l)$ and $\alpha(\tau_l)$ solve the ODEs

$$\dot{\beta}(\tau_l) = \frac{1}{2}\sigma_r^2 \beta^2(\tau_l) + \left(\tilde{u} g(\tau_l; t)\sigma_r^2 - \varkappa_r - \sigma_r^2 h(\tau_l; t)\right)\beta(\tau_l) + \frac{1}{2}(\tilde{u} g(\tau_l; t)\sigma_r)^2,$$

$$\dot{\alpha}(\tau_l) = \varkappa_r \bar{r} \beta(\tau_l),$$

subject to the boundary conditions $\beta(0) = 0$ and $\alpha(0) = 0$, and where $g(\tau_l; t) \equiv \left(\sum_{j=l}^{L} q_j(0) B(t, T_j)\right)$ and $h(\tau_l; t) \equiv \left(\sum_{j=l+1}^{L} \Delta_{j-1} B(t, T_j) \frac{D(0,T_j)}{P_{l+1,L}(0)}\right)$.

From this it follows that the extended transform of $\tilde{X}_t$ is given by

$$\begin{aligned}\tilde{\Psi}^{Q_{l+1,L}}&(\tilde{w} \equiv (\tilde{w}\;\; 0)', \tilde{u}, \tilde{X}_t, 0, T_l)\\ &= \tilde{\psi}^{Q_{l+1,L}}(\tilde{u}, \tilde{X}_t, 0, T_l)(Y(\tau_l) + \tilde{w} y_{l,L}(0) + Z_2(\tau_l) r_0),\end{aligned}$$

where $Z_2(\tau_l)$ and $Y(\tau_l)$ solve the ODEs

$$\begin{aligned}\dot{Z}_2(\tau_l) &= \left[-\left(\varkappa_r + \sigma_r^2 h(\tau_l; t)\right) + \tilde{u} g(\tau_l; t)\sigma_r^2 + \sigma_r^2 \beta(\tau_l)\right] Z_2(\tau_l)\\ &\quad + \left[\tilde{u}\,(g(\tau_l; t)\sigma_r)^2 + g\,(\tau_l; t)\,\sigma_r^2 \beta(\tau_l)\right]\tilde{w},\end{aligned}$$

$$\dot{Y}(\tau_l) = \varkappa_r \bar{r} Z_2(\tau_l),$$

subject to the boundary conditions $Z_2(0) = 0$ and $Y(0) = 0$.

7.4.3.2 Two-Factor CIR Model

The two-factor CIR (2FCIR) model defines the short rate $r_t = x_{1,t} + x_{2,t}$, where

$$d\begin{bmatrix} x_{1,t} \\ x_{2,t} \end{bmatrix} = \begin{bmatrix} \varkappa_1 & 0 \\ 0 & \varkappa_2 \end{bmatrix} \left(\begin{bmatrix} \bar{x}_1 \\ \bar{x}_2 \end{bmatrix} - \begin{bmatrix} x_{1,t} \\ x_{2,t} \end{bmatrix} \right) dt + \begin{bmatrix} \sigma_1 \sqrt{x_{1,t}} & 0 \\ 0 & \sigma_2 \sqrt{x_{2,t}} \end{bmatrix} d \begin{bmatrix} W_{1,t}^{Q} \\ W_{2,t}^{Q} \end{bmatrix}.$$

Bond prices are given by $D(t,T) = \exp(A(t,T) - B_1(t,T)x_{1,t} - B_2(t,T)x_{2,t})$, where $A(t,T) \equiv \ln[A_1(t,T)A_2(t,T)]$ and, for $i = 1, 2$,

$$A_i(t,T) \equiv \left(\frac{2\delta_i e^{(\delta_i + \varkappa_i)\tau/2}}{(\delta_i + \varkappa_i)(e^{\delta_i \tau} - 1) + 2\delta_i} \right)^{2\varkappa_i \bar{x}_i / \sigma_i^2}$$

and

$$B_i(t,T) \equiv \frac{2(e^{\delta_i \tau} - 1)}{(\delta_i + \varkappa_i)(e^{\delta_i \tau} - 1) + 2\delta_i}.$$

Let $\tilde{X}_t \equiv [y_{l,L}(t) \;\; x_{1,t} \;\; x_{2,t}]'$ and note that the associated transform of $\tilde{X}_t$ is given by

$$\tilde{\psi}^{Q_{l+1,L}}(\tilde{u} \equiv (\tilde{u} \;\; 0 \;\; 0)', \tilde{X}_t, 0, T_l) = \exp[\alpha(\tau_l) + \tilde{u} y_{l,L}(0) + \beta_1(\tau_l)x_{1,0} + \beta_2(\tau_l)x_{2,0}],$$

where the functions $\beta_1(\tau_l), \beta_2(\tau_l)$, and $\alpha(\tau_l)$ solve the ODEs

$$\dot{\beta}_1(\tau_l) = \frac{1}{2}\sigma_1^2 \beta_1^2(\tau_l) + \left(\tilde{u} g_1(\tau_l;t)\sigma_1^2 - \varkappa_1 - \sigma_1^2 h_1(\tau_l;t) \right) \beta_1(\tau_l) + \frac{1}{2}\left(\tilde{u} g_1(\tau_l;t)\sigma_1\right)^2,$$

$$\dot{\beta}_2(\tau_l) = \frac{1}{2}\sigma_2^2 \beta_2^2(\tau_l) + \left(\tilde{u} g_2(\tau_l;t)\sigma_2^2 - \varkappa_2 - \sigma_2^2 h_2(\tau_l;t) \right) \beta_2(\tau_l) + \frac{1}{2}\left(\tilde{u} g_2(\tau_l;t)\sigma_2\right)^2,$$

$$\dot{\alpha}(\tau_l) = \varkappa_1 \bar{x}_1 \beta_1(\tau_l) + \varkappa_2 \bar{x}_2 \beta_2(\tau_l),$$

subject to the boundary conditions $\beta_1(0) = \beta_2(0) = \alpha(0) = 0$, and $g_i(\tau_l; t) \equiv \sum_{j=l}^{L} q_j(0) B_i(t, T_j)$ and $h_i(\tau_l; t) \equiv \sum_{j=l+1}^{L} \Delta_{j-1} B_i(t, T_j) \frac{D(0,T_j)}{P_{l+1,L}(0)}$, for $i = 1, 2$.

In a similar manner, the extended transform $\tilde{\Psi}^{Q_{l+1,L}}(\tilde{w}, \tilde{u}, \tilde{X}_t, 0, T_l) \equiv \tilde{\Psi}^{Q_{l+1,L}}((\tilde{w}\ 0\ 0)', (\tilde{u}\ 0\ 0)', \tilde{X}_t, 0, T_l)$ is given by the following general form:

$$\tilde{\Psi}^{Q_{l+1,L}}(\tilde{w}, \tilde{u}, \tilde{X}_t, 0, T_l) = \tilde{\psi}^{Q_{l+1,L}}(\tilde{u}, \tilde{X}_t, 0, T_l)$$

$$\times (Y(\tau_l) + \tilde{w} y_{l,L}(0) + Z_2(\tau_l) x_{1,0} + Z_3(\tau_l) x_{2,0}),$$

where $Z_2(\tau_l)$, $Z_3(\tau_l)$, and $Y(\tau_l)$ solve the ODEs

$$\dot{Z}_2(\tau_l) = \left[- \left(\varkappa_1 + \sigma_1^2 h_1(\tau_l; t)\right) + \tilde{u} g_1(\tau_l; t)\sigma_1^2 + \sigma_1^2 \beta_1(\tau_l)\right] Z_2(\tau_l)$$

$$+ \left[\tilde{u} g_1^2(\tau_l; t)\sigma_1^2 + g_1(\tau_l; t)\sigma_1^2 \beta_1(\tau_l)\right] \tilde{w},$$

$$\dot{Z}_3(\tau_l) = \left[- \left(\varkappa_2 + \sigma_2^2 h_2(\tau_l; t)\right) + \tilde{u} g_2(\tau_l; t)\sigma_2^2 + \sigma_2^2 \beta_2(\tau_l)\right] Z_3(\tau_l)$$

$$+ \left[\tilde{u} g_2^2(\tau_l; t)\sigma_2^2 + g_2(\tau_l; t)\sigma_2^2 \beta_2(\tau_l)\right] \tilde{w},$$

$$\dot{Y}(\tau) = \varkappa_1 \bar{x}_1 Z_2(\tau_l) + \varkappa_2 \bar{x}_2 Z_3(\tau_l),$$

subject to the boundary conditions $Z_2(0) = 0$, $Z_3(0) = 0$, and $Y(0) = 0$.

7.4.3.3 Time-Dependency Approximation

Again, time-dependency approximations can be introduced by means of replacing the time-dependent functions $B(t, T_j)$ and $B_i(t, T_j), i = 1, 2$, by the date-zero values $B(0, T_j)$ and $B_i(0, T_j)$ in the approximate, affine dynamics of the swap rate and model variables as defined under the swap measure. Computational efficiencies and error are discussed in the next section.

7.4.4 Computational Analysis

The purpose of this section is to provide a comprehensive discussion of the computational results and speeds of the FFT-based swaption pricing methodology proposed here. For this a distinction is made between the following alternative integral transform-based swaption pricing implementations: (a) the FFT-based swaption pricing methodology, under which is assumed the Schrager and Pelsser [21] affine approximation only, herein labeled the FFT-SP methodology; (b) the FFT-based swaption pricing methodology, under which is assumed the Schrager and Pelsser [21] affine approximation and time-dependency approximation, herein labeled the FFT-SP-TDA methodology; and (c) the transform-based and extended transform-based swaption pricing methodology proposed by Schrager and Pelsser [21], under which is assumed the Schrager and Pelsser [21] affine approximation and time-dependency approximation, herein labeled the T-SP-TDA methodology. Computational results are generated under each of the alternative swaption pricing methodologies and under each of the alternative affine interest rate models considered in the previous sections; see Tables 7.4.1–7.4.7. The swaption contracts priced are assumed to be 1-year forward-payer swaptions written on underlying 4-year swap contracts that pay annual fixed payments, i.e., 1×4 swaption contracts. Note that, for the FFT-based swaption pricing methodologies, a high resolution of $N = 2^{16}$ is used, along with a strike price spacing of 39 basis points.

Note further that, to benchmark the pricing effect of the original Schrager and Pelsser [21] approximation, swaption prices are also reported in Tables 7.4.1–7.4.7 that are generated using Monte Carlo simulations (MC-SIM) of the nonaffine exact swap measure dynamics related to the alternative affine model specifications considered, which represent specializations of the general exact swap measure dynamics given in equations (7.2.2) and (7.2.3). For each of the Monte Carlo simulation implementations, note that 200,000 simulations and 100 time steps per simulation are used in generating the results.

TABLE 7.4.1 Swaption Prices: 1FV Model

	FFT-SP-TDA Run Time: 1.1720 s	T-SP-TDA Avg. Run Time: 0.0160 s	FFT-SP Run Time: 952.8 s	MC-SIM Run Time: 2516.4 s
Strikes	Prices	Prices	Prices	Prices
0.0134	0.0727	0.0727	0.0732	0.0730
0.0173	0.0605	0.0605	0.0612	0.0610
0.0212	0.0491	0.0491	0.0500	0.0498
0.0251	0.0388	0.0388	0.0399	0.0396
0.0290	0.0298	0.0298	0.0309	0.0307
ATMF: 0.0329	0.0221	0.0221	0.0232	0.0231
0.0368	0.0158	0.0158	0.0169	0.0169
0.0407	0.0109	0.0109	0.0119	0.0119
0.0446	0.0072	0.0072	0.0080	0.0081
0.0485	0.0045	0.0045	0.0052	0.0053
0.0524	0.0027	0.0027	0.0033	0.0033

Parameters: $\varkappa_r = 0.1$; $\bar{r} = 0.045$; $\sigma_r = 0.02$; $r_t = 0.03$.

TABLE 7.4.2 Swaption Prices: 2FV Model

	FFT-SP-TDA Run Time: 1.1880s	T-SP-TDA Avg. Run Time: 0.0329 s	FFT-SP Run Time: 1406.0 s	MC-SIM Run Time: 5939.8 s
Strikes	Prices	Prices	Prices	Prices
0.0014	0.0756	0.0756	0.0761	0.0762
0.0053	0.0628	0.0628	0.0635	0.0635
0.0092	0.0508	0.0508	0.0516	0.0517
0.0131	0.0399	0.0399	0.0409	0.0410
0.0171	0.0304	0.0304	0.0315	0.0316
ATMF: 0.0210	0.0223	0.0223	0.0235	0.0236
0.0249	0.0158	0.0158	0.0169	0.0171
0.0288	0.0107	0.0107	0.0117	0.0119
0.0327	0.0069	0.0069	0.0078	0.0080
0.0366	0.0043	0.0043	0.0050	0.0051
0.0405	0.0025	0.0025	0.0031	0.0032

Parameters: $\varkappa_r = 0.1$; $\mathcal{K}_{\bar{r}} = 0.25$; $\bar{\theta} = 0.035$; $\sigma_r = 0.02$; $\sigma_{\bar{r}} = 0.022$; $\rho = -0.2$; $r_t = 0.03$; $\bar{r}_t = 0.05$.

TABLE 7.4.3 Swaption Prices: 1FGV Model

	FFT-SP-TDA Run Time: 1.2340 s	T-SP-TDA Avg. Run Time: 0.0470 s	FFT-SP Run Time: 935.8 s	MC-SIM Run Time: 2945.8 s
Strikes	Prices	Prices	Prices	Prices
0.0134	0.0727	0.0727	0.0732	0.0732
0.0173	0.0605	0.0605	0.0612	0.0612
0.0212	0.0491	0.0491	0.0500	0.0500
0.0251	0.0388	0.0388	0.0399	0.0399
0.0290	0.0298	0.0298	0.0309	0.0309
ATMF: 0.0329	0.0221	0.0221	0.0232	0.0233
0.0368	0.0158	0.0158	0.0169	0.0169
0.0407	0.0109	0.0109	0.0119	0.0119
0.0446	0.0072	0.0072	0.0080	0.0081
0.0485	0.0045	0.0045	0.0052	0.0053
0.0524	0.0027	0.0027	0.0033	0.0033

Parameters: $\varkappa_1 = 0.1$; $\sigma_1 = 0.02$; $r_t = 0.03$; $x_{1,t} = -0.015$.

TABLE 7.4.4 Swaption Prices: 2FGV Model

	FFT-SP-TDA Run Time: 1.2500 s	T-SP-TDA Avg. Run Time: 0.0620 s	FFT-SP Run Time: 1193.0 s	MC-SIM Run Time: 5826.2 s
Strikes	Prices	Prices	Prices	Prices
0.0255	0.0663	0.0663	0.0663	0.0664
0.0294	0.0532	0.0532	0.0533	0.0534
0.0333	0.0405	0.0405	0.0407	0.0408
0.0372	0.0287	0.0287	0.0291	0.0291
0.0411	0.0185	0.0185	0.0190	0.0191
ATMF: 0.0450	0.0106	0.0106	0.0112	0.0112
0.0489	0.0053	0.0053	0.0058	0.0059
0.0528	0.0022	0.0022	0.0026	0.0026
0.0567	0.0008	0.0008	0.0010	0.0010
0.0606	0.0002	0.0002	0.0003	0.0003
0.0645	0.0001	0.0001	0.0001	0.0001

Parameters: $\varkappa_1 = 0.1$; $\varkappa_2 = 0.2$; $\sigma_1 = 0.01$; $\sigma_2 = 0.005$; $\delta = 0.025$; $\rho_{1,2} = -0.2$; $x_{1,t} = 0.01$; $x_{2,t} = 0.02$.

TABLE 7.4.5 Swaption Prices: 3FGV Model

	FFT-SP-TDA Run Time: 1.2660 s	T-SP-TDA Avg. Run Time: 0.0630 s	FFT-SP Run Time: 1857.6 s	MC-SIM Run Time: 8658.2 s
Strikes	Prices	Prices	Prices	Prices
0.0472	0.0630	0.0630	0.0630	0.0629
0.0511	0.0506	0.0506	0.0507	0.0506
0.0550	0.0386	0.0386	0.0388	0.0387
0.0589	0.0275	0.0275	0.0278	0.0277
0.0628	0.0179	0.0179	0.0183	0.0182
ATMF: 0.0667	0.0104	0.0104	0.0109	0.0108
0.0706	0.0053	0.0053	0.0057	0.0057
0.0745	0.0023	0.0023	0.0026	0.0026
0.0784	0.0008	0.0008	0.0010	0.0011
0.0823	0.0003	0.0003	0.0004	0.0004
0.0862	0.0001	0.0001	0.0001	0.0001

Parameters: $\varkappa_1 = 0.1$; $\varkappa_2 = 0.2$; $\varkappa_3 = 0.5$; $\sigma_1 = 0.01$; $\sigma_2 = 0.005$; $\sigma_3 = 0.002$; $\delta = 0.06$; $\rho_{1,2} = -0.2$; $\rho_{1,3} = -0.1$; $\rho_{2,3} = 0.3$; $x_{1,t} = 0.01$; $x_{2,t} = 0.005$; $x_{3,t} = -0.02$.

TABLE 7.4.6 Swaption Prices: 1FCIR Model

	FFT-SP-TDA Run Time: 495.1 s	T-SP-TDA Avg. Run Time: 2.0620 s	FFT-SP Run Time: 2073.3 s	MC-SIM Run Time: 3445.2 s
Strikes	Prices	Prices	Prices	Prices
0.0108	0.0699	0.0699	0.0727	0.0724
0.0147	0.0573	0.0573	0.0626	0.0625
0.0186	0.0472	0.0472	0.0541	0.0539
0.0225	0.0390	0.0390	0.0463	0.0465
0.0264	0.0322	0.0322	0.0399	0.0401
ATMF: 0.0303	0.0268	0.0268	0.0339	0.0345
0.0342	0.0223	0.0223	0.0292	0.0297
0.0381	0.0188	0.0188	0.0246	0.0255
0.0420	0.0158	0.0158	0.0212	0.0220
0.0459	0.0137	0.0137	0.0177	0.0189
0.0499	0.0117	0.0117	0.0152	0.0163

Parameters: $\varkappa_r = 0.1$; $\bar{r} = 0.045$; $\sigma_r = 0.2$; $r_t = 0.03$.

TABLE 7.4.7 Swaption Prices: 2FCIR Model

	FFT-SP-TDA Run Time: 583.6 s	T-SP-TDA Avg. Run Time: 1.8590 s	FFT-SP Run Time: 4090.3 s	MC-SIM Run Time: 6977.4 s
Strikes	Prices	Prices	Prices	Prices
0.0272	0.0617	0.0617	0.0663	0.0692
0.0311	0.0475	0.0475	0.0597	0.0585
0.0350	0.0350	0.0350	0.0538	0.0491
0.0389	0.0248	0.0248	0.0422	0.0408
0.0428	0.0167	0.0167	0.0372	0.0340
ATMF: 0.0467	0.0108	0.0108	0.0285	0.0277
0.0506	0.0066	0.0066	0.0240	0.0226
0.0545	0.0037	0.0037	0.0180	0.0185
0.0584	0.0019	0.0019	0.0149	0.0150
0.0623	0.0009	0.0009	0.0108	0.0119
0.0662	0.0003	0.0003	0.0090	0.0098

Parameters: $\varkappa_1 = 0.2$; $\varkappa_2 = 0.1$; $\bar{x}_1 = 0.025$; $\bar{x}_2 = 0.03$; $\sigma_1 = 0.2$; $\sigma_2 = 0.1$; $x_{1,t} = 0.02$; $x_{2,t} = 0.025$.

In the comparison of the FFT-SP and Monte Carlo simulation swaption prices reported, it is first noted that the former methodology effectively prices for all contracts under the various Vasicek and generalized Vasicek models considered. In particular, it is noted for the in-the-money-forward (ITMF) and at-the-money-forward (ATMF) swaptions that the approximation introduced by Schrager and Pelsser [21] is excellent, with pricing errors relative to the corresponding Monte Carlo simulation prices of less than 1%. For the case of the 1FCIR model, the effect of the approximation introduced by Schrager and Pelsser [21] is again noted to be quite effective. For the case of the ITMF and ATMF swaption contracts, pricing errors relative to the corresponding Monte Carlo simulation prices are observed to be less than 2%. However, for the out-of-the-money-forward (OTMF) swaption contracts, it is noted that the approximation is not as effective compared with the Vasicek and generalized Vasicek models, with pricing errors relative to the corresponding Monte Carlo simulation prices in the range 1.7–6.7%. However, for the 2FCIR model, it is noted that the Schrager and Pelsser [21] approximation is much less effective for all moneyness-category swaptions, with pricing errors relative to the corresponding Monte Carlo simulation prices in the range 2.1–9.5%.

With the pricing results of the FFT-SP and Monte Carlo simulation methodologies compared, focus is now turned to the computational run times of the alternative methodologies. It is first noted for the 1FV and 2FV models that the computational run time of the FFT-SP methodology is significantly lower than the Monte Carlo simulation approach, with reductions on the order of, approximately, 62% and 76%, respectively. Likewise, for the generalized Vasicek models, the FFT-SP pricing methodology results in significantly lower computational run times relative to the Monte Carlo simulation approach, with reductions on the order of, approximately, 68, 80, and 79% for the 1FGV, 2FGV, and 3FGV models, respectively. Finally, it is similarly observed that the FFT-SP swaption pricing methodology results in significant computational run-time efficiencies for the 1FCIR and 2FCIR models, with reductions on the order of, approximately, 40% for both models.

Now, comparing the computational results and speeds under the alternative FFT-SP-TDA and T-SP-TDA methodologies for all models, it is noted that the former methodology generates equivalent results to the latter while offering substantial computational efficiencies, particularly given the high resolution used for the FFT implementations. To better understand the pricing error introduced by the time-dependency approximation under both swaption pricing methodologies, swaption prices are compared with the results generated by the FFT-SP methodology. It is most notable that, for the various Vasicek and generalized Vasicek models considered, the time-dependency approximations made are quite effective in the pricing of ITMF and ATMF swaption contracts. In particular, for these swaption contracts, it can be observed that pricing errors relative to the corresponding FFT-SP prices are all less than 5.5%, with the level of relative pricing error decreasing as the ITMF swaption contracts become more in-the-money. However, for the OTMF contracts, the effectiveness of the time-dependency approximations made is not as good, with pricing errors relative to the corresponding FFT-SP prices observed in the range 6.5–33%. In fact, it is generally observed that the level of relative pricing error increases as the OTMF swaption contracts become more out-the-money. Hence, from a pricing perspective, the time-dependency approximations made are quite effective in the pricing of ITMF and ATMF swaption contracts for all the Vasicek and generalized Vasicek models considered.

From the perspective of computational run time, it is notable that, for the various Vasicek and generalized Vasicek models, the FFT-SP-TDA pricing methodology results in highly significant reductions in run time compared with the FFT-SP methodology, with reductions on the order of over 99%. Such dramatic reductions in run time reflect the fact that, under the time-dependency approximations made for the various Vasicek and generalized Vasicek models, the required transforms and extended transforms for the implementations are derived in analytic form, eliminating the need for any numerical evaluations. Hence, in summary of the above discussions, it is concluded that the FFT-SP-TDA swaption-pricing methodology quite effectively prices ITMF and ATMF swaption contracts, but does so at much greater computational speeds when compared with the FFT-SP pricing methodology.

Turning now to the CIR-type models, it is first noted for the 1FCIR model that the time-dependency approximation made in the implementation of the FFT-SP-TDA and T-SP-TDA pricing methodologies is actually quite ineffective in the pricing of all swaption contracts, showing consistent evidence of significant underpricing relative to the corresponding FFT-SP prices. In particular, it is noted for the ATMF swaptions contract that the relative pricing error is approximately 21%, with pricing errors decreasing from this as the ITMF swaption contracts become more in-the-money. In the case of the 2FCIR model, it is noted that the ineffectiveness of the time-dependency approximation is even more pronounced, with the FFT-SP-TDA and T-SP-TDA pricing methodologies significantly underpricing relative to the FFT-SP methodology, particularly in the case of the OTMF swaption contracts.

7.5 MODEL TESTING USING EURIBOR SWAPTIONS DATA

7.5.1 Implied Estimation and In-Sample Pricing Performance

Using an extensive EURIBOR payer-swaptions data set, this section presents an implied parameter estimation of the 1FV, 2FV, 1FGV, 2FGV, 3FGV, and 1FCIR models and evaluates the performance of these alternative models on the basis of in-sample pricing performance. The data set comprises swaption quotes on a selection of weekly observations over the period 1/1/2001–12/31/2004, obtained from Datastream and equating to 209 trade dates. The swaption quotes represent at-the-money-forward (ATMF) quotes for 44 different contract types,[4] and are given with the typical market convention of Black-implied volatilities. Hence, the entire data set consists of a total of 9196 swaption quotes. Note also that, by market convention, the swaptions are written on swap contracts that pay six-month EURIBOR against a fixed annual payment; see, for instance, Muck [19].

The associated structural parameters and latent-state variables, as defined under each of the competing model specifications, are estimated according to the following daily nonlinear ordinary least squares (NL-OLS) specification:

$$\min_{\{\theta\}} \sum_{i=1}^{N_t} [C_{i,t} - \hat{C}_{i,t}(\theta)]^2,$$

where $\theta \equiv \{\theta, \vartheta\}$ and where θ and ϑ are, respectively, the date-t vectors of structural parameters and latent state variables; $C_{i,t}$ is the market price corresponding to swaption quote i on date t; $\hat{C}_{i,t}\,(\theta)$ is the theoretical model-based price of the swaption with the same contract details as swaption quote i on date t; and N_t is the total number of swaptions observed on date t, which in this case is equal to 44 for each trade date. The theoretical swaption prices are generated using the FFT-SP-TDA swaption pricing methodology, as detailed and discussed in the previous section. However, this swaption pricing methodology is effective in the pricing of ATMF swaption contracts for the Vasicek and generalized Vasicek models. The second motivation for this choice of pricing methodology is the dramatic computational run-time efficiencies it offers over the FFT-SP pricing methodology, which in turn naturally reduce the computational intensity of the required optimization under the NL-OLS

[4] The EURIBOR swaptions data set comprises 6-month, 1-year, 2-year, 3-year, 4-year, and 5-year forward contracts written on underlying swap contracts with a variety of maturities. Quotes are provided for nine 6-month forward, nine 1-year forward, eight 2-year forward, seven 3-year forward, six 4-year forward, and five 5-year forward contracts.

TABLE 7.5.1 Minimized Loss Function Values of Candidate Models

Model	Minimized Loss Function Value
1FV	0.00243
2FV	0.00093
1FGV	0.00715
2FGV	0.00572
3FGV	0.00135
1FCIR	0.00156

parameter estimation process. However, it is important to point out that, although the FFT-SP-TDA pricing methodology is not particularly effective for the pricing of ATMF swaption contracts under the 1FCIR model, the decision is made here to trade off this pricing error for the significant computational efficiencies it offers. Note also for this implementation that the 1-month EURIBOR rate observed on date t is used as a proxy for the unobservable short rate r_t.[5]

Table 7.5.1 provides the average of the daily minimized loss function values obtained for each model over the sample period. It is notable from Table 7.5.1 that, according to the average minimized loss function values, the 2FV model provides the best overall fit to the EURIBOR swaptions data set under consideration, followed in order by the 3FGV, 1FCIR, 1FV, 2FGV, and 1FGV models. The results obtained suggest that the 2FV model outperforms the higher dimensional 3FGV model, and the 1FV model similarly outperforms the higher dimensional 2FGV model. However, within the Vasicek type, it is interesting to note that the higher dimensional 2FV model, which extends the 1FV model by allowing the long-run mean short-rate level to follow a correlated mean-reverting Gaussian process, improves considerably on the in-sample fit of the 1FV model. In a similar manner, it is noted that, for the generalized Vasicek model type, the in-sample fit also improves with increased dimensionality. Furthermore, of the one-factor models, the results indicate that the 1FCIR model offers the best in-sample fit relative to the 1FV and 1FGV model specifications.

In addition to the above observations, note that, despite the equivalence of the 1FV and 1FGV models as established in section 7.4.2.1, it is the 1FV model representation, which specifies the short rate as an observable state variable,

[5] Note that, under the specifications of the one-, two-, and three-factor generalized Vasicek models, the parameter δ is not estimated, given that it is bound by the model constraint $\delta = r_t - \sum_i x_{i,t}$.

that is reported to provide a better fit relative to the alternative 1FGV model representation. In a similar way, despite the equivalence of the 2FV and 2FGV models the results report that the 2FV model representation, which again specifies the short rate as an observable state variable, is the one that provides a better fit over the 2FGV model representation. Hence, the evidence suggests that even within the same affine-model equivalence class, as made formal by Dai and Singleton [11], the specific representation of an assumed model is an important factor in optimizing in-sample pricing performance. Indeed, it appears that those models that specify the short rate as an observable state variable offer a superior fit relative to those equivalent models that define the short rate by an affine relation of latent state variables.

In using the NL-OLS parameter estimation framework, no formal goodness-of-fit tests can be used to compare alternative model specifications. The following discussion therefore provides an informal discussion of the overall in-sample pricing performance of the alternative affine interest rate models based on average absolute swaption pricing errors. Note first that high absolute pricing errors are reported for each of the affine interest-rate models, which provides clear evidence of significant misspecification in each case. Despite this important qualification, the 2FV model reports the lowest absolute errors for the majority of swaption contracts and hence provides the best in-sample fit. This is in line with the overall pricing performance of the 2FV model, as reported by the minimized loss function values in Table 7.5.1. It is also interesting to note that, for the majority of swaption contracts in the sample data, the 2FV model significantly outperforms the 1FV model. Hence, for the Vasicek-type models, it is clear that, for the most part, in-sample pricing performance improves with increased dimension. A similar observation is also made for the generalized Vasicek-type models, with the in-sample pricing performance again improving with increased dimension. In particular, it is notable that the 3FGV model considerably improves on the in-sample pricing performance of the 2FGV and 1FGV models for the majority of swaption contracts.

In addition to the above observations, it is noted that, of the one-factor models, the 1FGV model is reported to consistently outperform the 1FV and 1FCIR models for swaptions written on 1-year swap contracts. For swaptions written on the longer-term 6-, 7-, 8-, and 9-year swap contracts, the 1FCIR model is shown to outperform both the 1FV and 1FGV models. Mixed results are observed for the remaining swaptions written on 2-year up to 5-year swap contracts. Furthermore, of the two-factor models, the 2FV model outperforms the equivalent 2FGV model for the majority of swaption contracts. It is only for the swaptions written on 1-year swap contracts that the

TABLE 7.5.2 Implied Estimates : 1FV and 2FV Models

	1FV			2FV		
θ	$\varkappa_r$	0.217	(0.7)	$\varkappa_r$	0.210	(2.4)
	$\bar{r}$	0.074	(7.7)	$\varkappa_{\bar{r}}$	0.259	(2.7)
	σ_r	0.014	(0.8)	$\bar{\theta}$	0.075	(0.2)
				σ_r	0.025	(0.03)
				$\sigma_{\bar{r}}$	0.018	(0.2)
				$\rho_{r,\bar{r}}$	−0.074	(3.2)
ϑ				$\bar{r}_t$	0.020	(0.2)

2FGV model is reported to outperform the 2FV model. In fact, as can be observed from the reported absolute pricing errors, the 2FV model consistently outperforms the higher dimensional 3FGV model for all swaptions written on swap contracts with a maturity greater than 3 years. This observation is consistent with the overall pricing performance of these models, as reported in Table 7.5.1.

7.5.1.1 *Analysis of Implied Estimates*

Tables 7.5.2, 7.5.3, and 7.5.4, respectively, report the average of the daily estimates obtained for the Vasicek, the generalized Vasicek, and the CIR models.

From the results for the Vasicek-type models reported in Table 7.5.2, it is first noted that the speed of mean reversion of the short rate in the case of the 1FV model is 0.217, which corresponds to a half-life of interest-rate shocks of approximately 38 months. The constant long-run mean level to which the

TABLE 7.5.3 Implied Estimates : GV Models

	1FGV			2FGV			3FGV		
θ	$\varkappa_1$	0.451	(1.2)	$\varkappa_1$	0.572	(3.6)	$\varkappa_1$	0.884	(2.2)
	σ_1	0.003	(1.0)	$\varkappa_2$	0.510	(3.4)	$\varkappa_2$	0.524	(1.6)
				σ_1	0.015	(0.2)	$\varkappa_3$	0.455	(2.1)
				σ_2	0.015	(0.2)	σ_1	0.017	(0.5)
				$\rho_{1,2}$	−0.754	(2.1)	σ_2	0.017	(0.7)
							σ_3	0.031	(0.5)
							$\rho_{1,2}$	0.286	(3.3)
							$\rho_{1,3}$	−0.140	(3.2)
							$\rho_{2,3}$	0.560	(4.0)
ϑ	$x_{1,t}$	0.013	(0.01)	$x_{1,t}$	0.017	(1.6)	$x_{1,t}$	0.031	(1.9)
				$x_{2,t}$	−0.009	(1.1)	$x_{2,t}$	−0.059	(1.6)
							$x_{3,t}$	−0.021	(3.8)

TABLE 7.5.4 Implied Estimates : 1FCIR Model

		1FCIR	
θ	$\varkappa_r$	0.221	(0.04)
	$\bar{r}$	0.088	(2.16)
	σ_r	0.098	(0.03)

short rate reverts is estimated to be 7.4%, which is considerably higher than the average 1-month EURIBOR rate of 2.65%. For the 2FV model, the speed of mean reversion of the short rate is almost identical to the case of the 1FV model at 0.210 (corresponding to a half-life of approximately 40 months), with the speed of mean reversion of the stochastic long-run mean short rate reported at a similar value of 0.259 (corresponding to a half-life of approximately 32 months). Interestingly, the average daily estimate of the latent long-run mean short rate in the 2FV model is reported to be 2%, with the level to which it reverts reported at a much higher rate of 7.5%. Also of importance in this analysis of the 2FV model is the weak negative correlation reported between the short-rate and long-run mean short-rate processes. Note finally from the results of Table 7.5.2 that the volatility coefficient of the short-rate process σ_r is estimated to be 1.7% under the 1FV model, which is lower than that estimated under the 2FV model at 2.5%.

For the generalized Vasicek models considered in the study, note first from Table 7.5.3 that, for the 1FGV model, the speed of mean reversion associated with the latent-state variable $x_{1,t}$ is estimated at 0.451 (corresponding to a half-life of 18 months). The average estimate of $x_{1,t}$ over the sample period is 1.3%. Recall from the definition of the 1FGV model that $r_t = \delta + x_{1,t}$, for constant δ, and as noted by Cortazar, Schwartz, and Naranjo [8], δ represents the long-run mean short rate to which the short rate reverts. Calculating an estimate of δ for each trade date in the sample period leads to an average figure for the entire sample period of 1.7%. Further, it is noted that the volatility parameter σ_1 is reported at a very low level of 0.3%.

Under the equivalence of the 1FV and 1FGV model specifications as established in section 7.4.2.1, the speeds of mean-reversion and volatility parameters are equivalent. It is interesting to note from the above results, however, that the 1FGV-model estimates of the speeds of mean reversion and volatility parameters are significantly different from the equivalent parameters under the 1FV-model specification. It is first noted that the speed of mean reversion under the 1FGV is reported to be less than half that estimated for the 1FV model. This implies that, relative to the 1FV model, the 1FGV reports a much lower half-life of interest-rate shocks, indicating that interest-rate shocks are far less persistent. It is next noted that the volatility parameter under the 1FGV model is estimated to be over five times lower than that reported for the case of the 1FV model. Hence, this indicates a much less volatile stochastic process relative to the 1FV model.

In a similar way, an estimate of the long-run mean level of the short-rate parameter δ under the 1FGV model is reported at 1.7%. This result is significantly less than the equivalent long-run mean level of the short rate $\bar{r}$ of 7.4% as estimated under the 1FV model. Hence, this and the above observations provide evidence that, despite the equivalence of the 1FV and 1FGV models, the parameter-estimation process does not result in similar findings for the two model specifications.

In the case of the 2FGV model, the mean-reversion rates associated with the two factors $x_{1,t}$ and $x_{2,t}$ are reported to be 0.572 and 0.510, respectively (corresponding to half-lives of approximately 15 and 16 months, respectively). Therefore, it is clear that the two factors in the 2FGV model revert to the long-run mean level of the factors of 0% at similar rates. Also of note is that the average estimates of the two factors $x_{1,t}$ and $x_{2,t}$ are 1.7% and −0.9%, respectively. Similar to previously stated results, note from the definition of the 2FGV model that the short rate $r_t = \delta + x_{1,t} + x_{2,t}$, and so over the entire sample period the average estimate of the long-run mean level of the short rate δ works out to be 2.14%. Also of note is that the results report a very strong negative correlation between the two underlying factors $x_{1,t}$ and $x_{2,t}$.

For the 3FGV model, the three underlying factors $x_{1,t}, x_{2,t}$, and $x_{3,t}$ have associated mean-reversion rates of 0.884, 0.524, and 0.455, respectively (corresponding to half-lives of 9, 16, and 18 months, respectively). Note also that the estimates obtained for the latent-state variables $x_{1,t}, x_{2,t}$, and $x_{3,t}$ are, respectively, 3.1%, −5.9%, and −2.1%. Under the structure of the 3FGV model, the short rate is given by $r_t = \delta + x_{1,t} + x_{2,t} + x_{3,t}$, and so the average estimate of δ over the sample period gives a long-run mean level of the short rate of 7.9%, which is considerably higher than the corresponding estimates for δ obtained under the 1FGV and 2FGV models. Finally, note from the correlation-parameter estimates that there is a considerable positive correlation between the underlying factors $x_{1,t}$ and $x_{2,t}$ and between $x_{2,t}$ and $x_{3,t}$. However, the results suggest a moderately negative correlation between the underlying factors $x_{1,t}$ and $x_{3,t}$.

Turning now to the parameter estimates under the 1FCIR model, it is first noted from Table 7.5.4 that the speed of mean reversion $\varkappa_r$, the long-run mean level of the short rate $\bar{r}$, and the volatility coefficient σ_r are reported to be 0.221, 8.8%, and 9.8%, respectively. Comparing these results with the parameter counterparts under the 1FV model, it is noted that the speed of mean reversion and long-run mean short-rate estimates are quite similar under both models. However, it is notable that the volatility coefficient parameter is estimated to be seven times higher under the 1FCIR model.

To conclude the discussion, the focus is now turned to the time-series properties of the implied structural parameter and latent-state variable estimates.[6] For all of the affine interest-rate models tested, the majority of structural-parameter and, where applicable, latent-state variable estimates exhibit substantial instability over the sample period. When viewed over the entire sample period or over identifiable subperiods of the sample period, this parameter instability manifests itself in the form of considerable oscillatory behavior in the weekly estimates or shifts in the level of parameter estimates. Such observations can be interpreted as providing evidence of model misspecification. However, another possible explanation for the observation of shifts in the level of parameter estimates is that the optimization process is alternating between local minima.

7.5.2 Out-of-Sample Pricing Performance

The candidate models are now examined to evaluate the out-of-sample pricing performance of each. An out-of-sample pricing evaluation is conducted for each affine interest-rate model according to the following loss-function specification:

$$\sum_{i=1}^{N_t} [C_{i,t} - \hat{C}_{i,t}(\theta_{lag})]^2,$$

where $\theta_{lag} = \{\theta_{lag}, \vartheta_{lag}\}$, and θ_{lag} and ϑ_{lag} are, respectively, the vectors of structural parameters and latent-state variables as estimated on the last trade date previous to date t. Table 7.5.5 reports the average of the daily loss-function values calculated. It is interesting to note that the ranking of the candidate models based on out-of-sample pricing performance is not entirely consistent with the ranking established in the previous section for the in-sample fit of the models. In particular, the 3FGV model is shown to provide the best out-of-sample pricing performance, followed in order by the 2FV and 1FCIR models.

[6] An extensive range of time-series graphs for the implied structural parameter and latent-state variable estimates exists. To conserve on space, these graphs are not presented here but are readily available upon request.

TABLE 7.5.5 Out-of-Sample Loss Function Values of Candidate Models

Model	Out-of-Sample Loss Function Value
1FV	0.00103
2FV	0.00088
1FGV	0.00711
2FGV	0.00707
3FGV	0.00052
1FCIR	0.00096

However, it is noted from the results that the 2FV model outperforms the 3FGV model, and the 1FV model outperforms the 2FGV model, which is in line with the observations on in-sample fit made in the previous section.

7.6 CONCLUSIONS AND FUTURE RESEARCH

This chapter outlines how the transform-inversion approach of Carr and Madan [4] has been independently adapted by the authors to allow the use of FFT technology to price European-exercise interest-rate swaptions. Under a range of modeling assumptions, the FFT is shown to be a superfast and accurate engine to price a range of swaption contracts around forward-at-the-money strikes. Future research should investigate extensions that will ensure (a) term-structure consistency for the class of approximate-affine models investigated in this chapter and (b) the validity of replacing low-variance martingales with their time-zero conditional expected values under an affine jump-diffusion framework. Finally, it is proposed to investigate the potential of a recursive quadrature-based FFT routine in pricing swaption contracts with exotic or early-exercise features (e.g., Bermudan swaptions) under multifactor affine model dynamics for the term structure.

REFERENCES

1. Andersen, L. and J. Andreasen (1999). Factor dependence of Bermudan swaption prices: fact or fiction? *Journal of Financial Economics 62*, 3–37.
2. Black, F. (1976). The pricing of commodity contracts. *Journal of Financial Economics 3*, 167–179.
3. Brace, A., M. Gatarek, and M. Musiela (1997). The market model of interest rate dynamics. *Mathematical Finance 7*, 127–155.

4. Carr, P. and D. Madan (1999). Option valuation using the fast Fourier transform. *Journal of Computational Finance 3*, 463–520.
5. Chen, R. and L. Scott (1992). Pricing interest rate options in a two-factor Cox-Ingersoll-Ross Model of the term structure. *The Review of Financial Studies 5*, 613–636.
6. Collin-Dufresne, P. and R.S. Goldstein (2001). Stochastic correlation and the relative pricing of caps and swaptions in a generalized-affine framework. Working paper.
7. Collin-Dufresne, P. and R.S. Goldstein (2002). Pricing swaptions within an affine framework. *Journal of Derivatives 10*, 1–18.
8. Cortazar, G., E.S. Schwartz, and L. Naranjo (2003). Term structure estimation in low-frequency transaction markets: a Kalman filter approach with incomplete panel data. Finance Working Paper.
9. Cox, J.C., J. Ingersoll, and S. Ross (1979). Duration and the measurement of basis risk. *Journal of Business 52*, 51–61.
10. Cox, J.C., J. Ingersoll, and S. Ross (1985). A theory of the term structure of interest rates. *Econometrica 53*, 385–407.
11. Dai, Q. and K. Singleton (2000). Specification analysis of affine term structure models. *Journal of Finance 55*, 1943–1978.
12. d'Aspremont, A. (2003). Interest rate model calibration using semidefinite programming. *Applied Mathematical Finance 10*, 183–213.
13. Driessen, J., P. Klaassen, and B. Melenberg (2003). The performance of multi-factor term structure models for pricing and hedging caps and swaptions. *Journal of Financial and Quantitative Analysis 38*, 635–672.
14. Duffie, D. and R. Kan (1996). A yield-factor model of interest rates. *Mathematical Finance 4*, 379–406.
15. Duffie, D., J. Pan, and K. Singleton (2000). Transform analysis and option pricing for affine jump-diffusions. *Econometrica 68*, 1343–1376.
16. Geman, H., N. El Karoui, and J.C. Rochet (1995). Changes of numeraire, changes of probability measure and option pricing. *Journal of Applied Probability 32*, 443–458.
17. Hibbert, J., P. Mowbray, and C. Turnbull. (2001). A stochastic asset model and calibration for long-term financial planning purposes. Technical report, Barrie & Hibbert Ltd., Edinburgh.
18. Longstaff, F.A., P. Santa-Clara, and E.S. Schwartz (2001). The relative valuation of caps and swaptions: theory and empirical evidence. *Journal of Finance 56*, 2067–2109.
19. Muck, M. (2003). Do surprising central bank moves influence interest rate derivatives prices? Empirical evidence from the European caps and swaptions market. Working paper.
20. Munk, C. (1999). Stochastic duration and fast coupon bond option pricing in multifactor models. *Review of Derivatives Research 3*, 157–181.
21. Schrager, D.F. and A.A.J. Pelsser (2004). Pricing swaptions and coupon bond options in affine term structure models. Working Paper.

22. Singleton, K.J. and L. Umantsev (2002). Pricing coupon-bond options and swaptions in affine term structure models. *Mathematical Finance 12*, 427–446.
23. Wei, J. (1997). A simple approach to bond option pricing. *Journal of Futures Markets 17*, 131–160.
24. Vasicek, O.A. (1977). An equilibrium characterization of the term structure. *Journal of Financial Economics 5*, 177–188.

CHAPTER 8

Empirical Testing of Local Cross Entropy as a Method for Recovering Asset's Risk-Neutral PDF from Option Prices

Vladimír Dobiáš
University College Dublin, Dublin, Ireland

Contents

Abstract: In this chapter we test the performance of the local cross entropy as a method for extracting the underlying asset price distribution implied by option prices. The fast Fourier transform algorithm is used to generate the "true" risk-neutral distribution as well as theoretical option prices consistent with several known models, and the local cross-entropy criterion is applied to recover this distribution given a small number of option prices. The resulting distribution is compared with the "true" distribution in terms of entropy pseudodistance and error between the theoretical and implied option prices. We also investigate the effect of noise added to the theoretical option prices. Finally, the local cross entropy is applied to market data, and the risk-neutral distribution underlying the S&P 500 and NASDAQ 100 index options is obtained at various maturities.

8.1 INTRODUCTION

One of the most interesting pieces of information one can gain from derivative markets is the expected distribution of asset prices of a risk-neutral agent.[1] It may be useful to policy makers for pricing other (exotic) derivatives or for checking that a no-arbitrage condition holds across different derivative markets. In recent years several methods for extracting the risk-neutral probability distribution function (RN-PDF) have been suggested. Jackwerth [1] provides a good classification of these methods. We focus on one promising branch of nonparametric methods that draws insight from information theory and employs the concept of entropy. We need to address the issue of incomplete markets because, in reality, we are dealing with a situation where there are more states of nature than linearly independent securities. For this reason an extra condition is needed to represent the missing information.

One possibility is the maximum entropy approach (ME) pioneered by Edelman and Buchen [2] and Buchen and Kelly [3], which solves this problem by maximizing the randomness of the RN-PDF. Such a distribution would be the least prejudiced with respect to the missing information. Unfortunately, this method is numerically unstable. The practical drawbacks of ME could

[1] We are concentrating exclusively on a risk-neutral distribution. To obtain a real-world distribution, one must assume a particular utility function. Alternatively, we can assume that the risk aversion of market participants does not change over time, and then we could infer the evolution of the real-world distribution based on the evolution of the risk-neutral counterpart.

be overcome by minimizing the Kullback-Leibler pseudodistance to some reference (prior) distribution minimizing Kullback-Leibler (MKL). However, the resulting solution suffers from arbitrariness in the choice of the prior distribution. Finally, as the latest development, Edelman [4] proposed a minimum local cross-entropy (MLCE) criterion that minimizes the entropy pseudodistance of the distribution to a slightly smoothed version of itself and thus effectively removes the subjectivity imposed by the reference distribution.

An empirical testing of the MLCE method on both simulated and market data is the main focus of this work. The chapter is organized in the following way: in section 8.2 the methodology for a simulated experiment is described. Section 8.3 presents the results of the MLCE method on option prices generated from various models as well as on market data using S&P 500 and NASDAQ 100 index options. Section 8.4 concludes the chapter.

8.2 METHODOLOGY

This section outlines the methodology used in the simulated experiment. First we examine the theoretical background of the recovery of the implied distributions, with special emphasis on local cross entropy as one of the methods for extracting the "true" distribution in an incomplete market setting. Then we shift focus to the selection of models and ways to generate data, namely the terminal distribution and option prices. Lastly, we describe the performance of the MLCE method.

8.2.1 Recovery of the Implied RN-PDF

The goal is to recover a risk-neutral distribution $q(S_T)$ that would be consistent with a set of M linearly independent assets: $M - 2$ options, an underlying asset (or a forward contract), and a riskless bond. In other words, the market prices $\{C^*_{j=1,2,\ldots,M-2;t}, S^*_t, B^*_t\}$ should coincide with theoretical prices $\{C_{j=1,2,\ldots,M-2;t}, S_t, B_t\}$ that are implied by the terminal distribution $q(S_T)$ as given by the risk-neutral pricing formula:

$$V(S_t) = e^{-r(T-t)} \int_{\mathbb{R}} \text{payoff}(S_T)\, q(S_T)\, \mathrm{d}S_T. \tag{8.2.1}$$

Consider a discrete setting where there are N states of nature $\{S_{i,T}\}_{i=1,2,\ldots,N}$, that are equally spaced with a step size δS_T. Equation (8.2.1) can be approximated for a particular payoff function as follows:

- For (call) options:

$$C_{j;t}(S_t) = e^{-r(T-t)} \sum_{i=1}^{N} \max(S_{i,T} - X_j, 0)\, q(S_{i,T})\, \delta S_T \tag{8.2.2}$$

- For underlying asset (or forward):

$$\begin{aligned} S_t &= e^{-r(T-t)} \sum_{i=1}^{N} S_{i,T}\, q(S_{i,T})\, \delta S_T \\ \Rightarrow F_t(S_t) &= \sum_{i=1}^{N} S_{i,T}\, q(S_{i,T})\, \delta S_T = S_t\, e^{r(T-t)} \end{aligned} \tag{8.2.3}$$

- For riskless bond with face value 1:

$$B_t = e^{-r(T-t)} \sum_{i=1}^{N} q(S_{i,T})\, \delta S_T = e^{-r(T-t)} \tag{8.2.4}$$

Put differently, the discrete probabilities $q^{'}(S_{i,T}) = q(S_{i,T})\, \delta S_T$ must sum to 1.

Essentially there are two possibilities:

1. **Complete markets setting** ($N = M$): the number of securities equals the number of unknown variables $q(S_{i,T})$. To ensure that for each state of nature in the future there exists an Arrow-Debreu security, the option strike prices $X_j, j = 1, 2, \ldots, M-2$ must coincide with the states of nature $S_{i,T}, i = 2, \ldots, N-1$. Then all we need to do to get a unique solution is to solve a system of linear equations (see Figure 8.2.1a):

$$e^{-r(T-t)} \begin{bmatrix} \max(S_{1,T} - X_1, 0) & \ldots & \max(S_{T,N} - X_1, 0) \\ \ldots & \ldots & \ldots \\ \max(S_{1,T} - X_{M-2}) & \ldots & \max(S_{T,N} - X_{M-2}) \\ S_{1,T} & \ldots & S_{T,N} \\ 1 & \ldots & 1 \end{bmatrix} \begin{bmatrix} q'(S_{1,T}) \\ q'(S_{2,T}) \\ \ldots \\ q'(S_{N,T}) \end{bmatrix}$$

$$= \begin{bmatrix} C^*_{1,t} \\ C^*_{2,t} \\ \ldots \\ C^*_{M-2,t} \\ S^*_t \\ B^*_t \end{bmatrix}.$$

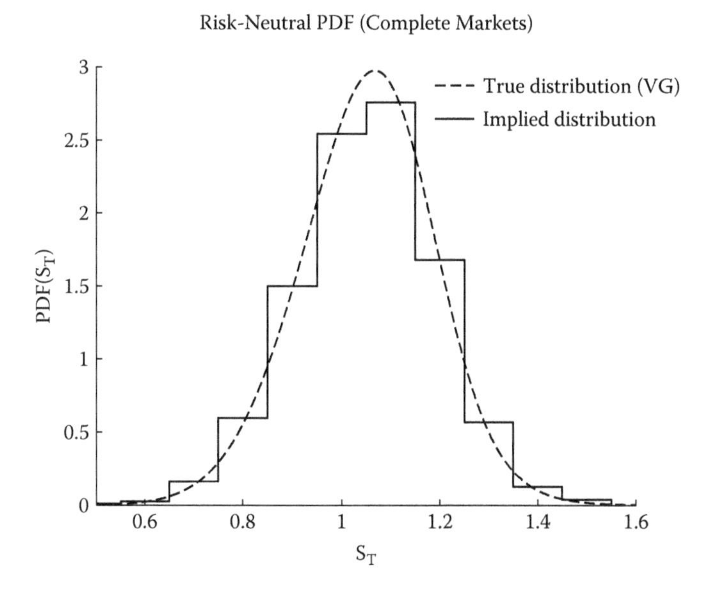

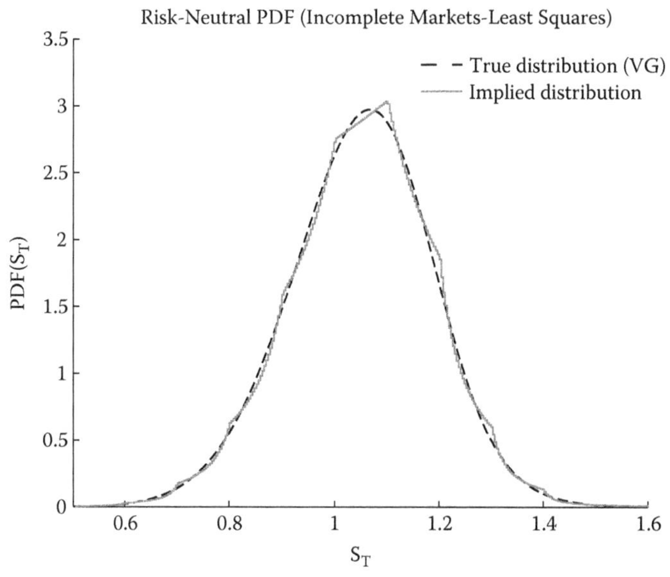

FIGURE 8.2.1 (a) Complete markets and (b) incomplete markets with no smoothing condition (solution in the least squares sense).

2. **Incomplete markets setting** ($N > M$): we have more unknowns than securities (and thus equations), which means that we are dealing with an underdetermined system. We could solve the system in the least-square sense (Figure 8.2.1b), but it is more theoretically sound to add

an extra condition (the MLCE criterion) that accounts for the lack of information and effectively serves as a smoothing condition. Equations (8.2.2), (8.2.3), and (8.2.4) define a region of feasible solutions, and they serve as constraints in an optimization problem. The MLCE criterion ensures uniqueness of the solution.

The optimization problem that enables the true distribution to be recovered from the set of option prices is defined as follows [4]:

$$\text{(MLCE)} \quad \min_{q(S_{i,T})} \mathcal{L}[q(S_T)] = \sum_{i=1}^{N} \frac{(q(S_{i,T})'')^2}{q(S_{i,T})} \tag{8.2.5}$$

subject to:

- Nonnegativity of probabilities: $q(S_{i,T}) \geq 0$
- $M - 2$ option price constraints—equation (8.2.2)
- Underlying asset constraint—equation (8.2.3)
- Riskless bond constraint (normalization condition)—equation (8.2.4)

8.2.2 Model Selection

Four models with different risk-neutral asset-price dynamics will be used to test the performance of the MLCE method.

8.2.2.1 Black–Scholes–Merton Model

The risk-neutral process for the stock price is the lognormal process [9]:

$$\frac{dS_t}{S_t} = r\,dt + \sigma\,dW_t.$$

Parameter r is a risk-neutral drift, σ a volatility of the diffusion, and W_t denotes a standard Wiener process.

8.2.2.2 Variance Gamma Model

The variance gamma model is based on a time-changed arithmetic Brownian motion X_t, where the random time change is given by a gamma process γ_t [10–12]:

$$X_t(\sigma, \nu, \theta) = \theta\,\gamma_t + \sigma\,W(\gamma_t).$$

The stock price is modeled as follows:

$$S_t = S_0\,e^{(r+\omega)t + X_t(\sigma,\nu,\theta)}.$$

The drift is modified by a compensator term ω, which is chosen such that the discounted stock price is a martingale. Parameter σ refers to a volatility of Brownian motion, ν to a variance of the gamma time change (kurtosis parameter), and θ to a drift of Brownian motion (skewness parameter).

8.2.2.3 Stochastic Volatility Model

The stochastic volatility is a generalization of the lognormal process, where volatility is modeled by a mean-reverting process [13]:

$$\frac{\mathrm{d}S_t}{S_t} = r\,\mathrm{d}t + \sqrt{V_t}\,\mathrm{d}W_t^{(1)},$$

$$\mathrm{d}V_t = \kappa\,(\theta - V_t)\,\mathrm{d}t + \sigma\,\sqrt{V_t}\,\mathrm{d}W_t^{(2)},$$

$$\mathrm{d}W_t^{(1)}\,\mathrm{d}W_t^{(2)} = \rho\,\mathrm{d}t,$$

where κ is the rate of mean reversion of the variance process, θ is the long-run variance, σ is the instantaneous volatility of the variance, and ρ is the correlation between the asset and variance processes.

8.2.2.4 Jump Diffusion Model

The stock price is modeled with a mixed jump diffusion process [14]:

$$\frac{\mathrm{d}S_t}{S_t} = (r + \omega)\,\mathrm{d}t + \sigma\,\mathrm{d}W_t + (e^J - 1)\,\mathrm{d}N_t.$$

The arrival of jumps is assumed to be driven by a Poisson process N_t with constant frequency of jumps per year λ. When the jump occurs, the logarithm of the jump size J is assumed to be normally distributed with mean jump size μ_j and jump volatility σ_j. Again, parameter ω refers to the compensator term and σ to the volatility of the diffusion part.

8.2.3 Data Generation

In this section we show how to generate the terminal distribution of asset prices that we will eventually try to recover as well as theoretical option prices that will carry information about the "true" distribution.

We rely on a one-to-one relationship between characteristic function (CF) and probability distribution function, where the link is a continuous Fourier transform:

$$q(x) = \frac{1}{2\pi} \int_{\mathbb{R}} e^{-i\theta x}\,\phi(\theta)\;\mathrm{d}\theta \tag{8.2.6}$$

and an inverse continuous Fourier transform, respectively:

$$\phi(\theta) = \mathbb{E}[e^{i\theta x}] = \int_{\mathbb{R}} e^{i\theta x} q(x) \, dx, \tag{8.2.7}$$

where $q(x)$ refers to the probability distribution function of a random variable x (usually log-asset price), and $\phi(\theta)$ is the corresponding characteristic function.

Theoretically, both transforms are possible, but in terms of practical usefulness, the forward transform of equation (8.2.6) is more interesting because the CF is readily available in closed form for a large number of popular models for which the PDF is not available in closed form. Models for which a closed-form CF exist include all of the above-mentioned models; mixed jump-diffusion stochastic volatility models, see Bates [18] and Duffie, Pan, and Singleton [19]; and a large number of Lévy processes, see Schoutens [20] and references therein for more information.

To derive the "true" terminal distribution, we will work in a discrete setting where the continuous Fourier transform is approximated by a discrete Fourier transform. Specifically, we evaluate the integral of equation (8.2.6) only at a finite number of points $\{\theta_j\}_{j=0,1,\ldots,N-1}$ and thus calculate the value of the PDF at a particular point x_k. We can harness the fast Fourier transform (FFT) algorithm developed by Cooley and Tukey [5] to substantially boost the computational efficiency.

To generate theoretical option prices from the model distributions, we follow Carr and Madan [6]. They show that the vanilla call option price can be computed as:

$$C_T(x) = \frac{e^{-\alpha x}}{2\pi} \int_{\mathbb{R}} e^{-ix\theta} \, \Psi(\theta) \, d\theta, \tag{8.2.8}$$

where x is the log-strike price, $\alpha > 0$ is a damping parameter that enforces square integrability, and

$$\Psi(\theta) = \frac{e^{-r(T-t)} \, \phi(\theta - (\alpha + 1)\, i)}{\alpha^2 + \alpha - \theta^2 + (2\alpha + 1)\theta \, i}. \tag{8.2.9}$$

Again, this integral can be discretized and quickly evaluated using the FFT algorithm. We obtain option prices for a variety of strike prices at the same time.

8.2.4 Performance Measurement

To measure the performance of the MLCE method in recovering the "true" distribution, the following criteria will be used:

1. The entropy pseudodistance to the true distribution—measured by Kullback–Leibler distance (KL):

$$\mathcal{K}[q(S_T), q^*(S_T)] = \sum_{i=1}^{N} q(S_{i,T}) \log\left(\frac{q(S_{i,T})}{q^*(S_{i,T})}\right),$$

   where $q(S_T)$ is the recovered risk-neutral distribution, and $q^*(S_T)$ is the true risk-neutral distribution.
2. The ability of the implied distribution to replicate the theoretical option prices—measured by root mean-squared error (RMSE):

$$\text{RMSE} = \sqrt{\frac{1}{N}\sum_{i=1}^{N}(C_{\text{theoretical}} - C_{\text{implied}})^2}.$$

   Call option prices C_{implied} are calculated via risk-netural pricing formula in equation (8.2.1), where $q(S_T)$ is the recovered distribution. The integral is evaluated numerically using Simpson's rule.
3. The difference between descriptive statistics (mean, variance, skewness, and excess kurtosis) of $\log(\frac{S_T}{S_0})$ of the implied and the true distribution.

8.3 RESULTS

This section explores the empirical performance of the MLCE method in three settings: on clean model data, on model data perturbed with noise, and finally on market data.

8.3.1 Simulated Data

First, using equation (8.2.6) and an FFT algorithm, we generate the terminal distribution of the four models described in section 8.2.2 for risk-free interest rate $r = 0.05$, initial asset price $S_0 = 1300$, time horizon $T = 6$ months, and parameters typical for the S&P 500 index [7, 8] (see Figure 8.3.1a):

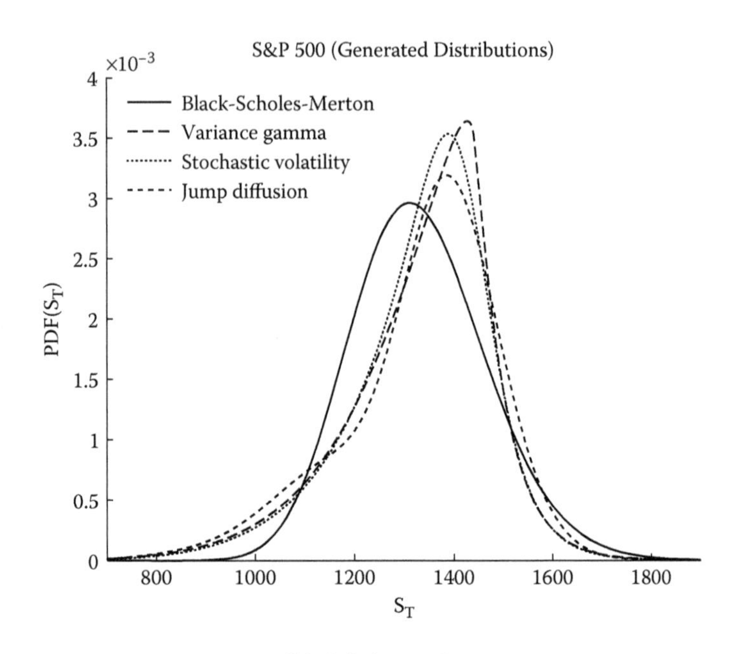

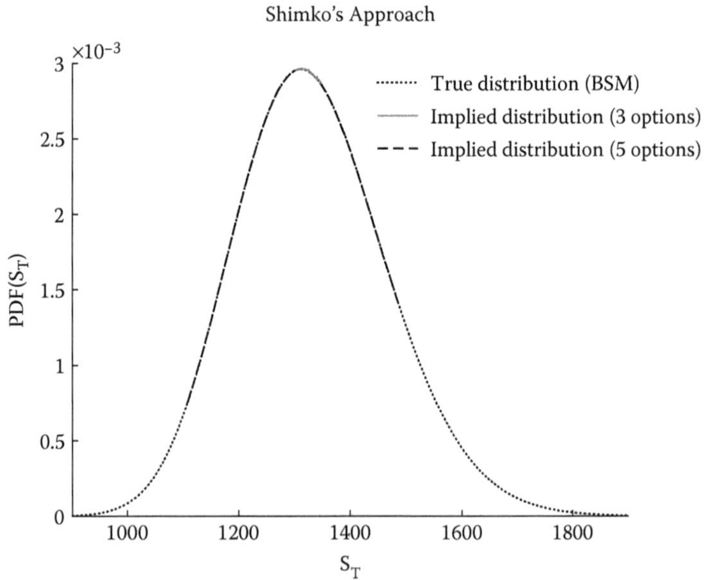

FIGURE 8.3.1 (a) Distribution of S&P 500 stock index generated by four different models; (b) Shimko's approach for recovering RN-PDF.

TABLE 8.3.1

Model	Parameter	Value
Black–Scholes–Merton	volatility	$\sigma = 0.14437$
Variance gamma	volatility of Brownian motion	$\sigma = 0.13503$
	variance of the time change	$\nu = 0.39608$
	drift of Brownian motion	$\theta = -0.16798$
Stochastic volatility	mean reversion rate	$\kappa = 1.15$
	long-run variance	$\theta = 0.04$
	volatility of variance	$\sigma = 0.39$
	correlation	$\rho = -0.64$
Jump diffusion	volatility	$\sigma = 0.09544$
	frequency of jumps	$\lambda = 0.77742$
	mean jump size	$\mu_j = -0.14899$
	jump volatility	$\sigma_j = 0.09411$

On each distribution we will perform three tests that vary in the amount of input information they utilize. The input information is represented by option prices at various strike prices. The three scenarios are:

1. One option with strike price $X = 1333$ (at-the-money-forward option)
2. Three options with strike prices $X_1 = 1300$, $X_2 = 1325$, and $X_3 = 1350$
3. Five options with strike prices $X_1 = 1100$, $X_2 = 1200$, $X_3 = 1300$, $X_4 = 1400$, and $X_5 = 1500$

The calculation of theoretical option prices is based on equation (8.2.8).

Next, we solve the MLCE optimization problem specified in equation (8.2.5). In practical implementation, the best performance was achieved with primal-dual interior point optimizers (KNITRO [15], IPOPT [16]) where, besides call options, we also include put counterparts calculated by the put-call parity. Another slightly less effective way would be to translate only out-of-the-money calls into in-the-money puts. Figure 8.3.2 summarizes all the possibilities of using the put-call parity: (a) translating all calls to puts, (b) translating only out-of-the-money calls, or (c) including puts while keeping all the calls. The graph indicates that call options tend to explain well the right side of the distribution, whereas puts influence the left side. The reason lies in the Jacobian of the option price constraint, which (for calls) is zero for stock prices below the strike price. This in turn gives little guidance to the optimization engine for the search direction in the left tail. Thus including put options yields great benefit, especially if we have only a small number of options. The opposite argument applies to put options and the right side of the distribution.

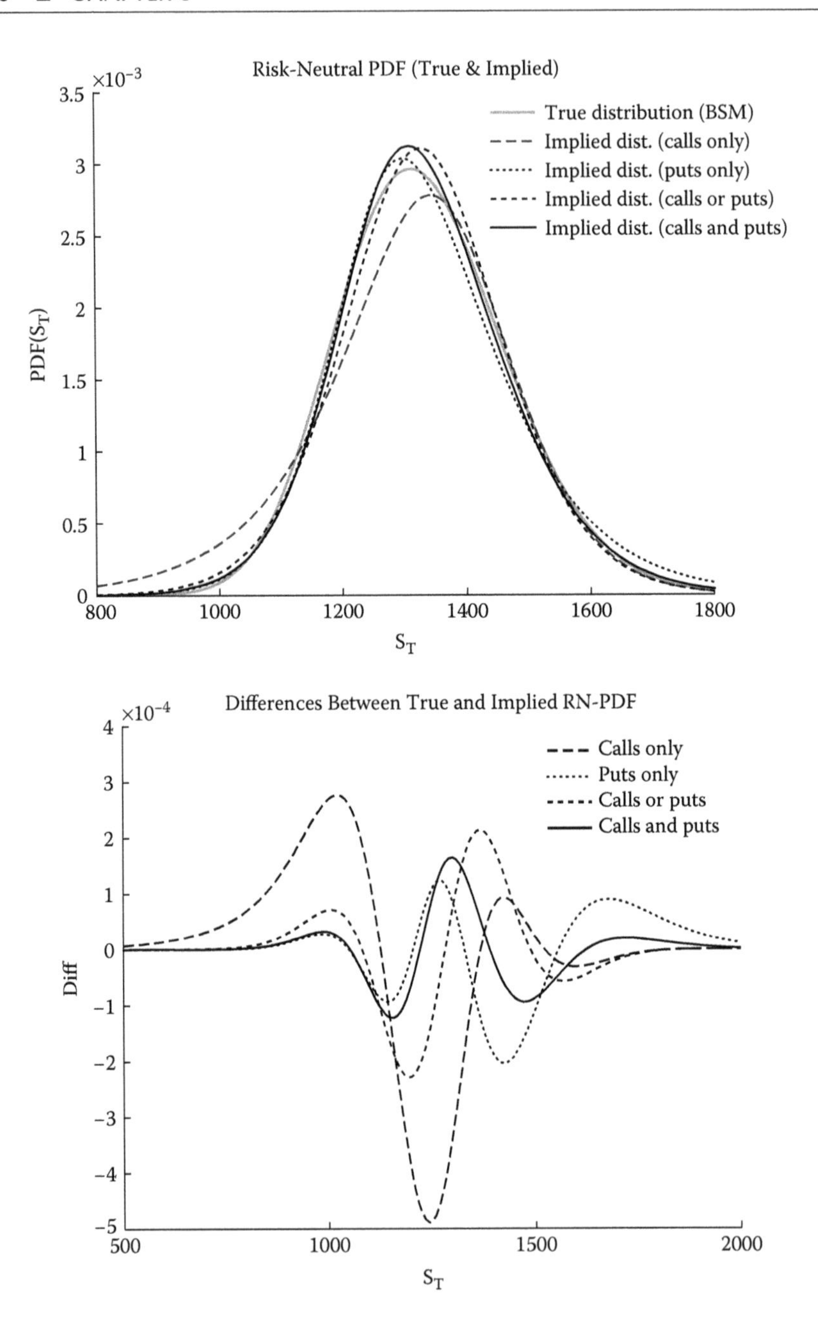

FIGURE 8.3.2 Benefits of including puts via put-call parity.

Any additional constraint is computationally expensive, so we can speed up the optimization if we omit the normalization condition of equation (8.2.4) and let the optimizer produce relative values that we subsequently renormalize to make sure that the integral of the PDF is equal to 1.

In the next step we visually compare the recovered distribution with the true one (Figures. 8.3.3–8.3.6). Clearly, the more information we input, the more closely the recovered distribution tracks the original one. The results are even more pronounced in the (σ, X) space, where σ refers to implied volatility and X to strike price.

Finally, analyzing the performance measures described in section 8.2.4 (see appendix A), we conclude that:

- In all four cases, the more options we include, the closer is the implied distribution to the true one as measured by KL distance.
- Comparing the error in call option prices across different models points out that the MLCE method works best for the Black–Scholes–Merton model, followed by stochastic volatility, variance gamma, and jump diffusion. It confirms the expected result that the more complicated the shape of the distribution is, the harder it is to discover.
- The distribution characteristics (mean, variance, skewness, excess kurtosis) converge to the true values as we add more options, though not necessarily in a monotone fashion.

8.3.2 Noisy Prices

We carry out a comparison test inspired by Buchen and Kelly [3]. They tested the performance of the maximum entropy criterion on two distributions—a discrete test distribution and a log-normal distribution. For the first distribution they also perturb the option prices with 7.5% white noise. Unfortunately, they do not apply the noise to the lognormal distribution. We will fill this gap and see if the MKL and (more importantly) the MLCE criterion succeeds even in this setting.

The lognormal distribution has the following parameters: initial stock price $S_t = 50$, volatility $\sigma = 25\%$, time to expiration $T - t = 60$ days, risk-free interest rate $r = 10\%$, and strike prices $X_1 = 45$, $X_2 = 50$, and $X_3 = 55$. The prior distribution for MKL is lognormal with volatility $\sigma = 35\%$. Before adding the noise, the recovered distribution using the MKL approach is a piecewise exponential function (Figure 8.3.7). The MLCE solution is smooth. After applying the noise, the situation does not change much (see Figure 8.3.8).

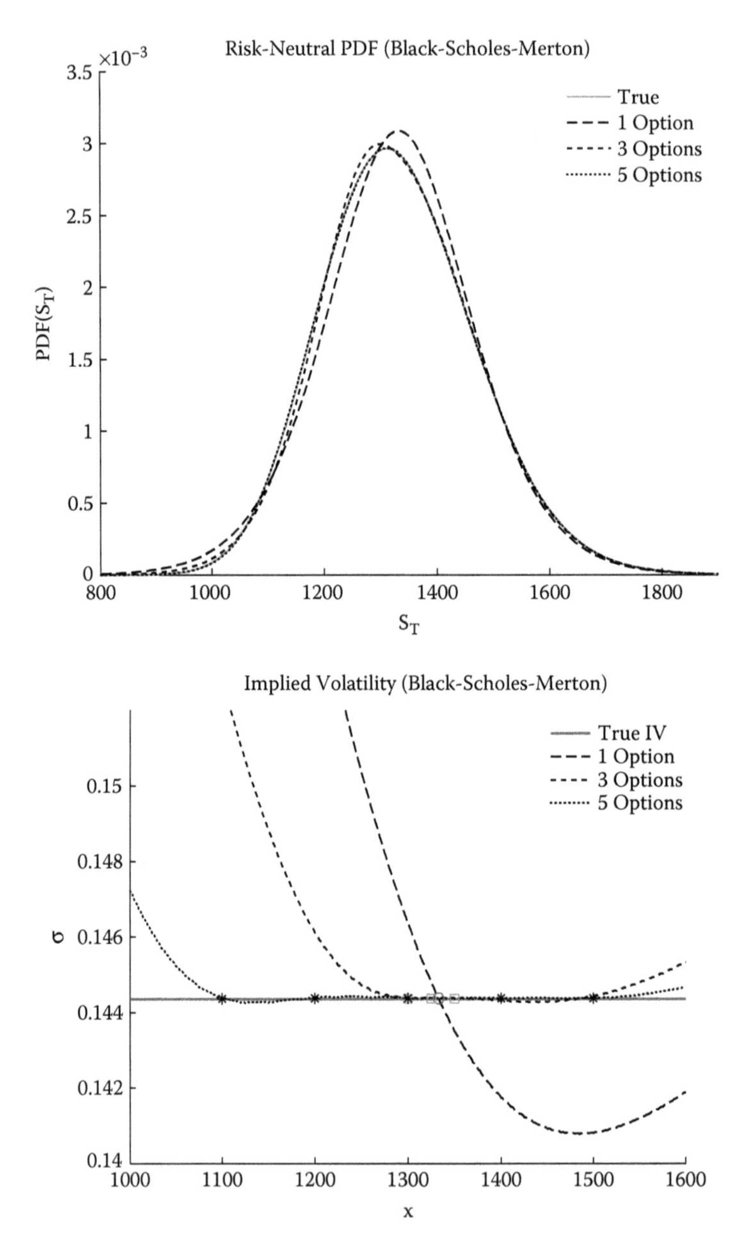

FIGURE 8.3.3 Recovered risk-neutral PDF and implied volatility for Black–Scholes–Merton (the circle, squares, and stars represent the strike prices of the option constraints).

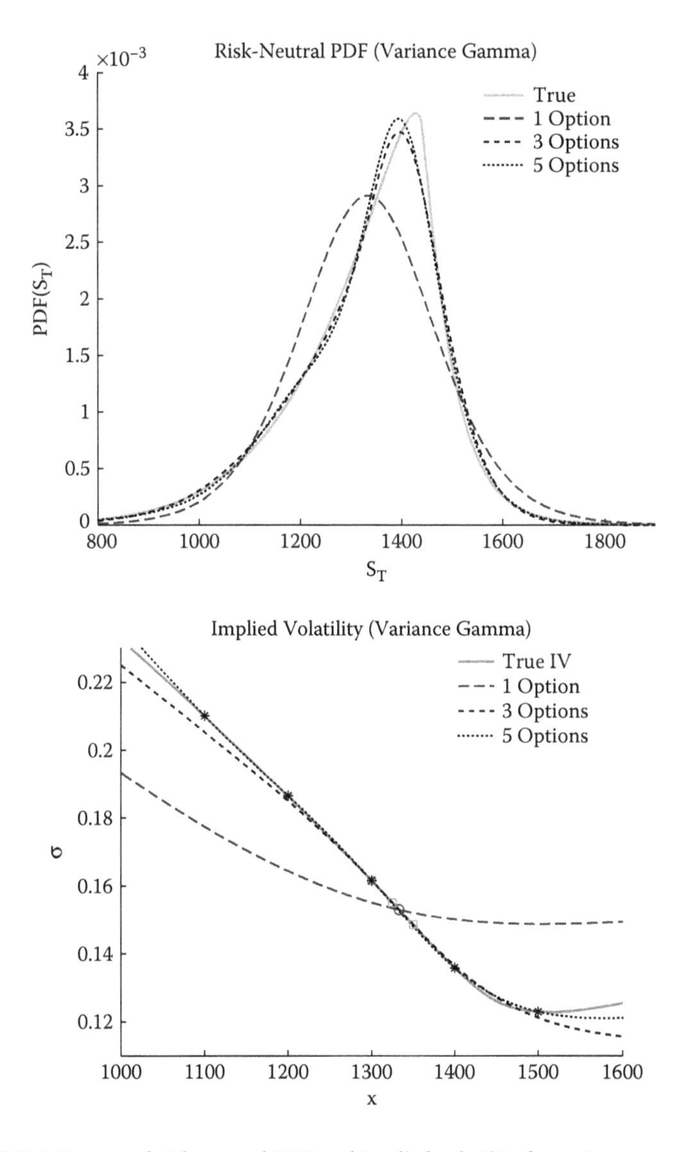

FIGURE 8.3.4 Recovered risk-neutral PDF and implied volatility for variance gamma.

If the perturbed option constraints still allow solution, the MKL would provide distribution with spikes at the strike prices, whereas MLCE yields a smooth distribution. The optimization fails if the call option violates the arbitrage constraint $S - Xe^{-r(T-t)}$ (parity puts would have negative values).

The characteristics of the two distributions resulting from MKL and MLCE are very similar, but as we have mentioned before, the MLCE method is superior to the MKL because it does not need the specification of the prior distribution, as this distribution is built into the method itself.

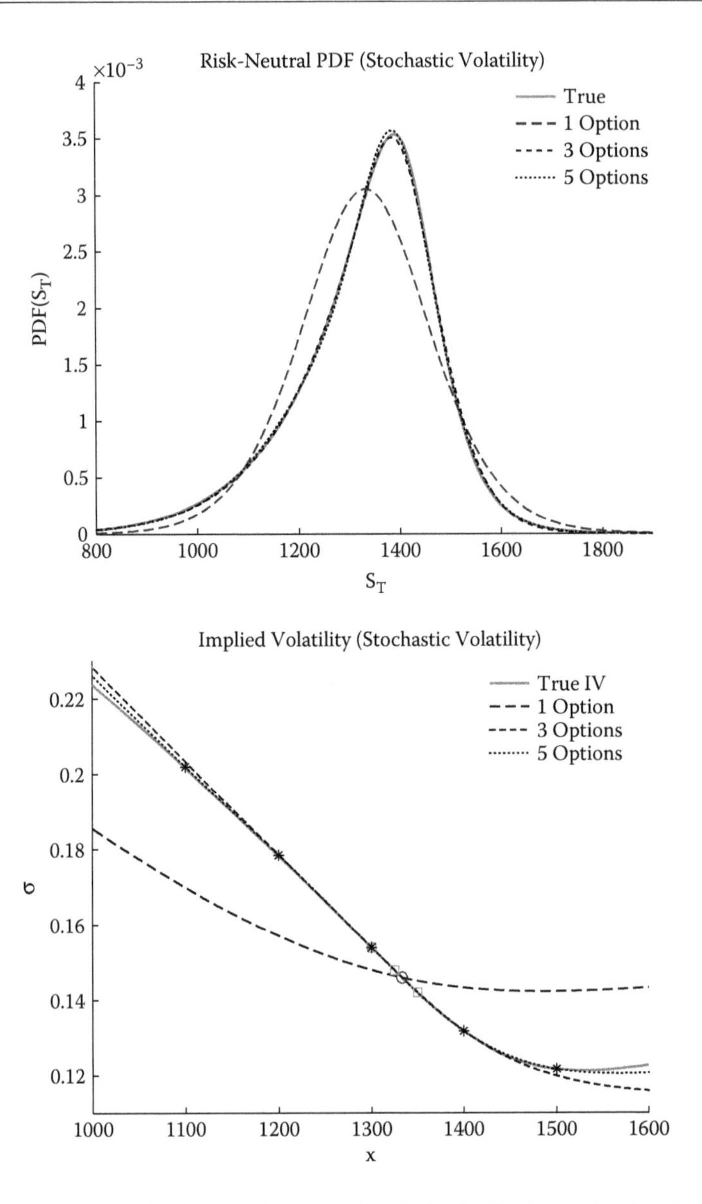

FIGURE 8.3.5 Recovered risk-neutral PDF and implied volatility for stochastic volatility.

8.3.3 Market Data

We can easily apply the same method to traded option prices. The only change is that each option equality constraint is now replaced with two inequality constraints. This ensures that the implied option price lies within the bid–ask spread.

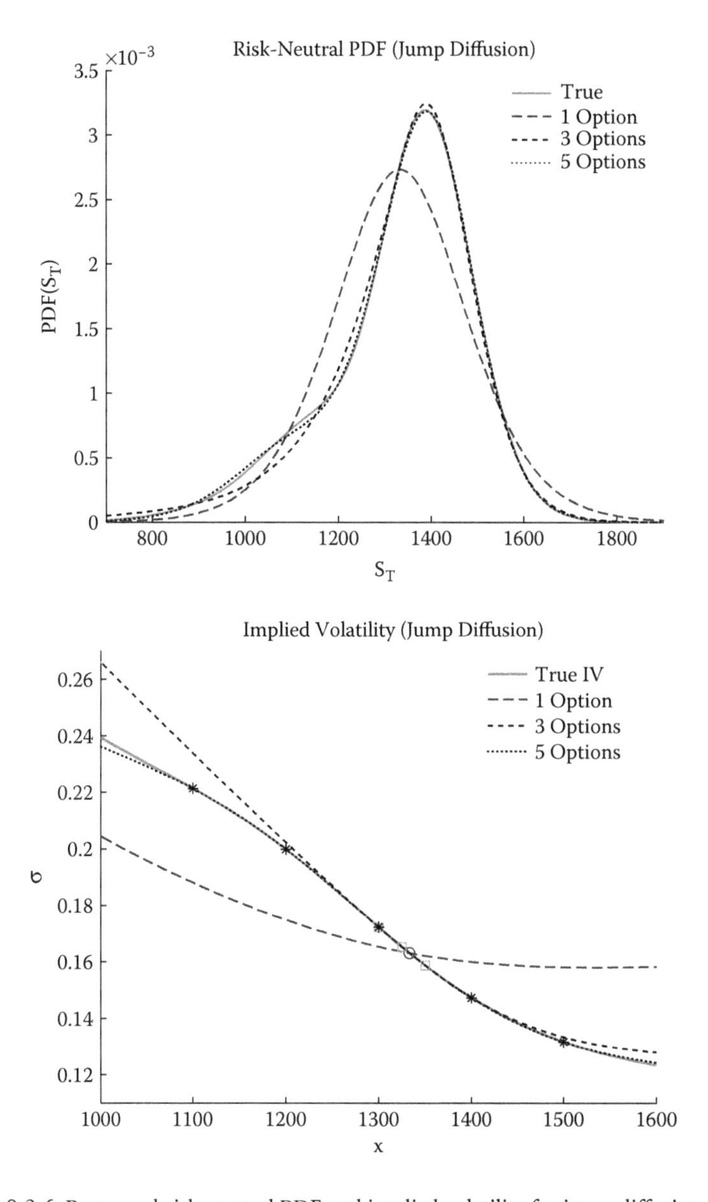

FIGURE 8.3.6 Recovered risk-neutral PDF and implied volatility for jump diffusion.

The resulting distributions for seven maturities of S&P 500 options (SPX) are shown in Figure 8.3.9a. The option data were downloaded from the CBOE Web site, and only the quotes at which a trade was made were used. The number of available strike prices varies quite substantially across different maturities, with a maximum of 45 and a minimum of 2 options. Nevertheless, as long

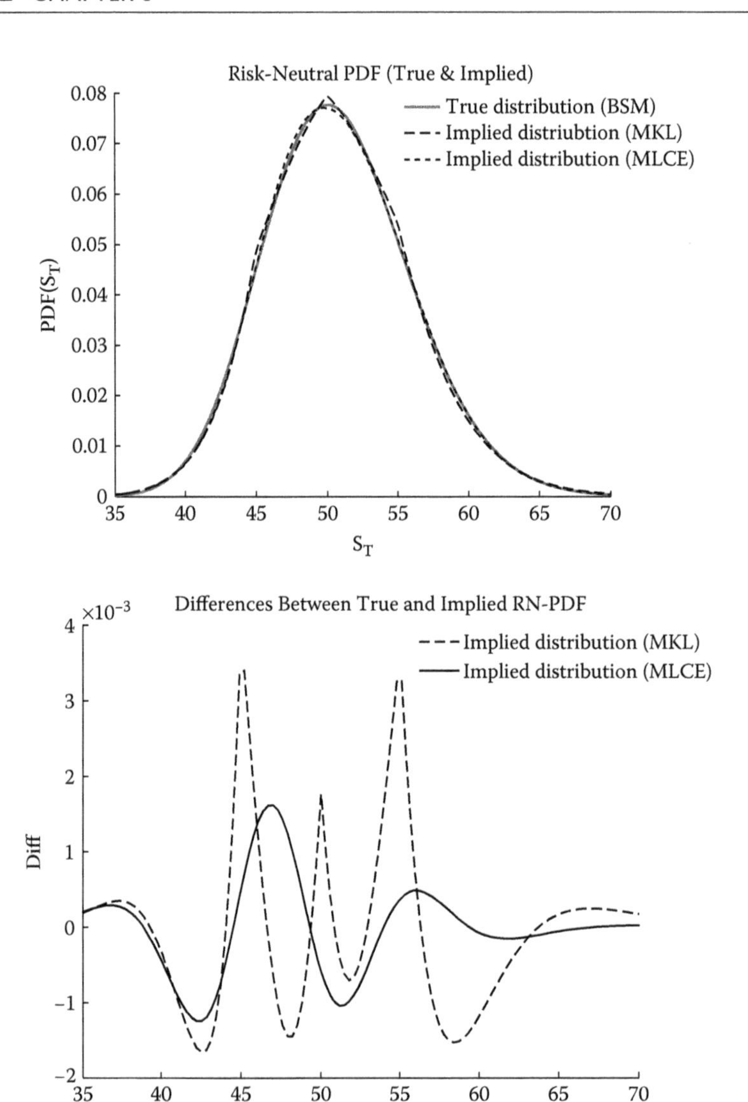

FIGURE 8.3.7 Buchen–Kelly test without noise.

as we have at least 2 options[2] it does not cause any problems for the MLCE method. The IPOPT optimizer converges to an acceptable solution (tolerance

[2] As we have alluded to before, it would be even sufficient to have one option, but then we would need to use put/call parity to obtain the corresponding put/call price, which would have to be included in the optimization problem as well.

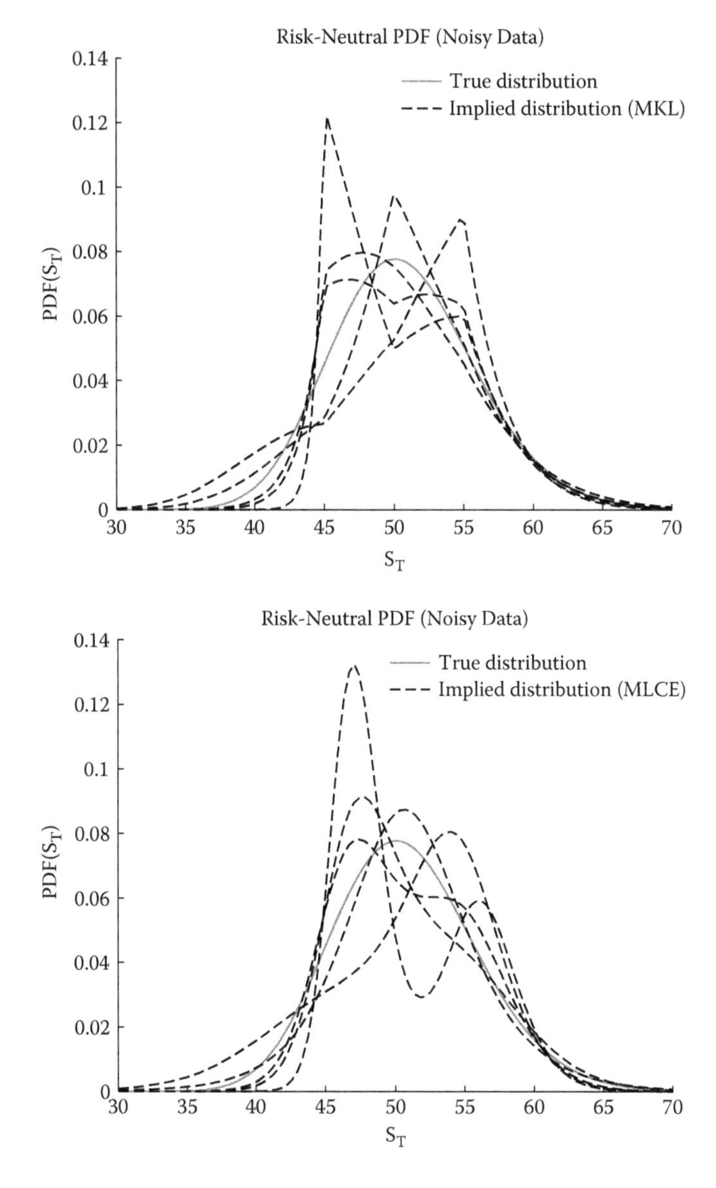

FIGURE 8.3.8 Buchen–Kelly test with noise for five random perturbations: (a) MKL, (b) MLCE.

level 10^{-16}) within 60 iterations, which takes less than 2 s on a computer with a Pentium 4 2.80 GHz running Debian Linux.

Figure 8.3.9b graphs distributions that were obtained for six maturities of NASDAQ 100 options (NDX). Because these options are less liquid and for the last three maturities we only have one option at which a trade was made,

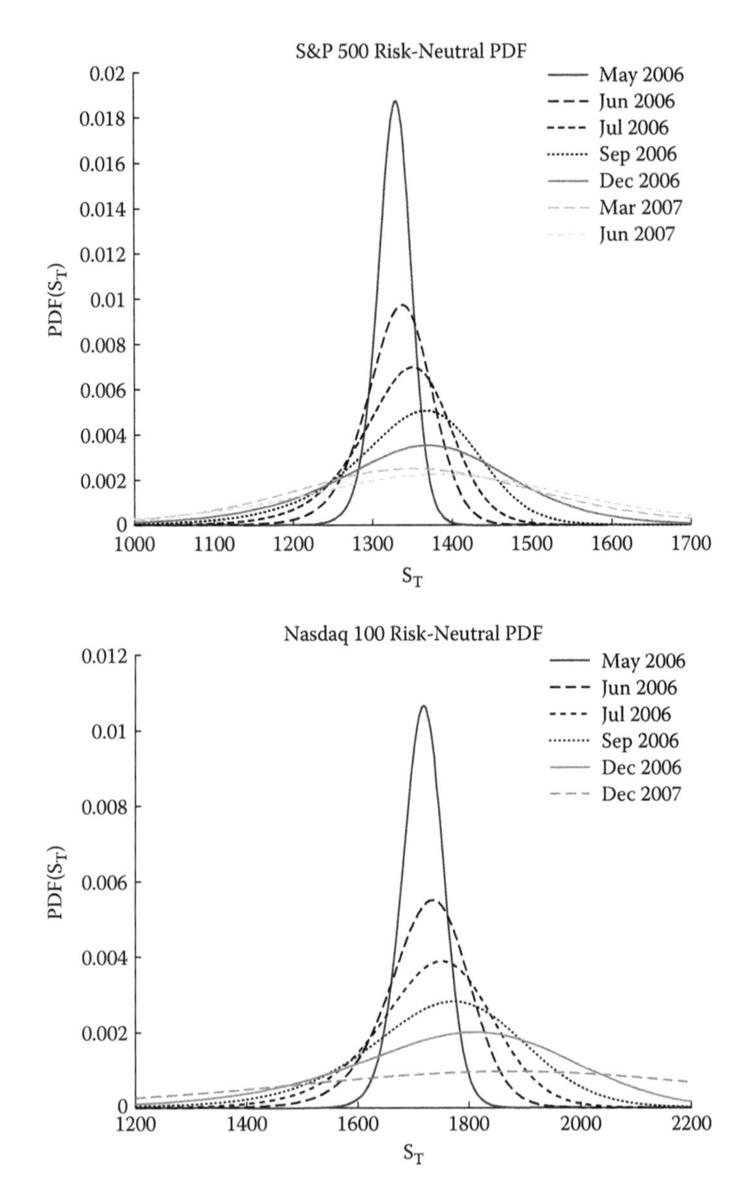

FIGURE 8.3.9 Risk-neutral distribution for (a) S&P 500 for seven maturities (May, June, July, September, December 2006, March and June 2007); (b) NASDAQ 100 for six maturities (May, June, July, September, December 2006, and December 2007) as expected on May, 9 2006.

we decided to take a different approach than in the S&P 500 options and to use all available quotes.

In some cases the optimization engine would report unfeasibility of the solution or would not converge. The perpetrators are deep out-of-the-money options that would be quoted at prices 0.05 or 0.10, even though the model price would be much smaller. It is caused by the unwillingness of market makers to sell any contract below these amounts. After removing one or two of these prices, the convergence is successful again. Moreover, the loss of information is small because the other type of option (call/put) at the same strike price usually remains part of the optimization.

8.4 CONCLUSION

We have tested the minimum local cross-entropy method for recovering the risk-neutral probability distribution function on a wide variety of simulated models—ranging from the usual benchmark Black–Scholes–Merton model, through variance gamma from the infinite activity Lévy-process family, to Merton's jump diffusion and Heston's stochastic volatility model. The MLCE method is successful in all four models. It quickly converges to the true distribution as we add more and more options. It even copes with more complicated distributions such as those produced by variance gamma and jump diffusion models.

Comparing the MLCE method with the classic implied volatility smoothing approach developed by Shimko [17], we can see that MLCE is superior in two aspects:

1. MLCE works for a small number of option prices, even for one option. Depending on the interpolation scheme, Shimko's approach needs a higher number of option prices to achieve satisfactory results.
2. MLCE provides the distribution for the whole range of stock prices, including the tails. Shimko's method provides the distribution only within the interval of available strike prices (see Figure 8.3.1b) and has to deal with the tails by extrapolation, e.g., by grafting-in lognormal tails. It contrasts with tails produced by MLCE that are not arbitrary but are based on a theoretical concept inspired by information theory.

The speed of convergence is fast. Industrial-strength optimizers find the solution within several seconds on a standard PC. For noisy data, if the option constraints do not violate the no-arbitrage conditions, the MLCE method still provides a smooth distribution. Moving away from simulated data, we tested MLCE on real market prices. The optimization is slightly different in the sense that we have a large number of inequality constraints (bid and ask prices). Nevertheless, the MLCE again quickly converges to a smooth distribution.

APPENDIX A

Tables in this section summarize the ability of the MLCE method to recover the true distribution in terms of Kullback–Leibler pseudodistance (KL), the error between theoretical and implied option prices (RMSE), and descriptive statistics of both distributions. Moreover, we include information about the local cross entropy (LCE), the number of iterations and elapsed time needed for the KNITRO optimizer to converge to a satisfactory solution given that we use 300 discrete points and the tolerance level is set to 10^{-16}.

ACKNOWLEDGMENTS

The author would like to kindly acknowledge the helpful comments and suggestions of David Edelman, Conall O'Sullivan, and an anonymous referee.

Research supported by GAČR 402/06/0209.

TABLE 8.4.1 Black–Scholes–Merton

	True	Implied Distribution Using MLCE		
	Distribution	(1 option)	(3 options)	(5 options)
KL	—	0.01811	0.00302	0.00011
LCE	7.330E − 9	5.049E − 9	6.130E − 9	7.149E − 9
RMSE	—	0.1776	0.19211	0.41267
Mean	0.01979	0.01941	0.01967	0.01976
Variance	0.01042	0.01142	0.01077	0.01054
Skewness	0	−0.42218	−0.10207	−0.00272
Kurtosis	−0.00157	0.99805	0.30551	−0.03352
Iterations	—	77	101	106
Time	—	0.15 s	0.25 s	0.32 s

TABLE 8.4.2 Variance Gamma

	True	Implied Distribution Using MLCE		
	Distribution	(1 option)	(3 options)	(5 options)
KL	—	0.05681	0.00410	0.00384
LCE	5.373E − 08	3.983E − 9	1.289E − 8	1.380E − 8
RMSE	—	0.1881	0.20606	0.43031
Mean	0.01804	0.01870	0.01823	0.01790
Variance	0.01471	0.01288	0.01433	0.01525
Skewness	−1.43671	−0.45180	−1.28855	−1.59633
Kurtosis	3.83464	1.06703	2.79155	4.97809
Iterations	—	78	106	105
Time	—	0.15 s	0.27 s	0.33 s

TABLE 8.4.3 Stochastic Volatility

	True	Implied Distribution Using MLCE		
	Distribution	(1 option)	(3 options)	(5 options)
KL	—	0.04139	0.00105	0.00044
LCE	1.20E − 8	4.834E − 9	1.117E − 8	1.156E − 8
RMSE	—	0.1795	0.19737	0.42600
Mean	0.01864	0.01928	0.018547	0.018576
Variance	0.01348	0.01168	0.013734	0.013642
Skewness	−1.36414	−0.42744	−1.47102	−1.4123
Kurtosis	3.63782	1.01055	4.18778	3.9586
Iterations	—	77	88	92
Time	—	0.15 s	0.21 s	0.29 s

TABLE 8.4.4 Jump Diffusion

	True	Implied Distribution Using MLCE		
	Distribution	(1 option)	(3 options)	(5 options)
KL	—	0.06562	0.01016	0.00055
LCE	8.958E − 9	3.065E − 9	7.732E − 9	8.674E − 9
RMSE	—	0.2004	0.21820	0.43690
Mean	0.01714	0.01782	0.01621	0.01716
Variance	0.01663	0.01470	0.01914	0.01660
Skewness	−1.31029	−0.48743	−1.81389	−1.22364
Kurtosis	2.66095	1.14291	5.5502	2.08009
Iterations	—	76	77	114
Time	—	0.15 s	0.19 s	0.35 s

REFERENCES

1. J. C. Jackwerth, "Option implied risk-neutral distributions and implied binomial trees: A literature review," *Journal of Derivatives*, pp. 66–92, September 1999.
2. D. Edelman and P. Buchen, "Garch or posterior volatility? An alternative approach to volatility smile modelling," *Australian National University Workshop on Stochastics and Finance, Financial Mathematics Research Report*, pp. 465–470, 1995.
3. P. W. Buchen and M. F. Kelly, "The maximum entropy distribution of an asset inferred from option price," *Journal of Financial and Quantitative Analysis*, vol. 31, pp. 143–159, March 1996.
4. D. Edelman, "Local cross-entropy," *Risk Magazine*, vol. 17, July 2004.
5. J. W. Cooley and J. W. Tukey, "An algorithm for the machine calculation of complex Fourier series," *Mathematics of Computation*, vol. 19, pp. 297–301, 1965.

6. P. Carr and D. B. Madan, "Option pricing and the fast Fourier transform," *Journal of Computational Finance*, vol. 2, no. 4, pp. 61–73, 1999.
7. K. Matsuda, "Calibration of Lévy option pricing models: Application to S&P 500 futures option," 2005.
8. G. Bakshi, C. Cao, and Z. Chen, "Empirical performance of alternative option pricing models," *Journal of Finance*, vol. 52, pp. 2003–2049, December 1997.
9. F. Black and M. Scholes, "The pricing of options and corporate liabilities.," *Journal of Political Economy*, vol. 81, pp. 637–654, May/June 1973.
10. D. B. Madan and E. Seneta, "The variance gamma (V.G.) model for share market returns," *Journal of Business*, vol. 63, pp. 511–524, October 1990.
11. D. B. Madan and F. Milne, "Option pricing with VG martingale components," *Mathematical Finance*, vol. 1, no. 4, pp. 39–56, 1991.
12. D. Madan, P. Carr, and E. Chang, "The variance gamma process and option pricing," *European Finance Review*, vol. 2, pp. 79–105, 1998.
13. S. L. Heston, "A closed-form solution for options with stochastic volatility with applications to bond and currency options," *Review of Financial Studies*, vol. 6, pp. 327–343, 1993.
14. R. C. Merton, "Option pricing when underlying stock returns are discontinuous," *Journal of Financial Economics*, vol. 3, pp. 125–144, 1976.
15. R. H. Byrd, M. E. Hribar, and J. Nocedal, "An interior point algorithm for large-scale nonlinear programming," *SIAM Journal on Optimization*, vol. 9, no. 4, pp. 877–900, 1999.
16. A. Wächter and L. T. Biegler, "On the implementation of an interior-point filter line-search algorithm for large-scale nonlinear programming," *Math. Program.*, vol. 106, no. 1, pp. 25–57, 2006.
17. D. Shimko, "Bounds of probability," *Risk*, pp. 33–37, 1993.
18. D. Bates, "Jump and stochastic volatility: exchange rate processes implicit in Deutschemark options," *Review of Financial Studies*, vol. 9, pp. 69–107, 1996.
19. D. Duffie, J. Pan, and K. Singleton, "Transform analysis and asset pricing for affine jump-diffusions," *Econometrica*, vol. 68, pp. 1343–1376, 2000.
20. W. Schoutens, "Lévy processes in finance," *Wiley Series in Probability and Statistics*, Wiley, New York, 2003.

CHAPTER 9

Using Intraday Data to Forecast Daily Volatility: A Hybrid Approach

David C. Edelman
University College Dublin, Dublin, Ireland
Francesco Sandrini
Pioneer Investments, Dublin, Ireland

Contents

Abstract: Despite significant advances in volatility forecasting over the last 20 years, since the first introduction of ARCH and GARCH models, comparatively little attention has been paid to issues such as the sensitivity of out-of-sample volatility forecasts to changes in structural components of the models. We propose and estimate a hybrid GARCH model where the parameters are Kalman filtered, and the model is assessed and compared with standard GARCH methods, using several measures of efficacy. The framework is then extended to cope with intraday data, and we investigate whether working with intraday summaries brings any informative advantage in comparison to working with daily data only. As this chapter does not concern itself with intraday volatility forecasting, but focuses on daily volatility forecasting, daily summaries are developed that allow for intraday microstructure. These summaries are then calibrated and used for the daily forecasting problem. Kalman-filtering methods are then applied to these, and the results are then shown to compare favorably with the methods of the earlier section.

9.1 INTRODUCTION

The first part of this chapter aims to provide a description of the framework employed, essentially, the evaluation of whether there is some out-of-sample performance advantage in pairing two approaches such as Kalman filtering and GARCH (generalized autoregressive conditional heteroscedastic) to forecast daily volatility. The origin of what later came to be the GARCH model dates back to 1982, when Robert Engle [9] proposed the famous ARCH (autoregressive conditional heteroscedastic) formulation to provide a solution to the heterosfedasticity problem that seems to affect the vast majority of financial time series. The idea of increasing the estimation efficiency of the linear regression model by adding a second equation in which the conditional variance of the process is modeled as a linear function of squared returns was then improved in 1986 by Bollerslev [5], who proposed a new, more parsimonious process, autoregressive in the conditional variance; the new GARCH framework was born. With great synthesis and using a very general formulation, the GARCH framework can be represented as follows:

$$y_t = \beta_0 + x_t^T \beta + \epsilon_t \tag{9.1.1}$$

$$\epsilon_t \mid \Psi_{t-1} = N(0, h_t) \tag{9.1.2}$$

$$h_t = \alpha_0 + \sum_{i=1}^{q} \alpha_i \epsilon_{t-i}^2 + \sum_{j=1}^{p} \beta_j h_{t-j}. \tag{9.1.3}$$

It should be noted that for $p = 0$ the GARCH(p, q) model is also called an ARCH(q) model. It is worthwhile to stress that one of the key requirements of the model is a zero-mean series of observed returns. A degree of filtering is hence required, and the first equation aims to achieve this goal. In what follows, we propose a filtering routine, exogenous to the GARCH model, based on the Kalman filter. All the analysis has been performed using a $p = 5$, $q = 5$ specification. Results are rather robust with respect to the choice of a different specification. The idea behind the Kalman filter (from the contribution of Kalman) is to express a dynamic system in a particular form called state–space representation. The Kalman filter (KF) is an algorithm for sequentially updating a linear projection of the system. Following the KF schema in Johnson and Sakoulis [14] (slightly simplified), the observation equation is given by

$$r_{t+1} = f'\beta_{t+1} + u_{t+1}, \tag{9.1.4}$$

where r is the observable, β a (slowly) time-varying parameter, u a mean-zero random error, and f a data input vector, and where typically

$$u_t \sim N(0, \sigma^2). \tag{9.1.5}$$

Next, we have the process equation

$$\beta_t = \beta_{t-1} + \upsilon_t, \tag{9.1.6}$$

where υ refers to the (mean-zero) "process" innovation, and where typically

$$\upsilon_t \sim N(0, Q) \tag{9.1.7}$$

$$P_t = P_{t-1} + Q,$$

where P denotes the variance-covariance matrix of the current β estimate.

This summarizes the specification of the filter, whereas the prediction part ("Kalman update" step) of the filter can be schematized as follows:

$$\beta_{t,t-1} = \beta_{(t-1)|(t-1)}, \tag{9.1.8}$$

where "|" refers to the convention in Kalman filtering of conditioning on information at the previous time step

$$P_{t|(t-1)} = P_{(t-1)|(t-1)} + Q \tag{9.1.9}$$

$$\eta_{t|(t-1)} = r_t - f'_{t-1}\beta_{t|(t-1)} \tag{9.1.10}$$

$$\xi_{t|(t-1)} = f_{t-1}P_{t|(t-1)}f_{t-1} + \sigma^2. \tag{9.1.11}$$

The update happens in the state–space framework, and in particular

$$\beta_{t|(t)} = \beta_{t|(t-1)} + K_t\eta_{t|(t-1)} \tag{9.1.12}$$

$$P_{t|(t)} = P_{t|(t-1)} - K_t f'_{t-1} P_{t|(t-1)}. \tag{9.1.13}$$

K is the so-called Kalman gain, and its formulation is as follows:

$$K_t = P_{t|(t-1)} f_{t-1} \xi_{t|(t-1)}^{-1}. \tag{9.1.14}$$

This Kalman filtering routine is paired to the GARCH model in an iterative fashion and thereby provides a different input to the model, the motivation being that if a GARCH-type relationship exists, the nature of the relationship (and hence the parameters) is likely to change somewhat over time.

The next section presents the specifics of combining Kalman filtering with GARCH, followed by a range of tests that assess the benefits of this approach. The succeeding section extends these results using daily summaries of intraday data.

9.2 THE HYBRID FRAMEWORK

The introduction to this chapter presented an overview of the two main components of out-of-sample back-testing. Some readers may find the following conceptual characterization of the previously described process helpful. To describe in words what will be presented graphically in the flow diagram below (Figure 9.2.1), one can imagine the system at time t.

The Kalman filter performs a linear filtering on the past returns available until time $t-1$. The forecast for the return at time $r_t \mid \Psi_{t-1}$ is then compared with the realized FTSE daily return at time t. In a mathematical formulation, $\epsilon_t = r_t - r^{KF}_{t|\Psi_{t-1}}$. This forecasting residual is then stored in an array called Φ, and this array is used as an input for the GARCH estimation, using a

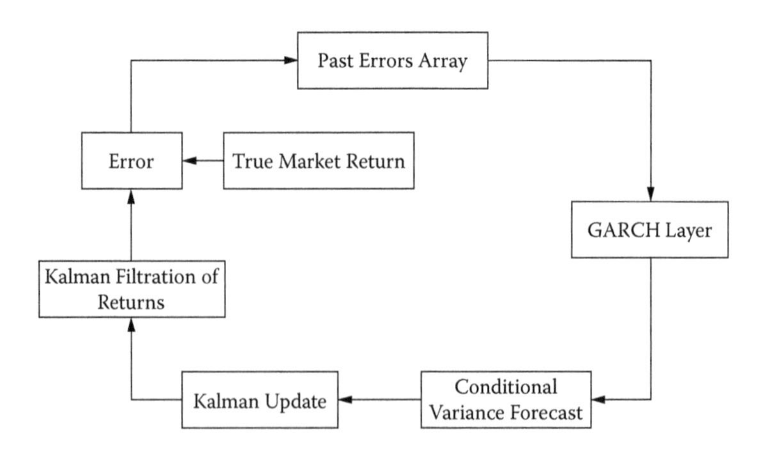

FIGURE 9.2.1 Interaction between the GARCH layer and the Kalman filter.

maximum-likelihood approach as follows. Given a parameterized family D_θ of probability distributions associated with either a known probability mass function (a discrete distribution) denoted by f_θ, we assume that we draw a sample, $x_1, \ldots, x_n$, of random variables from this distribution, and then using f_θ we can compute the probability density associated with our observed data

$$f_\theta(x_1, \ldots, x_n|\theta). \tag{9.2.1}$$

As a function of θ with $x_1, \ldots, x_n$, this is the likelihood function

$$L_\theta = f_\theta(x_1, \ldots, x_n|\theta) \tag{9.2.2}$$

When, as in this case for volatility, θ is not observable, the method of maximum likelihood searches for the value of θ, here the volatility h_t^2, that maximizes $L(\theta)$ as an estimate of h_t^2. The maximum likelihood estimator may not be unique, or indeed may not even exist. In the framework we set up, we shall use a continuous probability distribution, which is the Gaussian distribution

$$f(x|\mu, \sigma^2) = \frac{1}{\sqrt{2\pi\sigma^2}} e^{-\frac{(x-\mu)^2}{2\sigma^2}} \tag{9.2.3}$$

$$f(x_1, \ldots, x_n|\mu, \sigma^2) = \left(\frac{1}{\sqrt{2\pi\sigma^2}}\right)^n e^{-\sum \frac{(x_i-\mu)^2}{2\sigma^2}}. \tag{9.2.4}$$

The GARCH model produces a forecast for conditional volatility by searching in the parameter space for the level of h_t^2 in which k, the vector of past iteration errors (which have a mean equal to zero), is as "Gaussian" as possible. Obviously, the maximum-likelihood procedure does not work in the μ space; in other words, it assumes that the past iteration errors are already centered around a zero mean. This forecast for h_t^2 is then used via the Kalman update to provide a forecast for returns at time $t + 1$. This is a description of how the two methods interact to provide an out-of-sample strip of volatility forecasts. In Figure 9.2.2, we show the out-of-sample volatility forecast performed with KGARCH and GARCH.

As expected, the difference between the two volatility forecasts is very small, as seen in Figure 9.2.2, which also shows the difference between the two annualized standard deviations. Figures 9.2.3(a) and 9.2.3(b) concentrate on the details, extracting from the dataset some subperiods affected by surprisingly high returns in a low-volatility environment. When there is a surprise in

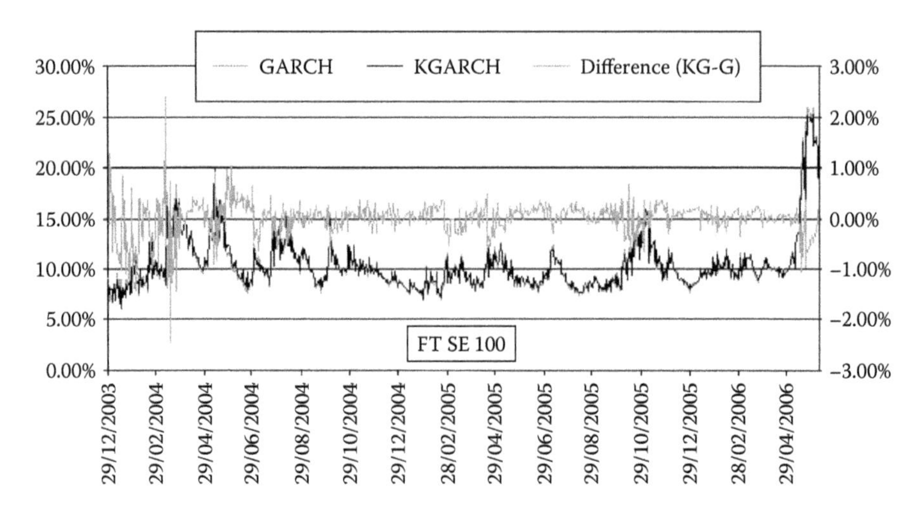

FIGURE 9.2.2 Out-of-sample volatility forecast FTSE 100 Stock Index, data sample from December 2002.

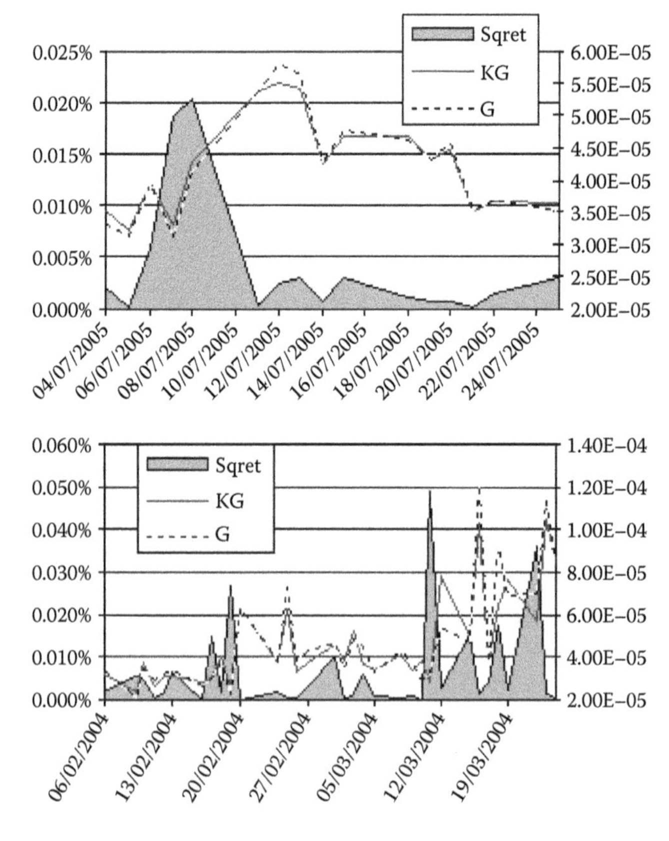

FIGURE 9.2.3 Particulars of the reactions of the two models to high squared return surprises: (a) subperiod number 1, example of return surprise; (b) subperiod number 2, example of return surprise.

terms of squared returns, the KF hybrid GARCH seems to react much more quickly, and then the reversion rate looks faster.

We now try to perform several more formal tests with a much larger data set. Using daily returns from Standard and Poor's 500 stock index starting from 1960 (more than 11,500 daily returns), the framework described above will be used to understand whether the hybrid between GARCH and Kalman filtering brings any advantage in forecasting daily volatility out of sample. Conditional volatility in financial time series is the most popular method for explaining why returns have thicker tails than those defined in the following equation:

$$f_{\theta}(x_1, \ldots, x_n|\theta). \tag{9.2.5}$$

The underlying philosophy in the GARCH class of models is that they reproduce (explain) the random variation in h and thus reverse this tail thickening. The fundamental question is what criteria one should use to judge the superiority of a volatility forecast. The literature describes different testing approaches, ranging from statistical-based evaluation, to utility (interval)-based evaluation, to profit-based (preference free) evaluation. See Knight and Satchell [16] for a detailed review of existing testing methodologies.

From the traditional statistical point of view, we wish to pick up the forecast that incorporates the maximum possible information or the minimum possible noise. The common feature in most of the papers—Cooper and Nelson [6], Diebold and Nason [8], and Hsieh and Tauchen [13]—is that the performance of the model is assessed on the basis of a traditional MSE (mean squared errors). This result relies on the used of squared returns as a proxy to the true, unobservable volatility. As Knight and Satchell [16] show, statistical measures probably underestimate the forecasting capability of GARCH models, especially when assessed out of sample. The approximation of the true volatility by the squared returns introduces substantial noise, which effectively inflates the estimated error statistics and removes any explanatory power of GARCH volatility forecasts with respect to the "true" volatility. See Knight and Satchell [16] for a formal proof and a quantification of the bias in a GARCH framework. A second class of measures concentrates on the forecast's unbiasedness by adapting the approach of Mincer and Zarnowitz [15]. The forecast for the volatility will be unbiased if the following regression:

$$h_{t+s} = \alpha + \beta \widehat{h}_{t+s} + \epsilon_{t+s} \tag{9.2.6}$$

returns a value for $\alpha = 0, \beta = 1$, and $E(\epsilon_{t+s}) = 0$. We apply this framework both to the FTSE and to the Standard and Poor's 500 data set.

TABLE 9.2.1 Results from Out-of-Sample Regression of Square FTSE Returns Against GARCH Forecast Variance (600 Out-of-Sample Points Used in the Regression)

Coefficient	Alpha 0	Beta 0	Residual Sum of Squares
Parameter value	1.94×10^{-6}	9.273×10^{-1}	0.433
Standard error	5.71×10^{-6}	9.9×10^{-2}	—

From Tables 9.2.1 and 9.2.2, can be seen that the β value in the KF-GARCH regression is closer to 1. Both of the regressions broadly accept the null hypothesis that the alpha coefficient is equal to zero; still, the hybrid approach provides a better parameter value. The R^2s of the two regressions were very similar.

A second testing framework was also implemented to check for omitted information, or, in other words, to understand whether a piece of information is captured by a model and not by the alternative specification. In the manner of Fair and Schiller [11], a first OLS (ordinary least squares) regression similar to the one introduced above was run. Next, the residuals from a regression between the volatility, as measured by the squared returns, and a model (here, GARCH) are saved, and then a second regression between the residuals and the variance, as forecast by the competing model, is run to assess if the latter contains some information not included by the former (if so, both regression coefficients should be different from zero). The out-of-sample regression on 600 returns from the FTSE stock index does not provide any clear indication in this sense. Both the regressions gave strong evidence for $\beta = 0$ and a negligible level of R^2. The same analysis performed on a broader data set of 4500 data points from Standard and Poor's 500 returns the same evidence.

Using the latter S&P data set, an alternative testing framework was applied to assess the superiority of one model to the other from a "coverage" point of view. In this setting, prediction intervals are produced from volatility forecasts, and coverage rates can be compared. The desired interval-coverage probability in this setting is a result of a utility theory analysis (obviously, there is a trade-off

TABLE 9.2.2 Results from Out-of-Sample Regression of Square FTSE Returns Against KF-GARCH Forecast Variance (600 Out-of-Sample Points Used in the Regression)

Coefficient	Alpha 0	Beta 0	Residual Sum of Squares
Parameter value	3.02×10^{-7}	0.9753	4.35×10^{-6}
Standard error	5.95×10^{-6}	0.1068	—

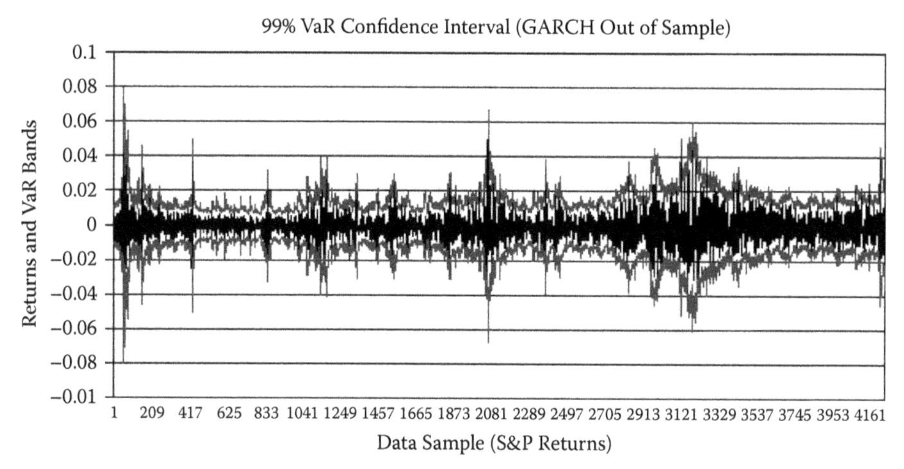

(a) Data sample 4500 US Equity Returns (random sample daily 1960–1978)

FIGURE 9.2.4 Out-of-sample confidence VaR intervals, 99 % probability on 4500 Standard and Poor's data points.

between interval width and coverage probability), but the analysis below will focus on several common values.

Figure 9.2.4 displays the value at risk (VaR) obtained by applying the out-of-sample daily volatility forecast to calculate, for every point, the maximum loss at 99 % probability. Here we assume that the only asset we hold in our portfolio is the Standard and Poor's 500 stock index. The problem setup obviously grows in complexity if more than one asset class is considered. This setup is important to count breaches of these VaR intervals and eventually to calculate the average or aggregated width of the intervals (the budget of risk that is allocated). The question we want to answer is whether the KF GARCH model is able to provide any improvement in this sense (i.e., superior coverage or narrower intervals). We repeat the procedure over different probability levels to gather evidence of this.

Table 9.2.3 shows the results obtained by changing the confidence level.

The evidence found over the Standard and Poor's data set is quite encouraging. The KF GARCH model is successful in managing the heavy tailedness that seems to affect financial-returns time series as compared with a traditional GARCH model. This conclusion is drawn using out-of-sample back-testing. The histogram in Figure 9.2.5 shows the distribution of the standardized variables as obtained by the two models. More-extreme events are localized on the GARCH distribution tags than in the KF version of the GARCH; the mean and standard deviation of the KF version are closer to 0 and 1, respectively.

TABLE 9.2.3 Coverage-Based Test Showing the Number of Breaches at Different Confidence Levels (Random Sample, 1960–1978 Standard and Poor's 500)

	Model	
Confidence Level (%)	KF GARCH	GARCH
90	0.209	0.211
91	0.1876	0.1923
92	0.1703	0.1732
93	0.1514	0.1547
94	0.1336	0.1355
95	0.1116	0.1161
96	0.0937	0.0954
97	0.0714	0.0720
98	0.0520	0.0528
99	0.0302	0.0321

A better approximation to a Gaussian distribution is visible in Figure 9.2.5. From the random sample we selected, it is also possible to look at this from a testing point of view.

The KF GARCH framework produces a distribution of standardized returns that is closer to Gaussian over the sample period. All of the statistics (Table 9.2.4) are better for KF than for GARCH; the Jarque–Bera test rejects

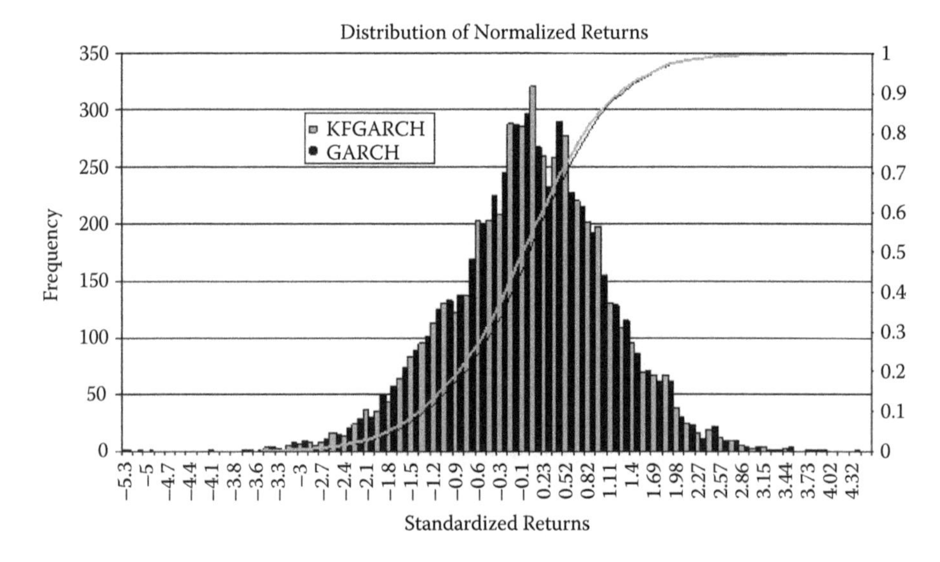

FIGURE 9.2.5 Standardized returns and cumulated probability for KF GARCH and GARCH models.

TABLE 9.2.4 Normality Test on Standardized Returns under the Two Competing Model Specifications (Random Sample, 1960–1978 Standard and Poor's 500)

	Model	
	KF GARCH	GARCH
Mean	−0.028	−0.10
Standard deviation	1.031	1.048
Skewness	−0.14	−0.15
Kurtosis	3.58	3.75
Jarque–Bera	79.73	124.49

the normality (Gaussian) assumption less strongly than for the GARCH specification. The Jarque–Bera test is a goodness-of-fit measure of departure from normality, based on the sample kurtosis and skewness

$$JB = \frac{n-k}{6}\left(S^2 + \frac{(K-3)^2}{4}\right), \tag{9.2.7}$$

where S is the skewness, K is the kurtosis, n is the number of available observations, and k is the number of coefficients used to create the series. The statistic has an asymptotic chi-squared distribution with two degrees of freedom and can be used to test the null hypothesis that the data are from a Gaussian distribution, as samples from a Gaussian distribution have an expected skewness of zero and an expected kurtosis of 3. As the equation shows, any deviation from this increases the JB statistic.

Figure 9.2.6 shows the tendency of the KF version of the model to react more quickly than the GARCH (out of sample) after big moves as measured by squared returns.

Thus it seems that, in a univariate environment and using daily data, the combination of a Kalman filter and the usage of a cumulative array of errors produces better results when assessed from an out-of-sample perspective.

9.3 ADDING INTRADAY DATA TO THE FRAMEWORK

The natural extension of this analysis is to understand whether this approach, which seems to be quite flexible, can be applied to forecasting daily volatility using intraday data. We used FTSE intraday data from December 2002 (Figure 9.3.1). The increasing availability of high-frequency data has made possible the detailed study of intraday statistics. We sampled data every half-hour from FTSE. Taking historical standard deviation on the intraday returns, we spot

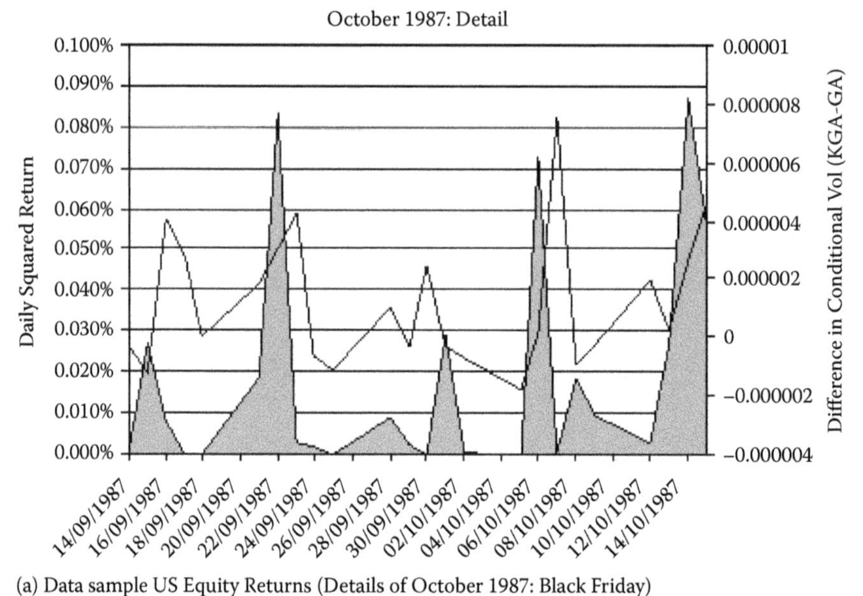

(a) Data sample US Equity Returns (Details of October 1987: Black Friday)

FIGURE 9.2.6 Evaluation of the difference between conditional variance of KF GARCH and GARCH in regime of high squared returns (squared returns shaded).

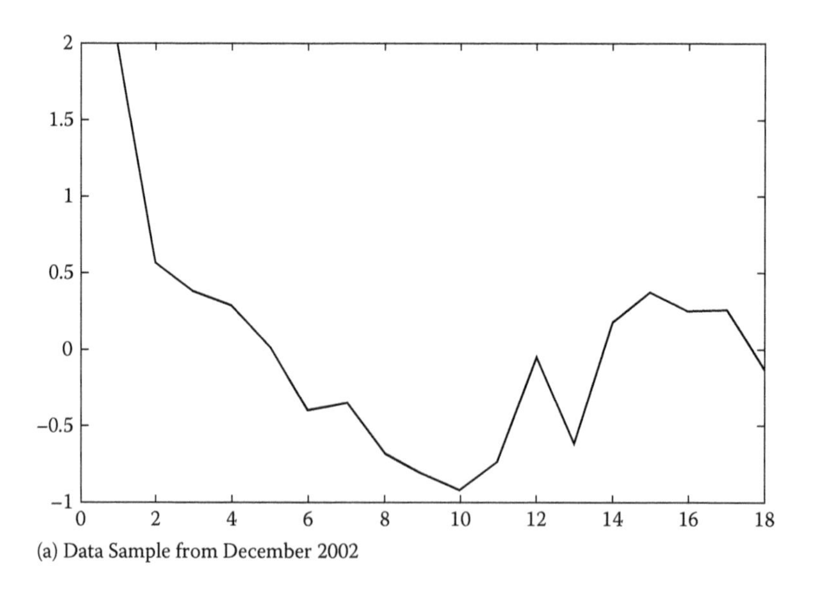

(a) Data Sample from December 2002

FIGURE 9.3.1 Presence of J-shape as explained by ANOVA filtration.

the presence of a U-shaped pattern; a more-refined analysis based on the use of dummy variables to separate daily from intraday effect gave similar results.

This is quite consistent with findings from a rather wide portion of the existing literature. A recent literature review by Daigler [7] shows the existence of intraday U-shaped curves in volatility (more precisely, over certain markets, a reverse J can be observed) as well as in volumes and bid-ask spreads. A stream of empirical research and several different theories try to explain these patterns.

For example, if we consider trading activity, we observe that trading often occurs at the opening because information flows in over the closed period. Moreover, it is pointed out by Acar and Petitdidier [1] (again, an accurate review of this stream of literature can be found in Knight and Satchell [16]) that traders might have different opinions at the opening stage, creating possibly high volatility and high trading volumes. The existence of these patterns can be easily observed in different markets, from New York to London and from Tokyo to Chicago. High-frequency data are subject to measurement errors, and very long time series cannot be easily found and, when available, are often not homogeneous from the point of view of opening and closing times of the underlying markets. Collectors of intraday data include Reuters, Telerates, etc.

In contrast, this chapter attempts to make a distinct and possibly more practical use of intraday data than in the prevailing literature so far; intraday price-change dynamics in the FTSE are investigated in an effort to achieve the ultimate goal, which is forecasting daily returns or using a generalization of volatility on lower frequencies.

Following this, and extension to the type of analysis explained in the previous chapter to intraday frequency (in particular to half-hourly FTSE price sampling), the goal is to understand whether a dynamic analysis of the volatility of prices within every time bucket (for example from 11 a.m. to 11:30 a.m.) can bring some information when properly aggregated at daily frequency.

From a technical point of view, what we perform is an out-of-sample back-testing (see Table 9.3.1) where, in two consecutive stages, we try to examine the available data sample (intradaily and daily) to extrapolate a volatility forecast at daily frequency. This is achieved by pairing the methodology presented above (at intraday level) with a second likelihood calibration using daily returns, in a rather nested way.

$$h_t = \alpha_h + \sum_{i=1}^{n} \beta_{h,i} h_{t,i} + \epsilon_t. \tag{9.3.1}$$

Here n is the number of intraday time buckets (in the analysis performed, every half hour), and i is the index identifying the bucket. The purpose of the

TABLE 9.3.1 Results from Out-of-Sample Regression of Square FTSE Returns Against Intraday KF-GARCH Forecast Variance (600 Out-of-Sample Points Used in the Regression)

Coefficient	Alpha 0	Beta 0	Residuals Sum of Squares
Parameter value	3.27×10^{-7}	1.32	4.21×10^{-6}
Standard error	5.4×10^{-6}	0.130	—

analysis that follows is to compare the performance of this nested intraday hybrid model with the corresponding daily hybrid model that we presented in the previous section of this chapter, in order to understand whether there is an improvement in performance by using a broader data set. Figure 9.3.2 illustrates the use of the KF GARCH methodology at the time-bucket level, and in particular we display the difference in the forecast volatility levels at 15.00, when the U.S. equity market is opening and new information is flowing into the system, and at 12.00, when the number of trades is traditionally very low.

It is possible to see that, even in a low-volatility environment such as the available intraday data set for the FTSE 100, the presence of spikes or innovations is more evident in the 2 p.m. pattern (Figure 9.3.2[b]) than in the 12 a.m. one (Figure 9.3.2.[a]). This is explainable and consistent with the reverse J shape theory discussed previously. Figure 9.3.3 shows the comparison among different daily volatility forecasts (out of sample on FTSE 100), adding to the previous picture of the intraday GARCH model filtered with the previously described Kalman-filtering technique. It is possible to see that it produces a steadily lower level of volatility in a low-volatility environment, such as the one used to evaluate the models.

We performed some tests, in the same fashion used before, to understand whether this procedure leads to some benefits in terms of daily volatility forecast.

From a first set of diagnostics, we observe a value for the β parameter that is not so close to 1, in exchange for a gain of R^2 of around 25% from a value around 11.3 for the two models at daily frequency to a value of 14. Again, establishing whether one model is better than another is far from easy. To further investigate this view, a simple test for omitted information was run (the same test that did not give significant evidence in assessing KF GARCH against GARCH at daily frequency). In the manner of Fair and Schiller [11], first an OLS regression between squared returns and GARCH conditional volatility is run out of sample. After saving residuals, a second regression performed between the residuals from the first OLS regression and the variance as forecast by the competing model (in this case the intraday KF GARCH) to

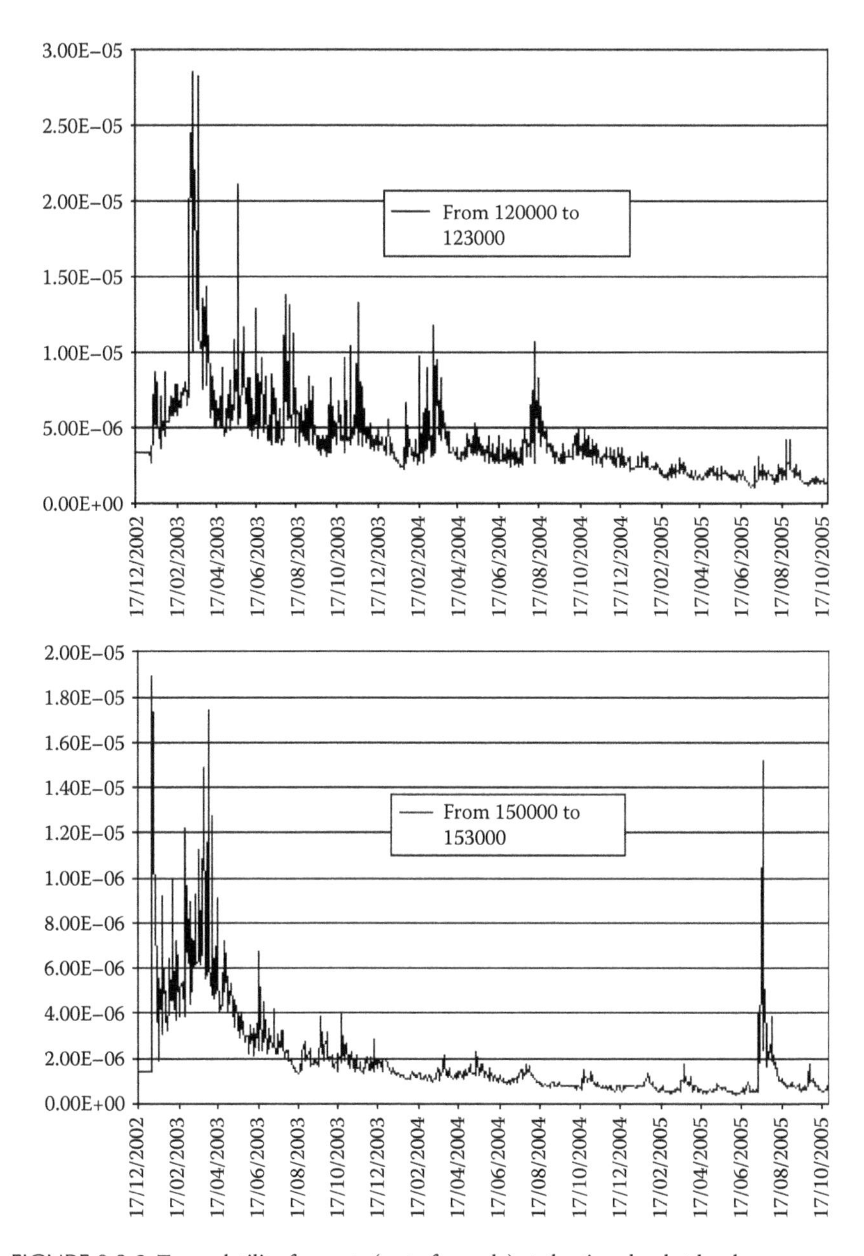

FIGURE 9.3.2 Two volatility forecasts (out of sample) at the time-bucket level.

assess whether it introduces any new, useful explanatory information; if so, both the regression coefficients should be different from zero (see Table 9.3.2).

The test returns an R^2 that is different from zero and a value for β that is also significantly different from zero. This did not happen when we performed the same test using two different specifications but the same data set (daily,

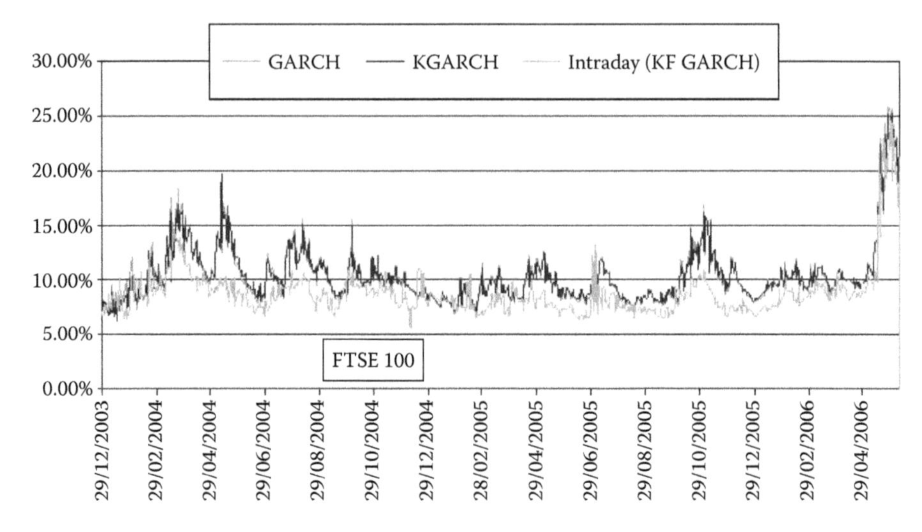

FIGURE 9.3.3 Different daily volatility out-of-sample forecasts, including the intraday model.

KF GARCH against GARCH). There is information not captured by a daily parametric model that is captured in this intraday framework. We think it could be interesting to compare the two out-of-sample forecasts with a measure of implied volatility. In financial mathematics, the implied volatility of a financial instrument is the volatility implied by the market price of a derivative based on a theoretical pricing model. We cannot invert the function (see, for example, the Black–Scholes equation) analytically; however we can find the unique value of the volatility that makes the Black–Scholes equation hold by using a root finding algorithm such as Newton's method.

$$C(S,T) = SN(d_1) - Ke^{-rT}N(d_2) \tag{9.3.2}$$

$$d_1 = \frac{ln(S/K) + (r + \sigma^2/2)T}{\sigma\sqrt{T}} \tag{9.3.3}$$

$$d_2 = d_1 - \sigma\sqrt{T}. \tag{9.3.4}$$

TABLE 9.3.2 Test for Omitted Information

Coefficient	Alpha 0	Beta 0	Residuals Sum of Squares
Parameter value	-1.22×10^{-5}	0.36	4.28×10^{-6}
Standard error	5.45×10^{-6}	0.120	—

Interestingly, the implied volatility of options rarely corresponds to the historical volatility (i.e., the volatility of a historical time series). This is because implied volatility includes future expectations of price movement, which are not reflected in historical volatility. The Chicago Board Options Exchange Volatility Index (VIX), measures how much of a premium investors are willing to pay for options as insurance to hedge their equity positions. The VIX is calculated using a weighted average of implied volatility of at-the-money and near-the-money striked in options on the Standard and Poor's 500 index futures. Because intraday data on the German Stock Market (DAX) are available, we choose to repeat the comparison we did for FTSE using the DAX index. The VDAX index represents the implied volatility of the German stock exchange assuming 45 days remaining until expiration of the DAX index options. Eight subindices are also calculated, each corresponding to the lifetime of the current DAX options.

A check was run for the correlations between the out-of-sample volatility forecast from GARCH, KF GARCH, and IKF GARCH, with the implied volatility measure plotted in Figure 9.3.4. The correlation between GARCH and the implied variance is very high, around 0.72. A similar number comes from the Kalman-filtered GARCH model. The correlation between the implied measure and the out-of-sample intraday forecast is lower, around 0.62. The Kalman-filtered GARCH estimated on daily data produces a slightly better

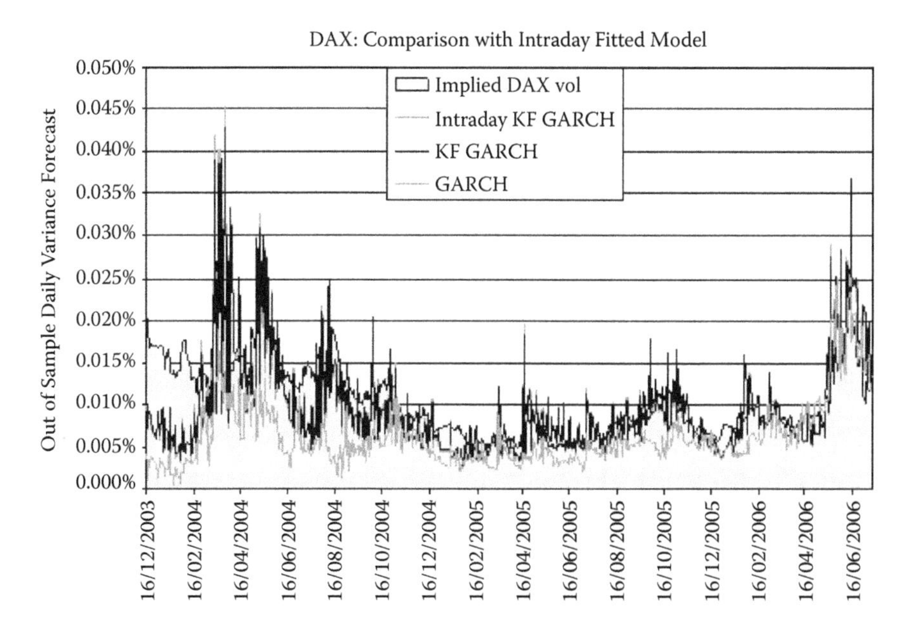

FIGURE 9.3.4 Different daily volatility out-of-sample forecasts, including the intraday model.

performance in forecasting implied volatility. In particular, we get a higher R^2 of 0.515 against 0.503 from the GARCH, and a β coefficient slightly closer to 1, 0.665 against 0.653 from the GARCH. Although this may seem a rather marginal gain, it deserves pointing out that the overall structure of the model is very similar, the only change being in the way the available information was filtered. When we consider the intraday data, however, Kalman-filtered GARCH results begin to show a distinct difference, a much better β equal to 0.9, but traded off against a lower R^2 of 0.435. The testing for omitted information in this case was inconclusive.

When squared returns were regressed on the out-of-sample volatility forecasts from GARCH or KF GARCH and then the residuals were regressed on the DAX options implied variance, it was found that no additional information appeared to be captured by the implied volatility measure (the relevant regression parameter in particular was not statistically different from zero). What one concludes from this is that the implied volatility (obtained by reversing the Black–Scholes formula starting from option prices) is more correlated with GARCH models estimated using only daily data. This leads us to believe that this approach, GARCH, is the one most broadly available to the financial community, and it is the one probably more used by dealers when pricing options for trading. If it is true that implied volatility includes more forward-looking information than an unconditional variance estimated over a long time series of past returns, it appears equally true that this does not hold when we replace the unconditional variance with a conditional one estimated with a GARCH model.

A slightly different framework is necessary to ascertain whether there is any marginal value in intraday data in forecasting daily volatility. Required is the specification of a less variable (and, ideally, unbiased) measure for daily volatility (we know that using squared returns involves a bias, see Knight and Satchell [16]). We feel that intraday data can be used to provide both a new measure to assess the performance of competing models and an input for different model specifications. Therefore, a framework for adjusting for microstructures will be developed here, making use of some constructs from numerical analysis, such as sparse matrices, to come to a summary of the information that can be useful in this sense.

Note first that if daily volatility consisted of a number t of equal intraday periods and the volatility through the day was constant, we could use the following formula:

$$\sigma_{\text{day}} = \sqrt{t}\sigma_{\text{intradaytimebucket}}. \tag{9.3.5}$$

From the analysis we performed, we showed how this does not match empirical evidence due to microstructure issues and opening adjustments. So it is not possible to use the following estimator:

$$\sqrt{t}\sqrt{\frac{\sum_{i=1}^{n}(y_i - \bar{y})^2}{n-1}}. \tag{9.3.6}$$

Instead, we propose a log ANOVA method as an alternative model for calibrating daily volatility. Notionally, we can write the relation existing between intraday and daily volatility as

$$\log\left(\sigma_{i,j}^2\right) = -\gamma + \log\left(\sigma_i^2\right) + a_j \tag{9.3.7}$$

$$\sigma_{i,j}^2 = \sigma_i^2 e^{a_j + \varepsilon_{i,j}} \tag{9.3.8}$$

for the day i and for the intraday bucket j. The result we want to achieve is a general index level for daily volatility, the square of which is proportional to the square of realized daily volatility. The calibration constant is computed using moments:

$$\frac{1}{n}\sum_{i=1}^{N}\frac{r_i^2}{\bar{\sigma}_i^2} = 1. \tag{9.3.9}$$

Let r_i denote mean-adjusted (filtered) logarithmic returns. Then $\bar{z}_i = r_i/\bar{\sigma}_i^2$ should have the following properties: (a) interval breaches at different confidence levels unbiased, (b) $\bar{z}_i$ and interval breaches uncorrelated, (c) histogram looks Gaussian and (d) $\frac{1}{n}\sum_{i=1}^{N}(\bar{z}_i)^m (m = 1, 2, 3, 4)$.

The proposed ANOVA filtration returns a time series of daily realized volatilities. This time series is slightly more correlated with the out-of-sample unconditional variances, as obtained using the set of models presented so far. This is quite evident looking at Table 9.3.3. Moreover, it is an unbiased estimator of the volatility of the data-generating process, in contrast to r^2. The approximation of the data-generating process volatility or "true" volatility by squared returns inflates the error statistics, hence reducing the explanatory power of GARCH specifications.

We use this vector of realized volatilities both as an input (possibly unbiased) to evaluate whether a superior performance in forecasting daily volatility can be achieved and as a measure to determine relative performance between competing models, in place of s^2. Figure 9.3.5 compares the realized daily volatility with the squared daily returns of the FTSE 100 stock index.

TABLE 9.3.3 Model Correlation with the Volatility Measure

Model	Corr. with Squared Ret. 0	Corr. with the New Realized Vol.
GARCH	0.47	0.48
KF GARCH	0.46	0.49
Intraday KF GARCH	0.37	0.48

With this new measure, we came to a new way to assess the performance of the models. What we obtained is a new benchmark that can be used in the same testing setup as before. We run simple linear regressions (OLS) using squared returns and the newly realized filtered measure to understand the performance of the model using intraday data.

The newly realized volatility measure, being slightly more correlated with the out-of-sample volatility forecasts, delivers not only a higher R^2, but does not change the conclusions that were obtained under the s^2 testing framework, which, as we recall, was a biased measure of volatility. Thus, it is clear that there is substantial advantage to be obtained by using the same model but with a different data set, or, in other words by moving from daily to intraday data as input. The evidence shown in Table 9.3.4 confirms that the use of intraday

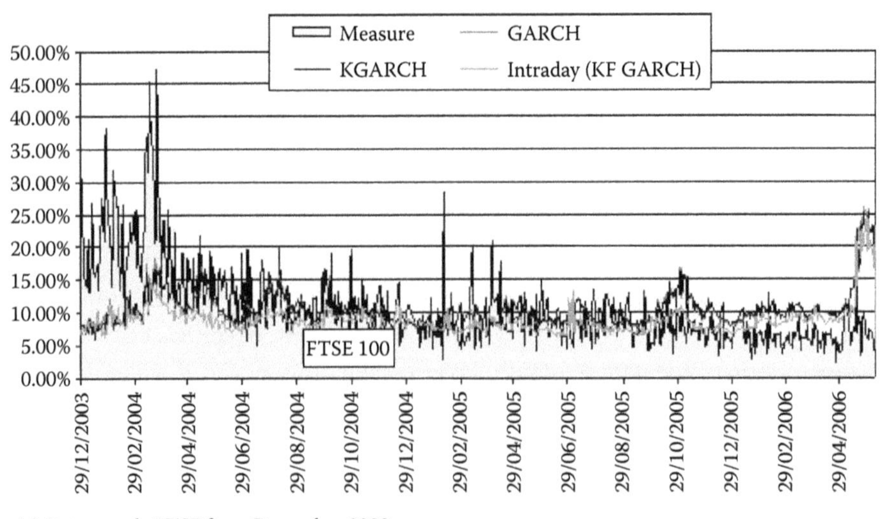

(a) Data sample FTSE from December 2003

FIGURE 9.3.5 Different daily volatility out-of-sample forecasts, including the intraday model.

TABLE 9.3.4 Reassessing Model Performance under the Newly Realized Volatility Measure

Model	R^2 Using Squared Ret.	R^2 Under New Realized Vol.
GARCH	0.12	0.18
KF GARCH	0.123	0.182
Intraday KF GARCH	0.142	0.192

data allows access to a richer information set than the traditional framework when forecasting volatility at the daily level.

9.4 CONCLUSION

The collection of studies presented here suggests a framework for testing the sensitivity of the out-of-sample performance of GARCH (and other volatility forecasting) models to some of their components such as, for example, a filtering procedure. It has been shown here that in pairing the GARCH econometric framework with Kalman-filtering techniques, results were obtained that appear to be quite convincing even though, in the absence of a unique framework to assess model out-of-sample performance, not all diagnostics were able to demonstrate clear superiority of the approach without further innovations (specifically, the inclusion of calibrated intraday summaries).

A much clearer picture was obtained by enhancing the model, first by adding intraday data to the proposed framework at the input, and then subsequently at the output level, thereby demonstrating that these measures can be used either way to boost the forecast of lower-frequency data (here, daily) or to provide newly realized volatility measures that can be used as an additional benchmark to assess the performance of competing models. The results presented here seem to suggest that intraday summaries may soon be used routinely in the analysis of daily data, both in forecasting and post hoc assessment of volatility forecasting methodology.

ACKNOWLEDGMENTS

The authors wish to thank in particular Andrey Demidov, senior financial engineer, Pioneer Investments, for valuable help in computationally expediting the work here, as well as for the very helpful comments on the approach.

REFERENCES

1. Acar, E. and Petididier, E. (2002), "Modelling slippage: an application to the Bund futures contract," in *Forecasting Volatility in the Financial Markets*, 2nd ed., Butterworth-Heinemann, Oxford.
2. Alexander, C. (2001), *Market Models: A Guide to Financial Data Analysis*, 2nd ed., John Wiley & Sons Ltd., New York.
3. Alexander, C. (1996), "Evaluating the use of RiskMetrics as a risk measure tool for your operation: what are its advantages and limitations?" *Derivatives: Use Trading and Regulations*, 2, 277–285.
4. Bauwens, Luc C.A.A and Giot, P. (1994), *Econometric Modelling of Stock Market Intraday Activity*, (vol. 38), *Advanced Studies in Theoretical and Applied Econometrics*, Springer, New York.
5. Bollerslev, T. (1966), "Generalized autoregressive conditional heteroschedasticity," *Journal of Econometrics*, 31, 307–327.
6. Cooper, J.P. and Nelson, C.R. (1975), "The ex ante prediction performance: of the St. Louis and FRB-MIT-PENN econometric models and some results on composite predictors," *Journal of Money, Credit and Banking*, 7, 1–32.
7. Daigler, R.T. (1997), "Intraday futures volatility and theories of market behaviour," *Journal of Futures Markets*, 17, 45–74.
8. Diebold, F.X. and Nason, J.A. (1990), "Non parametric exchange rate prediction," *Journal of International Economics*, 28, 315–332.
9. Engle, R.F. (1982), "Autoregressive conditional heteroscedasticity with estimates of the variance of United Kingdom inflation," *Econometrica, Econometric Society*, 50, 987–1007.
10. Engle, R.T. (1995), *ARCH: Selected Readings (Advanced Texts in Econometrics)*, Oxford University Press, Oxford.
11. Fair, R.C. and Schiller, R.J. (1990), "Comparing the information in forecasts from econometric models," *American Economic Review, American Economic Association*, 80, 375–389.
12. Fair, R.C. and Schiller, R.J. (1989), "The informational content of ex ante forecasts," *NBER Working Paper Series*, paper no. W2503, NBER, Cambridge, MA.
13. Hsieh, D.A. and Tauchen, G. (1997), "Estimation of stochastic volatility models with diagnostics," *Journal of Econometrics*, 81, 159–192.
14. Johnson, L. and Sakoulis, G. (2003), "Maximising equity market sector predictability in a Bayesian time varying parameter model," *SSRN Working Paper Series*, July II 2003, paper no. 396642.
15. Mincer, J. and Zarnowitz, V. (1969), "The evaluation of economic forecasts," in *Economic Forecasts and Expectations*, National Bureau of Economic Research, New York, pp. 83–111.
16. Knight, J. and Satchell, S. (2002), *Forecasting Volatility in the Financial Markets*, 2nd ed., Butterworth-Heinemann, Oxford.

CHAPTER 10

Pricing Credit from the Top Down with Affine Point Processes

Eymen Errais and Kay Giesecke
Stanford University, Stanford, CA
Lisa R. Goldberg
MSCI Barra, Inc., Berkeley, CA

Contents

10.1 EXTENDED ABSTRACT

A defaults cluster is illustrated in Figure 10.1.1, which shows the default history among Moody's rated U.S. issuers between 1970 and 2005. The clustering is driven by firm sensitivity to common economic factors, but it can also come from the feedback of an individual firm event to the aggregate level. A single default often results in the widening of credit spreads across the board. This phenomenon is empirically documented in Jorion and Zhang (2006) and exemplified in Figure 10.1.2, which shows the impact of Delphi's default and GM's announcement of a huge quarterly loss on the spread of the Dow Jones CDX North America Crossover Index swap.

In this chapter we develop a new *affine point process* framework to price and hedge a portfolio credit derivative such as the CDX index swap or an index tranche. In this top-down approach, a credit derivative is a path-dependent

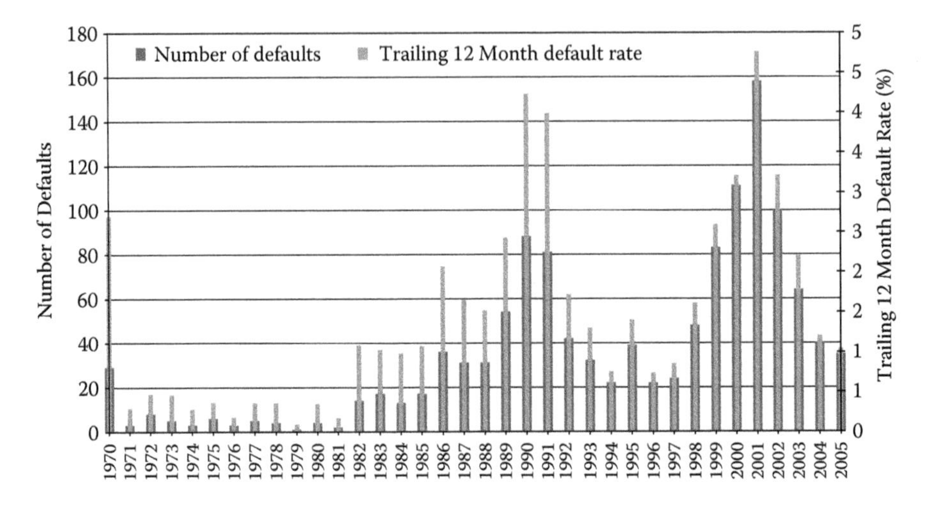

FIGURE 10.1.1 Number of Moody's rated U.S. issuers that defaulted in a given year between 1970 and 2005 (left scale) and the corresponding trailing 12-month default rate at the end of a year (right scale). The graph clearly indicates the clustering of defaults. (Source: Moody's.)

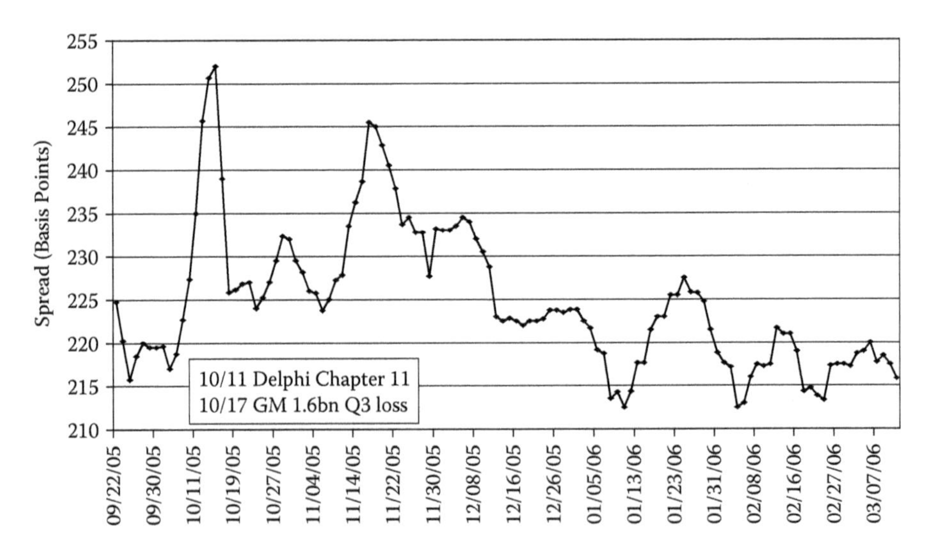

FIGURE 10.1.2 Dow Jones CDX.NA.XO 5-year credit swap index spread, which consists of 35 equally weighted credit default swaps referenced on BB and BBB rated issues. The time series shows the significant impact on the index spread of Delphi's default and GM's announcement of losses. Delphi and GM are not in the index. (Source: Credit Suisse.)

contingent claim on portfolio loss, which is modeled with a point process that records financial loss due to default. The magnitude of each loss is random. Default times are governed by an intensity driven by marketwide risk factors that follow an affine jump diffusion. The loss process itself is a risk factor, so that both the timing of past defaults and their recovery rates influence the future evolution of the portfolio loss. This *self-affecting* specification captures the feedback effects of events seen in the index spread (Figure 10.1.2). It also incorporates the negative correlation between default and recovery rates that is empirically documented in Altman et al. (2005) and many others. A straightforward extension of our specification includes a stochastic interest rate that is driven by the underlying risk factors. This extension accounts for the flight-to-quality effects often observed after a significant credit event.

An affine point process model leads to computationally tractable valuation, hedging, and calibration of credit derivatives. Based on the results of Duffie et al. (2000), we show that the conditional transform of an affine point process is an exponentially affine function of the risk factors with coefficients that satisfy ordinary differential equations. The transform determines the conditional distribution of future portfolio loss and the price of a contingent claim on portfolio loss. We implement a pricing calculator for affine point processes and provide numerical examples of term structures of index and tranche spreads implied by the Hawkes process (Hawkes, 1971). The Hawkes process is a self-affecting affine point process whose intensity increases at an event as a function of the realized loss. Between events, the intensity reverts to a time-varying level.

A portfolio credit derivative is sensitive to changes in the credit spread of the underlying names. The hedging of this exposure is important in practice. To price an instrument referenced on a constituent name and estimate hedge deltas, we use *random thinning* as proposed in Giesecke and Goldberg (2005). Here a fraction of the intensity is allocated to a given name. The resulting single-name intensities reflect the dependence structure in the portfolio. The prices of single-name instruments are exponentially affine functions of the risk factors, with coefficients determined by ordinary differential equations.

Model calibration is a two-step procedure. First an affine point process model is fit to standard index and tranche spreads. The resulting model can be used to consistently price tranches with nonstandard attachment points and maturities, and derivatives that depend on the loss dynamics such as forward starting tranches and options on indexes and tranches. Subsequently, a parametric thinning function is calibrated to each single-name swap curve. The parameters of the portfolio intensity are not affected, so the fit to the multiname market remains intact. The thinning function is used to estimate the delta hedge ratio for a portfolio constituent.

10.2 RELATED LITERATURE

Our contribution is to introduce a class of self-affecting models for portfolio loss that incorporate event feedback and the negative dependence between default and recovery rates. These dynamic models lead to computationally tractable valuation and hedging problems for portfolio credit derivatives. We follow the intensity-based top-down approach described in Giesecke and Goldberg (2005). Here the portfolio loss process is specified in terms of a default rate and a distribution for loss given default, and random thinning is used to estimate hedge ratios. Extant intensity specifications typically neglect the dependence between default and recovery rates. Examples include Davis and Lo (2001), who consider a homogeneous portfolio in which an event ramps up the intensity by a fixed factor for an exponential time. Ordinary differential equations govern the distribution of the default process. A more recent example is Giesecke and Tomecek (2005), where a stochastic time change is applied to a Poisson process to generate both affine and nonaffine self-affecting models of portfolio loss. As in this chapter, the price of a credit derivative is determined by the transform of the loss process. Ding et al. (2006) propose a class of self-affecting loss processes that are obtained by time-changing a birth process. The price of a credit derivative is given in terms of the Laplace transform of the time change. In Longstaff and Rajan (2006), defaults are driven by independent doubly stochastic processes that model idiosyncratic, sector-specific, and economy-wide events. Brigo et al. (2006) describe event arrivals by a mixed compound Poisson process and its doubly stochastic generalizations. Because they imply conditionally independent interarrival times, doubly stochastic models neglect the feedback from defaults. Failure to account for event feedback or the dependence between default and recovery rates may lead to a misrepresentation of the value and risk of a credit derivative position, since both these phenomena typically increase the likelihood of a large loss.

Some intensity-based top-down models have bottom-up counterparts. In a bottom-up model, the intensity of each portfolio constituent is the primitives. The dependence among firms must be built into the single-name models. An example is a doubly stochastic setting in which firm intensities are driven by common factors, as in Duffie and Garleanu (2001), Mortensen (2005), di Graziano and Rogers (2006), and Hurd and Kuznetsov (2006). Conditional on the factors, interarrival times are independent, so credit derivatives valuation is computationally convenient, but event feedback is ruled out. A second example is a setting with interaction, in which a default triggers a response in the intensities of the surviving firms. In Jarrow and Yu (2001) and Frey and Backhaus (2004), this response is exogenous. In the "frailty" models of Collin–Dufresne et al. (2003), Duffie et al. (2006), Giesecke (2004), Giesecke and Goldberg (2004), and Schonbucher (2004), the response is the result of learning from default. While these models are empirically plausible, they do not focus on the portfolio loss and so require substantial further steps to price credit derivatives. A third example is the static-copula approach that is popular in the industry. Here deterministic constituent intensities are fit to the single-name swap market and a fixed copula governs firm dependence. Although computationally convenient, a calibrated copula model does not generate a consistent set of tranche spreads across different attachment points and maturities. Thanks to specifying the time evolution of portfolio loss, our approach determines the term structure of tranche spreads for all attachment points simultaneously. This is a significant improvement, since the interpolation methods often used for copula models may generate arbitrage.

Affine point processes also have potential applications in the risk management of corporate debt portfolios, for which event feedback and the dependence between default and recovery rates are significant issues. For example, Das et al. (2005) test a bottom-up doubly stochastic model whose features are similar to those of the models endorsed by the regulatory authorities to estimate portfolio credit risk. They find evidence of historical default clustering in excess of that implied by the model they tested. This suggests that doubly stochastic models tend to underestimate risk capital. An affine point process model extends the doubly stochastic model to incorporate event feedback, which implies a more realistic degree of default clustering. It also accounts for the dependence between default and recovery rates, whose importance is emphasized in Basel II's pillar I guidelines, as seen in Basel Commission on Bank Regulation (2004). This body of rules and regulations requires banks to implement a "downturn" loss given default model, reflecting the fact that recovery rates vary with systematic risk factors.

REFERENCES

Edward Altman, Brooks Brady, Andrea Resti and Andrea Sironi. The Link between Default and Recovery Rates: Theory, Empirical Evidence and Implications. *Journal of Business*, 78(6): 2203–2227, 2005.

Basel Commission on Bank Regulation. International Convergence on Capital Measurement and Capital Standards. BIS, Basel, Switzerland, 2004.

Damiano Brigo, Andrea Pallavicini and Roberto Torresetti. Calibration of CDO Tranches with the Dynamical Generalized-Poisson Loss Model. Working paper, Banca IMI, Milan, 2006.

Pierre Collin-Dufresne, Robert Goldstein and Jean Helwege. Are Jumps in Corporate Bond Yields Priced? Modeling Contagion via the Updating of Beliefs. Working paper, Carnegie-Mellon University, Pittsburgh, PA, 2003.

Sanjiv Das, Darrell Duffie, Nikunj Kapadia and Leandro Saita. Common Failings: How Corporate Defaults Are Correlated, 2005. *Journal of Finance*, forthcoming.

Mark Davis and Violet Lo. Modeling Default Correlation in Bond Portfolios. In Carol Alexander, ed., *Mastering Risk*, vol. 2: *Applications*. Prentice Hall, New York, 2001, pp. 141–151.

Giuseppe di Graziano and Chris Rogers. A Dynamic Approach to the Modelling of Correlation Credit Derivatives Using Markov Chains. Working paper, University of Cambridge, U.K., 2006.

Xiaowei Ding, Kay Giesecke and Pascal Tomecek. Time-Changed Birth Processes and Multi-Name Credit. Working paper, Stanford University, Stanford, CA, 2006.

Darrell Duffie, Andreas Eckner, Guillaume Horel and Leandro Saita. Frailty Correlated Default. Working paper, Stanford University, Stanford, CA 2006.

Darrell Duffie and Nicolae Garleanu. Risk and Valuation of Collateralized Debt Obligations. *Financial Analysts Journal*, 57(1): 41–59, 2001.

Darrell Duffie, Nikunj Kapadia and Leandro Saita. Common Failings: How Corporate Defaults are Correlated. *Journal of Finance*, V62, 93–117, 2007.

Darrell Duffie, Jun Pan and Kenneth Singleton. Transform Analysis and Asset Pricing for Affine Jump-Diffusions. *Econometrica*, 68: 1343–1376, 2000.

Rudiger Frey and Jochen Backhaus. Portfolio Credit Risk Models with Interacting Default Intensities: a Markovian Approach. Working paper, Department of Mathematics, University of Leipzig, Germany, 2004.

Kay Giesecke. Correlated Default with Incomplete Information. *Journal of Banking and Finance*, 28: 1521–1545, 2004.

Kay Giesecke and Lisa Goldberg. Sequential Defaults and Incomplete Information. *Journal of Risk*, 7(1): 1–26, 2004.

Kay Giesecke and Lisa Goldberg. A Top Down Approach to Multi-Name credit. Working paper, Stanford University, Stanford, CA, 2005.

Kay Giesecke and Pascal Tomecek. Dependent Events and Changes of Time. Working paper, Stanford University, Stanford, CA, 2005.

Alan G. Hawkes. Spectra of Some Self-Exciting and Mutually Exciting Point Processes. *Biometrika*, 58(1): 83–90, 1971.

Thomas Hurd and Alexey Kuznetsov. Fast CDO Computations in the Affine Markov Chain Model. Working paper, McMaster University, Hamilton, ON, 2006.

Robert A. Jarrow and Fan Yu. Counterparty Risk and the Pricing of Defaultable Securities. *Journal of Finance*, 56(5): 555–576, 2001.

Philippe Jorion and Gaiyan Zhang. Good and Bad Credit Contagion: Evidence from Credit Default Swaps, 2006. *Journal of Financial Economics*, forthcoming.

Francis Longstaff and Arvind Rajan. An Empirical Analysis of Collateralized Debt Obligations. Working paper, UCLA, Los Angeles, 2006.

Allan Mortensen. Semianalytical Valuation of Basket Credit Derivatives in Intensity-Based Models. *The Journal of Derivatives*, summer 2006, V13, 93–117.

Philipp Schonbucher. Information-Driven Default Contagion. Working paper, ETH Zurich, 2004.

Philipp Schonbucher. Portfolio Losses and the Term Structure of Loss Transition Rates: A New Methodology for Pricing of Portfolio of Credit Derivatives. Working paper, Department of Mathematics, ETH Zurich, 2005.

Jakob Sidenius, Vladimir Piterbarg and Leif Andersen. A New Framework for Dynamic Credit Portfolio Loss Modelling. Bank of America, 2005.

CHAPTER 11

Valuation of Performance-Dependent Options in a Black–Scholes Framework

Thomas Gerstner and Markus Holtz
Institut für Numerische Simulation, Universität Bonn, Bonn, Germany
Ralf Korn
Fachbereich Mathematik, TU Kaiserslautern, Kaiserslautern, Germany

Contents

Abstract: In this chapter, we introduce performance-dependent options as the appropriate financial instrument for a company to hedge the risk arising from the obligation to purchase shares as part of a bonus scheme for their executives. We determine the fair price of such options in a multidimensional Black–Scholes model that results in the computation of a multidimensional integral whose dimension equals the dimension of the underlying Brownian motion. The integrand is typically discontinuous, though, which makes accurate solutions difficult to achieve by numerical approaches. As a remedy, we derive a pricing formula that only involves the evaluation of smooth multivariate normal distributions. This way, performance-dependent options can efficiently be priced as it is shown by numerical results.

Keywords: option pricing, multivariate integration, Black–Scholes model

11.1 INTRODUCTION

Today, long-term incentive and bonus schemes often form a major part of the wages of the executives of companies. One widespread form of such schemes consists in giving the participants a conditional award of shares. More precisely, if the participant stays with the company for at least a prescribed time period, he or she will receive a certain number of shares of the company at the end of the period. The exact amount of shares is usually linked to the success of the company measured via a performance criterion such as the company's gain over the period or its ranking among comparable firms.

It is now a huge risk for a company to leave the resulting positions unhedged. As the purchase of vanilla call options on the maximum number of possibly needed shares binds too much capital, the appropriate financial instruments in this situation are so-called performance-dependent options. These options are financial derivatives whose payoff depends on the performance of one asset in comparison to a set of benchmark assets. Thereby, we assume that the performance of an asset is determined by the relative increase of the asset price over the considered period of time. The performance of the asset is then compared with the performances of the benchmark assets. For each possible outcome of this comparison, a different payoff of the derivative can be realized.

We use a multidimensional Black–Scholes model, see, for example, Karatzas [5] or Korn and Korn [7], for the dynamics of all asset prices required for the performance ranking. The martingale approach then yields a fair price of the performance-dependent option as a multidimensional integral whose dimension equals the dimension of the underlying Brownian motion. The integrand is typically discontinuous, however, which makes accurate numerical solutions difficult to achieve.

The main aim of this chapter is to demonstrate that the combination of a closed-form solution to the pricing problem for performance-dependent options with suitable numerical integration methods clearly outperforms standard numerical approaches. The derived formula only involves the evaluation of smooth multivariate normal distributions that can be computed quickly and robustly by numerical integration. In various numerical results, we illustrate the efficiency of this approach and its possibility to evaluate high-dimensional normal distributions in a superior way.

11.2 PERFORMANCE-DEPENDENT OPTIONS

Bonus schemes whose payoff depends on certain success criteria are a way to provide additional incentives for the executives of a company. Often, the executives obtain a conditional amount of shares of the company. The exact number depends on the ranking of the company's stock price increase in comparison with other (benchmark) companies. Such schemes induce uncertain future costs for the company, though. The appropriate way to hedge these risks are options that include the performance criteria in the definition of their payoff function, so-called performance-dependent options. In the following, we aim to derive pricing formulas for the fair price of these options.

We assume that there are n assets involved in total. The asset of the considered company gets assigned label 1, and the $n-1$ benchmark assets are labeled from 2 to n. The price of the ith asset varying with time t is denoted by $S_i(t), 1 \leq i \leq n$. All stock prices at time t are collected in the vector $\mathbf{S}(t) = (S_1(t), \ldots, S_n(t))$.

11.2.1 Payoff Profile

First, we need to define the payoff of a performance-dependent option at time T. To this end, we denote the relative price increase of stock i over the time interval $[0, T]$ by

$$\Delta S_i = S_i(T)/S_i(0).$$

We save the performance of the first asset in comparison to a given strike price K (typically, $K = S_1(0)$) and in comparison to the benchmark assets at time T in a ranking vector $\mathbf{Rank}(\mathbf{S}(T)) \in \{+,-\}^n$ that is defined by

$$\text{Rank}_1(\mathbf{S}(T)) = \begin{cases} + & \text{if } S_1(T) \geq K, \\ - & \text{else} \end{cases} \quad \text{and } \text{Rank}_i(\mathbf{S}(T)) = \begin{cases} + & \text{if } \Delta S_1 \geq \Delta S_i, \\ - & \text{else} \end{cases}$$

for $i = 2, \ldots, n$. For each possible ranking $\mathbf{R} \in \{+,-\}^n$, a bonus factor $a_{\mathbf{R}} \in \mathbb{R}^+$ defines the payoff of the performance-dependent option. For explicit examples of such bonus factors, see Section 11.3. In all cases we define $a_{\mathbf{R}} = 0$ if $\mathbf{R}_1 = -$.

The payoff of the performance-dependent option at time T is then defined by

$$V(\mathbf{S}(T), T) = a_{\mathbf{Rank}(\mathbf{S}(T))} \left(S_1(T) - K\right). \tag{11.2.1}$$

In the following, we aim to determine the fair price $V(\mathbf{S}(0), 0)$ of such an option at the current time $t = 0$.

11.2.2 Multivariate Black–Scholes Model

We assume that the stock price dynamics are given by

$$dS_i(t) = S_i(t) \left(\mu_i dt + \sum_{j=1}^{n} \sigma_{ij} dW_j(t) \right) \tag{11.2.2}$$

for $i = 1, \ldots, n$, where μ_i denotes the drift of the ith stock, σ the $n \times n$ volatility matrix of the stock price movements, and $W_j(t), 1 \leq j \leq n$, an n-dimensional Brownian motion. The matrix $\sigma\sigma^T$ is assumed to be positive definite.

The explicit solution of the stochastic differential equation (11.2.2) is then given by

$$S_i(T) = S_i(\mathbf{X}) = S_i(0) \exp\left(\mu_i T - \bar{\sigma}_i + X_i\right) \tag{11.2.3}$$

for $i = 1, \ldots, n$ with

$$\bar{\sigma}_i := \frac{1}{2} \sum_{j=1}^{n} \sigma_{ij}^2 T$$

and

$$X_i := \sum_{j=1}^{n} \sigma_{ij} W_j(T).$$

Hence, $\mathbf{X} = (X_1, \ldots, X_n)$ is a $N(\mathbf{0}, \mathbf{\Sigma})$-normally distributed random vector with $\mathbf{\Sigma} = \sigma\sigma^T T$.

11.2.3 Martingale Approach

In the above multidimensional Black–Scholes setting, the option price $V(\mathbf{S}(0), 0)$ is given by the discounted expectation

$$V(\mathbf{S}(0), 0) = e^{-rT} E\left[V(\mathbf{S}(T), T)\right] \tag{11.2.4}$$

of the payoff under the unique equivalent martingale measure, i.e., the drift μ_i in equation (11.2.3) is replaced by the riskless interest rate r for each stock i. Plugging in the density function $\varphi_{0,\Sigma}$ of the random vector $\mathbf{X}$, we get that the fair price of a performance-dependent option with payoff, shown in equation (11.2.1), is given by the n-dimensional integral

$$V(\mathbf{S}(0), 0) = e^{-rT} \int_{R^n} \sum_{\mathbf{R}\in\{+,-\}^n} a_{\mathbf{R}}(S_1(T) - K)\, \chi_{\mathbf{R}}(\mathbf{S}(T))\varphi_{0,\Sigma}(\mathbf{x})\, d\mathbf{x}. \tag{11.2.5}$$

Thereby, the expectation runs over all possible rankings $\mathbf{R}$, and the characteristic function $\chi_{\mathbf{R}}(\mathbf{S}(T))$ is defined by

$$\chi_{\mathbf{R}}(\mathbf{S}(T)) = \begin{cases} 1 & \text{if } \mathbf{Rank}(\mathbf{S}(T)) = \mathbf{R} \\ 0 & \text{else} \end{cases}.$$

11.2.4 Pricing Formula

We will now derive an analytical expression for the solution of equation (11.2.5) in terms of smooth functions. We denote the Gauss kernel by

$$\varphi_{\mu,\Sigma}(\mathbf{x}) := \frac{1}{(2\pi)^{n/2}(\det \Sigma)^{1/2}}\, e^{-\frac{1}{2}(\mathbf{x}-\mu)^T\Sigma^{-1}(\mathbf{x}-\mu)}$$

and denote the multivariate normal distribution corresponding to $\varphi_{0,\Sigma}$ with mean zero and covariance matrix Σ and the integral limits

$$c_i = \begin{cases} b_i & \text{if } R_i = + \\ -\infty & \text{else} \end{cases} \quad \text{and} \quad d_i = \begin{cases} \infty & \text{if } R_i = + \\ b_i & \text{else} \end{cases} \quad \text{for } i = 1, \ldots, n$$

which are depending on the ranking $\mathbf{R} \in \{+,-\}^n$ by

$$\Phi_{\mathbf{R}}(\Sigma, \mathbf{b}) := \int_{c_1}^{d_1} \ldots \int_{c_n}^{d_n} \varphi_{0,\Sigma}(\mathbf{x})d\mathbf{x}.$$

Furthermore, we define the comparison relation $\mathbf{x} \geq_{\mathbf{R}} \mathbf{y}$ for two vectors $\mathbf{x}, \mathbf{y} \in \mathbb{R}^n$ with respect to the ranking $\mathbf{R}$ by

$$\mathbf{x} \geq_{\mathbf{R}} \mathbf{y} \;:\Leftrightarrow\; R_i(x_i - y_i) \geq 0 \quad \text{for} \quad 1 \leq i \leq n.$$

To prove our main theorem we need the following two lemmas.

LEMMA 11.2.1
Let $\mathbf{b}$, $\mathbf{q} \in \mathbb{R}^n$, $\mathbf{A} \in \mathbb{R}^{n\times n}$ *with full rank, and* $\mathbf{\Sigma} \in \mathbb{R}^{n\times n}$ *symmetric and positive definite. Then*

$$\int_{\mathbf{Ax}\geq_{\mathbf{R}}\mathbf{b}} e^{\mathbf{q}^T\mathbf{x}}\varphi_{0,\mathbf{\Sigma}}(\mathbf{x})d\mathbf{x} = e^{\frac{1}{2}\mathbf{q}^T\mathbf{\Sigma}\mathbf{q}}\Phi_{\mathbf{R}}(\mathbf{A\Sigma A}^T, \mathbf{b} - \mathbf{A\Sigma q}).$$

PROOF A simple computation shows

$$e^{\mathbf{q}^T\mathbf{x}}\varphi_{0,\mathbf{\Sigma}}(\mathbf{x}) = e^{\frac{1}{2}\mathbf{q}^T\mathbf{\Sigma}\mathbf{q}}\varphi_{\mathbf{\Sigma q},\mathbf{\Sigma}}(\mathbf{x})$$

for all $\mathbf{x} \in \mathbb{R}^n$. Using the substitution $\mathbf{x} = \mathbf{A}^{-1}\mathbf{y} + \mathbf{\Sigma q}$ we obtain

$$\begin{aligned}\int_{\mathbf{Ax}\geq_{\mathbf{R}}\mathbf{b}} e^{\mathbf{q}^T\mathbf{x}}\varphi_{0,\mathbf{\Sigma}}(\mathbf{x})d\mathbf{x} &= e^{\frac{1}{2}\mathbf{q}^T\mathbf{\Sigma}\mathbf{q}} \int_{\mathbf{Ax}\geq_{\mathbf{R}}\mathbf{b}} \varphi_{\mathbf{\Sigma q},\mathbf{\Sigma}}(\mathbf{x})d\mathbf{x} \\ &= e^{\frac{1}{2}\mathbf{q}^T\mathbf{\Sigma}\mathbf{q}} \int_{\mathbf{y}\geq_{\mathbf{R}}\mathbf{b}-\mathbf{A\Sigma q}} \varphi_{0,\mathbf{A\Sigma A}^T}(\mathbf{y})d\mathbf{y}\end{aligned} \tag{11.2.6}$$

and thus the assertion.

LEMMA 11.2.2
We have $\mathbf{Rank}(\mathbf{S}(T)) = \mathbf{R}$ *exactly if* $\mathbf{AX} \geq_{\mathbf{R}} \mathbf{b}$ *with*

$$\mathbf{A} := \begin{pmatrix} 1 & 0 & \dots & 0 \\ 1 & -1 & \ddots & \vdots \\ \vdots & 0 & \ddots & 0 \\ 1 & 0 & 0 & -1 \end{pmatrix} \quad \textit{and} \quad \mathbf{b} := \begin{pmatrix} \ln\frac{K}{S_1(0)} - rT + \bar{\sigma}_1 \\ \bar{\sigma}_1 - \bar{\sigma}_2 \\ \vdots \\ \bar{\sigma}_1 - \bar{\sigma}_n \end{pmatrix}$$

where $\mathbf{A} \in \mathbb{R}^{n\times n}$ *and* $\mathbf{b} \in \mathbb{R}^n$.

PROOF Using equation (11.2.3) we see that $\text{Rank}_1 = +$ is equivalent to

$$S_1(T) \geq K \Longleftrightarrow X_1 \geq \ln\frac{K}{S_1(0)} - rT + \bar{\sigma}_1$$

which yields the first row of the system $\mathbf{AX} \geq_{\mathbf{R}} \mathbf{b}$. Moreover, for $i = 2, \dots, n$ the outperformance criterion $\text{Rank}_i = +$ can be written as

$$\frac{S_1(T)}{S_1(0)} \geq \frac{S_i(T)}{S_i(0)} \Longleftrightarrow X_1 - X_i \geq \bar{\sigma}_1 - \bar{\sigma}_i$$

which yields rows 2 to n of the system.

Now we can state the following pricing formula which, in a slightly more special setting, can be found in Korn [6].

THEOREM 11.2.1
In our market setting determined by the price model shown in equation (11.2.2), the price of a performance-dependent option with payoff, as shown in equation (11.2.1), is given by

$$V(\mathbf{S}(0), 0) = \sum_{\mathbf{R}\in\{+,-\}^n} a_{\mathbf{R}}(S_1(0)\,\Phi_{\mathbf{R}}(\mathbf{C}, \mathbf{d}) - e^{-rT}\,K\,\Phi_{\mathbf{R}}(\mathbf{C}, \mathbf{b}))$$

where $\mathbf{C} := \mathbf{A}\Sigma\mathbf{A}^T$ *and* $\mathbf{d} := \mathbf{b} - \mathbf{A}\Sigma\mathbf{e}_1$, *with* $\mathbf{A}$ *and* $\mathbf{b}$ *defined as in lemma 11.2.2 and with* $\mathbf{e}_1$ *being the first unit vector.*

PROOF The characteristic function $\chi_{\mathbf{R}}(\mathbf{S}(T))$ in the integral given in equation (11.2.5) can be eliminated using lemma 11.2.2, and we get

$$V(\mathbf{S}(0), 0) = e^{-rT} \sum_{\mathbf{R}\in\{+,-\}^n} a_{\mathbf{R}} \int_{\mathbf{Ax}\geq\mathbf{Rb}} (S_1(T) - K)\varphi_{0,\Sigma}(\mathbf{x})d\mathbf{x}. \qquad (11.2.7)$$

By equation (11.2.3), the integral term can be written as

$$S_1(0)e^{rT-\bar{\sigma}_1} \int_{\mathbf{Ax}\geq_{\mathbf{R}}\mathbf{b}} e^{x_1}\,\varphi_{0,\Sigma}(\mathbf{x})d\mathbf{x} - K \int_{\mathbf{Ax}\geq_{\mathbf{R}}\mathbf{b}} \varphi_{0,\Sigma}(\mathbf{x})d\mathbf{x}.$$

Application of lemma 11.2.1 with $\mathbf{q} = \mathbf{e}_1$ shows that the first integral equals

$$e^{\frac{1}{2}\mathbf{e}_1^T\Sigma\mathbf{e}_1}\,\Phi_{\mathbf{R}}(\mathbf{A}\Sigma\mathbf{A}^T, \mathbf{b} - \mathbf{A}\Sigma\mathbf{e}_1) = e^{\bar{\sigma}_1}\,\Phi_{\mathbf{R}}(\mathbf{C}, \mathbf{d}).$$

By a further application of lemma 11.2.1 with $\mathbf{q} = 0$, we obtain that the second integral equals $K\,\Phi_{\mathbf{R}}(\mathbf{C}, \mathbf{b})$, and thus the assertion holds.

Note that the price of a performance-dependent option does not depend on the stock prices $S_2(0), \ldots, S_n(0)$ of the benchmark companies, but only on the joint volatility matrix Σ. The pricing formula of theorem 11.2.1 allows an efficient valuation of performance-dependent options in the case of moderate-sized benchmarks. It requires the computation of up to 2^n many n-dimensional normal distributions. The actual number of integrals equals twice the number of nonzero bonus factors $a_{\mathbf{R}}$. In the case of large benchmarks, the complexity and dimensionality of the pricing formula can prevent its efficient application, though. These problems can be circumvented by using a reduced Black–Scholes model and suitable tools from computational geometry; for details see Gerstner and Holtz [3].

11.3 NUMERICAL RESULTS

In this section, we present numerical examples to illustrate the use of the pricing formula of theorem 11.2.1. In particular, we compare the efficiency of our algorithm to the standard pricing approach (denoted by STD) of quasi-Monte Carlo simulation of the expected payoff, shown in equation (11.2.4), based on Sobol point sets; see, for example, Glasserman [4]. Monte Carlo instead of quasi-Monte Carlo simulation led to significantly less accurate results in all our experiments. We systematically compare the use of our pricing formula with

- Quasi-Monte Carlo integration based on Sobol point sets (QMC)
- Product integration based on the Clenshaw–Curtis rule (P)
- Sparse-grid integration based on the Clenshaw–Curtis rule (SG)

for the evaluation of the multivariate cumulative normal distributions (see Genz [1]). The sparse-grid approach is based on the work of Gerstner and Griebel [2]. All computations were performed on a dual Intel® Xeon™ CPU 3.06-GHz processor.

We consider a Black–Scholes market with $n = 5$ assets. Thereby, we investigate the following three choices of bonus factors $a_{\mathbf{R}}$ in the payoff function shown in equation (11.2.1):

EXAMPLE 11.1
Linear ranking-dependent option:

$$a_{\mathbf{R}} = \begin{cases} m/(n-1) & \text{if } R_1 = + \\ 0 & \text{else.} \end{cases}$$

Here, m denotes the number of outperformed benchmark assets. The payoff depends on the rank of our company among the benchmark assets. If the company ranks first, there is a full payoff $(S_1(T) - K)^+$. If it ranks last, the payoff is zero. In between, the payoff increases linearly with the number of outperformed benchmark assets.

EXAMPLE 11.2
Outperformance option:

$$a_{\mathbf{R}} = \begin{cases} 1 & \text{if } \mathbf{R} = (+, \dots, +) \\ 0 & \text{else.} \end{cases}$$

A payoff only occurs if $S_1(T) \geq K$ and if all benchmark assets are outperformed.

EXAMPLE 11.3

Linear ranking-dependent option combined with an outperformance condition:

$$a_R = \begin{cases} m/(n-1) & \text{if } R_1 = + \text{ and } R_2 = + \\ 0 & \text{else.} \end{cases}$$

The bonus depends linearly on the number m of outperformed benchmark companies, as in Example 11.1. However, the bonus is only paid if company two is outperformed. Company two could, for example be the main competitor of our company.

In all cases, we use the model parameters $K = 100$, $S_1(0) = 100$, $T = 1$, $r = 5\%$, and as volatility matrix

$$\sigma = \begin{pmatrix} 0.1515 & 0.0581 & 0.0373 & 0.0389 & 0.0278 \\ 0.0581 & 0.2079 & 0.0376 & 0.0454 & 0.0393 \\ 0.0373 & 0.0376 & 0.1637 & 0.0597 & 0.0635 \\ 0.0389 & 0.0454 & 0.0597 & 0.1929 & 0.0540 \\ 0.0278 & 0.0393 & 0.0635 & 0.0540 & 0.2007 \end{pmatrix}.$$

The computed option prices and discounts compared with the price of the corresponding plain vanilla option given by 9.4499 are displayed in the second and third column of Table 11.3.1. The number of normal distributions (No. Int) that have to be computed is shown in the last column.

The convergence behavior of the four different approaches (STD, QMC, P, SG) to price the performance-dependent options from Examples 11.1 to 11.3 are shown in Figure 11.3.1, which displays the time that is needed to obtain a given accuracy. One can see that the standard approach (STD) and the product integration approach (P) perform worst for all accuracies. The convergence rates are clearly lower than 1 in all examples. The integration scheme STD suffers under the irregularity of the integrand, which is highly discontinuous and not of bounded variation. The product-integration approach suffers under

TABLE 11.3.1 Option Prices and Discounts Compared with the Corresponding Plain Vanilla Option and Number of Computed Normal Distributions

Example	$V(S_1, 0)$	Discount	No. Int
11.1	6.2354	34.02%	30
11.2	3.0183	68.06%	2
11.3	4.5612	51.73%	16

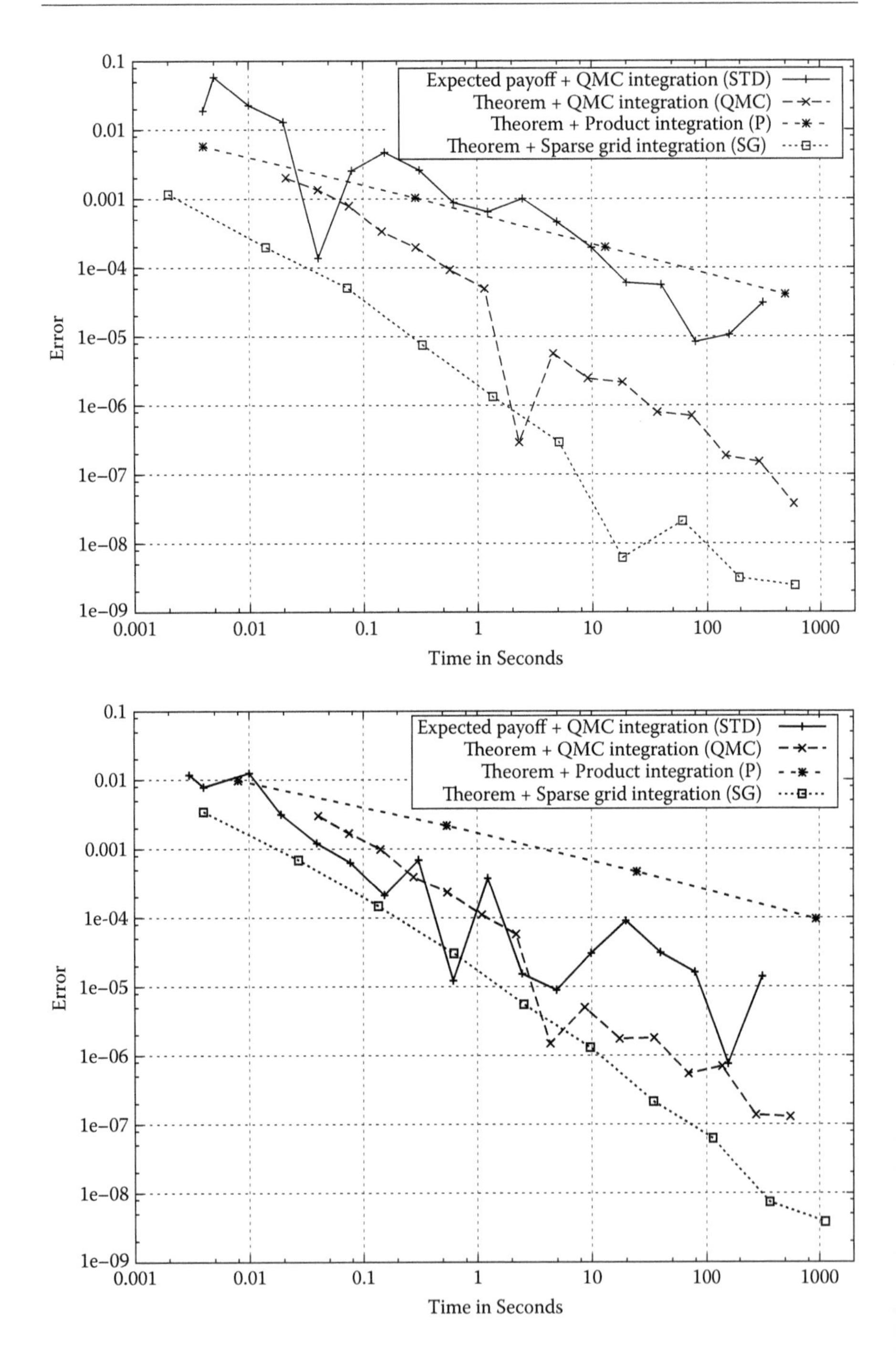

FIGURE 11.3.1 Errors and timings of the different numerical approaches to price the performance-dependent options of Examples 11.1 (top), 11.2 (bottom), and 11.3 (following page).

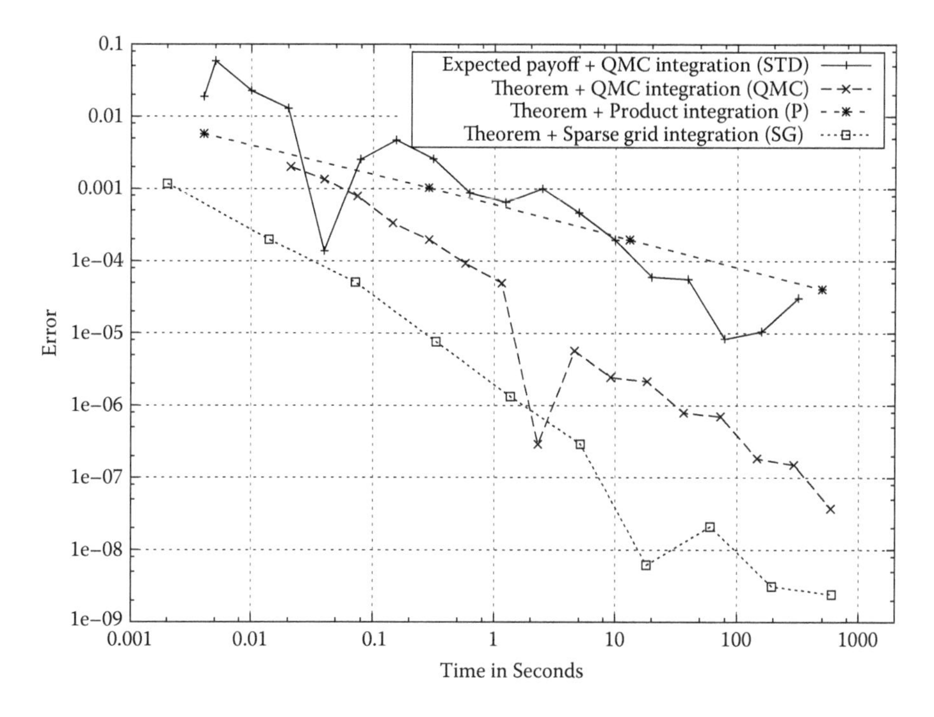

FIGURE 11.3.1 (*Continued*).

the curse of dimension. The use of the pricing formula from Theorem 11.2.1 combined with QMC or SG integration clearly outperforms the STD approach in terms of efficiency in all considered examples. The QMC scheme exhibits a convergence rate of about 1 independent of the problem. The combination of sparse-grid integration with our pricing formula (SG) leads to the best overall accuracies and convergence rates in all cases. Even very high accuracy demands can be fulfilled in less than a few seconds.

REFERENCES

1. Genz, A. Numerical computation of multivariate normal probabilities. *J. Comput. Graph. Statist.*, 1, 141–150, 1992.
2. Gerstner, T. and Griebel, M. Numerical integration using sparse grids. *Numerical Algorithms*, 18, 209–232, 1998.
3. Gerstner, T. and Holtz, M. Valuation of performance-dependent options. *Appl. Math. Finance*, to appear, 2007.
4. Glasserman, P. *Monte Carlo Methods in Financial Engineering*. Springer: New York, 2004.

5. Karatzas, I. *Lectures on the Mathematics of Finance*, vol. 8 of *CRM Monograph Series*. American Mathematical Society: Providence, RI, 1997.
6. Korn, R. A valuation approach for tailored options. Working paper, Fachbereich Mathematik, Universität Kaiserslautern, Germany 1996.
7. Korn, R. and Korn, E. *Option Pricing and Portfolio Optimization*, vol. 31 of *Graduate Studies in Mathematics*. American Mathematical Society: Providence, RI, 2001.

CHAPTER 12

Variance Reduction through Multilevel Monte Carlo Path Calculations

Michael B. Giles
Oxford University Computing Laboratory, Oxford, United Kingdom

Contents

Abstract: In this chapter we show that when estimating an expected value arising from a stochastic differential equation using Monte Carlo path simulations, the computational cost to achieve an accuracy of $O(\epsilon)$ can be reduced from $O(\epsilon^{-3})$ to $O(\epsilon^{-2}(\log \epsilon)^2)$ through the use of a multilevel approach. The analysis is supported by numerical results showing significant computational savings.

12.1 INTRODUCTION

In financial Monte Carlo path simulations, one is interested in the expected value of a quantity arising from the solution of a stochastic differential equation. To be specific, consider the vector stochastic differential equation (SDE) with general drift and volatility terms,

$$dS(t) = a(S,t)\,dt + b(S,t)\,dW(t), \quad 0 < t < T, \tag{12.1.1}$$

and given initial data S_0. Suppose we want to compute the expected value of $f(S(T))$, where $f(S)$ is a scalar function with a uniform Lipschitz bound, i.e., there exists a constant c such that

$$|f(U) - f(V)| \leq c\, \|U - V\|, \quad \forall\, U, V. \tag{12.1.2}$$

A simple Euler discretization of this SDE with time step h is

$$\widehat{S}_{n+1} = \widehat{S}_n + a(\widehat{S}_n, t_n)\, h + b(\widehat{S}_n, t_n)\, \Delta W_n,$$

and the simplest estimate for $E[f(S_T)]$ is the mean of the payoff values $f(\widehat{S}_{T/h})$ from N independent path simulations,

$$\widehat{Y} = N^{-1} \sum_{i=1}^{N} f(\widehat{S}^{(i)}_{T/h}).$$

It is well established that, provided $a(S,t)$ and $b(S,t)$ satisfy certain conditions [1, 7, 10], the expected mean square error (MSE) in the estimate $\widehat{Y}$ is asymptotically of the form

$$\text{MSE} \approx c_1 N^{-1} + c_2 h^2,$$

where c_1, c_2 are positive constants. To make this $O(\epsilon^2)$, so that the r.m.s. error is $O(\epsilon)$, requires that $N = O(\epsilon^{-2})$ and $h = O(\epsilon)$, and hence the computational complexity (cost) is $O(\epsilon^{-3})$ [2].

In this chapter we show that the cost can be reduced to $O(\epsilon^{-2}(\log \epsilon)^2)$ through the use of a multilevel method that reduces the variance for a given cost. Unlike the recent work of Kebaier [6], who used just two levels, with time steps h and $O(h^{1/2})$, to reduce the cost to $O(\epsilon^{-2.5})$, our technique uses a geometric sequence of time steps $h_l = M^{-l}, l = 0, 1, \ldots, L$, for integer $M \geq 2$, with the smallest time step h_L corresponding to the original h that determines the size of the Euler discretization bias.

This chapter gives a condensed presentation of the multilevel approach. Further details, the proof of the main theorem, and additional numerical results are available from a recent working paper [3].

12.2 MULTILEVEL MONTE CARLO METHOD

Consider Monte Carlo path simulations with different time steps $h_l = M^{-l}\, T$, $l=0,1,\ldots,L$. For a given Brownian path $W(t)$, let P denote the payoff $f(S(T))$, and let $\widehat{S}_{l,M^l}$ and $\widehat{P}_l$ denote the approximations to $S(T)$ and P using a numerical discretization with time step h_l. It is clearly true that

$$E[\widehat{P}_L] = E[\widehat{P}_0] + \sum_{l=1}^{L} E[\widehat{P}_l - \widehat{P}_{l-1}].$$

The multilevel method independently estimates each of the expectations on the right-hand side in a way that minimizes the overall variance for a given cost.

Let $\widehat{Y}_0$ be an estimator for $E[\widehat{P}_0]$ using N_0 samples, and let $\widehat{Y}_l$ for $l > 0$ be an estimator for $E[\widehat{P}_l - \widehat{P}_{l-1}]$ using N_l paths. The simplest estimator is a mean of N_l independent samples, which for $l > 0$ is

$$\widehat{Y}_l = N_l^{-1} \sum_{i=1}^{N_l} \left(\widehat{P}_l^{(i)} - \widehat{P}_{l-1}^{(i)}\right). \tag{12.2.1}$$

The key point here is that the quantity $\widehat{P}_l^{(i)} - \widehat{P}_{l-1}^{(i)}$ comes from two discrete approximations with different time steps but the same Brownian path. The variance of this simple estimator is $V[\widehat{Y}_l] = N_l^{-1} V_l$, where V_l is the variance of a single sample. Combining this with independent estimators for each of the other levels, the variance of the combined estimator $\widehat{Y} = \sum_{l=0}^{L} \widehat{Y}_l$ is

$$V[\widehat{Y}] = \sum_{l=0}^{L} N_l^{-1} V_l.$$

The computational cost is proportional to

$$\sum_{l=0}^{L} N_l\, h_l^{-1}.$$

Treating the N_l as continuous variables, the variance is minimized for a fixed computational cost by choosing N_l to be proportional to $\sqrt{V_l\, h_l}$.

In the particular case of the Euler discretization and the Lipschitz payoff function, provided that $a(S,t)$ and $b(S,t)$ satisfy certain conditions [1, 7, 10], there is $O(h)$ weak convergence, so that

$$E[\widehat{P}_l - P] = O(h_l),$$

and $O(h^{1/2})$ strong convergence, so that

$$E[\,\|\widehat{S}_{l,M^l} - S(T)\|^2\,] = O(h_l).$$

From the latter, together with the Lipschitz property shown in equation (12.1.2), it follows that

$$V[\widehat{P}_l - P] \;\le\; E[(\widehat{P}_l - P)^2] \;\le\; c^2\, E[\|\widehat{S}_{l,M^l} - S(T)\|^2],$$

and therefore $V[\widehat{P}_l - P] = O(h_l)$. Hence for the simple estimator shown in equation (12.2.1), the single-sample variance V_l is $O(h_l)$, and the optimal choice for N_l is asymptotically proportional to h_l. Setting $N_l = O(\epsilon^{-2} L h_l)$, the variance of the combined estimator $\widehat{Y}$ is $O(\epsilon^2)$. If L is chosen such that

$$L = \frac{\log \epsilon^{-1}}{\log M} + O(1),$$

as $\epsilon \to 0$, then $h_l = M^{-L} = O(\epsilon)$, and so the bias error $E[\widehat{P}_l - P]$ is $O(\epsilon)$. Consequently, we obtain a MSE that is $O(\epsilon^2)$, with a computational complexity that is $O(\epsilon^{-2}L^2) = O(\epsilon^{-2}(\log \epsilon)^2)$.

This analysis is generalized in the following theorem:

THEOREM 12.2.1

Let P denote a functional of the solution of stochastic differential equation (12.1.1) for a given Brownian path $W(t)$, and let $\widehat{P}_l$ denote the corresponding approximation using a numerical discretization with time step $h_l = M^{-l}\,T$.

If there exist independent estimators $\widehat{Y}_l$ based on N_l Monte Carlo samples, and positive constants $\alpha \ge \frac{1}{2}, \beta, c_1, c_2, c_3$ such that

1. $E[\widehat{P}_l - P] \le c_1\, h_l^{\alpha}$
2. $E[\widehat{Y}_l] = \begin{cases} E[\widehat{P}_0], & l = 0 \\ E[\widehat{P}_l - \widehat{P}_{l-1}], & l > 0 \end{cases}$
3. $V[\widehat{Y}_l] \le c_2\, N_l^{-1} h_l^{\beta}$
4. *C_l, the computational complexity of $\widehat{Y}_l$, is bounded by*

$$C_l \le c_3\, N_l\, h_l^{-1},$$

then there exists a positive constant c_4 such that for any $\epsilon < e^{-1}$ there are values L and N_l for which the multilevel estimator

$$\widehat{Y} = \sum_{l=0}^{L} \widehat{Y}_l,$$

has a mean square error with bound

$$MSE \equiv E[(\widehat{Y} - E[P])^2] < \epsilon^2$$

with a computational complexity C with bound

$$C \leq \begin{cases} c_4\,\epsilon^{-2}, & \beta > 1, \\ c_4\,\epsilon^{-2}(\log\epsilon)^2, & \beta = 1, \\ c_4\,\epsilon^{-2-(1-\beta)/\alpha}, & 0 < \beta < 1. \end{cases}$$

PROOF See [3].

12.3 NUMERICAL RESULTS

The multilevel algorithm used for the numerical tests is as follows:

1. Start with $L = 0$.
2. Estimate V_L using an initial $N_L = 10^4$ samples.
3. Calculate optimal N_l, $l = 0, \ldots, L$ for target variance.
4. Evaluate extra samples at each level as needed for new N_l.
5. Test for convergence by estimating remaining bias.
6. If not converged, set $L := L + 1$ and go to 2.

The numerical results are all obtained using $M = 4$; further analysis [3] suggests that this is close to optimal.

12.3.1 Geometric Brownian Motion

Figure 12.3.1 presents results for a European call option with discounted payoff

$$P = \exp(-r)\max(0, S(1) - 1),$$

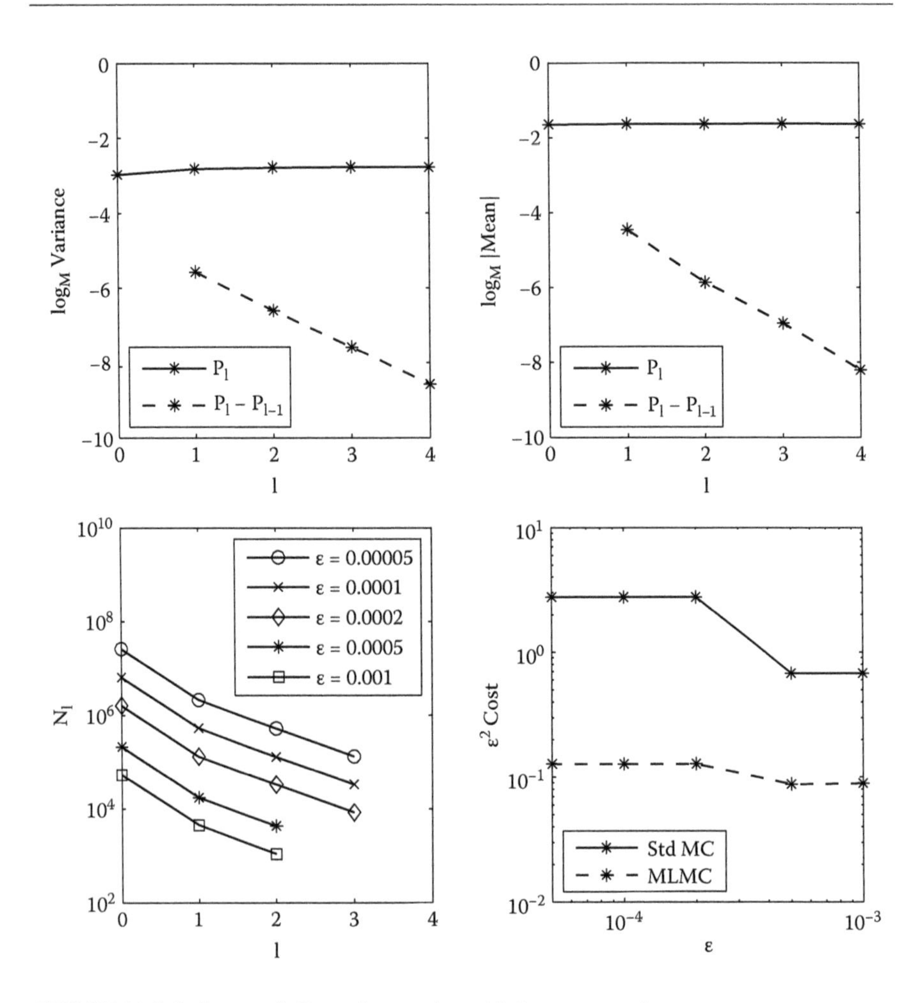

FIGURE 12.3.1 Geometric Brownian motion with European option.

with $S(t)$ based on simple geometric Brownian motion,

$$dS = r\, S\, dt + \sigma\, S\, dW, \quad 0 < t < 1,$$

with $S(0) = 1, r = 0.05$, and $\sigma = 0.2$.

The top left plot shows the behavior of the variance of both $\widehat{P}_l$ and $\widehat{P}_l - \widehat{P}_{l-1}$. The slope of the latter is approximately -1, showing that $V_l = V[\widehat{P}_l - \widehat{P}_{l-1}] = O(h)$. For $l = 4$, V_l is more than 1000 times smaller than the variance $V[\widehat{P}_l]$ of the standard Monte Carlo method with the same time step. The top right plot shows the $O(h)$ convergence of $E[\widehat{P}_l - \widehat{P}_{l-1}]$. Even at $l = 3$, the relative error $E[P - \widehat{P}_l]/E[P]$ is less than 10^{-3}. These two plots are both based on results from 10^6 paths for each value of the time step.

The bottom two plots have results from five multilevel calculations for different values of ϵ. Each line in the bottom left plot shows the values for $N_l, l = 0, \ldots, L$, with the values decreasing with l because of the decrease in both V_l and h_l. It can also be seen that the value for L, the maximum level of time-step refinement, increases as the value for ϵ decreases. The bottom right plot shows the variation with ϵ of $\epsilon^2 C$, where the computational complexity C is defined as

$$C = \sum_l N_l h_l^{-1}.$$

One line shows the results for the multilevel calculation, and the other shows the corresponding cost of a standard Monte Carlo simulation of the same accuracy, i.e., the same bias error corresponding to the same value for L, and the same variance. It can be seen that $\epsilon^2 C$ is a very slowly increasing function of ϵ^{-1} for the multilevel method, in agreement with the theory that predicts it to be proportional to $(\log \epsilon)^2$, whereas for the standard Monte Carlo method it is approximately proportional to ϵ^{-1}. For the most accurate case, $\epsilon = 5 \times 10^{-5}$, the multilevel method is more than 60 times more efficient than the standard method.

12.3.2 Heston Stochastic Volatility Model

Figure 12.3.2 presents results for the European call based on the Heston stochastic volatility model [5],

$$\mathrm{d}S = r\, S\, \mathrm{d}t + \sqrt{V}\, S\, \mathrm{d}W_1, \quad 0 < t < 1$$

$$\mathrm{d}V = \lambda\, (\sigma^2 - V)\, \mathrm{d}t + \xi\, \sqrt{V}\, \mathrm{d}W_2,$$

with $S(0) = 1$, $V(0) = 0.04$, $r = 0.05$, $\sigma = 0.2$, $\lambda = 5$, $\xi = 0.25$, and correlation $\rho = -0.5$.

In the discretization, $\sqrt{V}$ is replaced by $\sqrt{\max(V, 0)}$, but as $h \to 0$ the probability of the discrete approximation to the volatility becoming negative approaches zero for the chosen values of λ, σ, ξ. Consequently, the Euler discretization has the same order of weak and strong convergence as for the

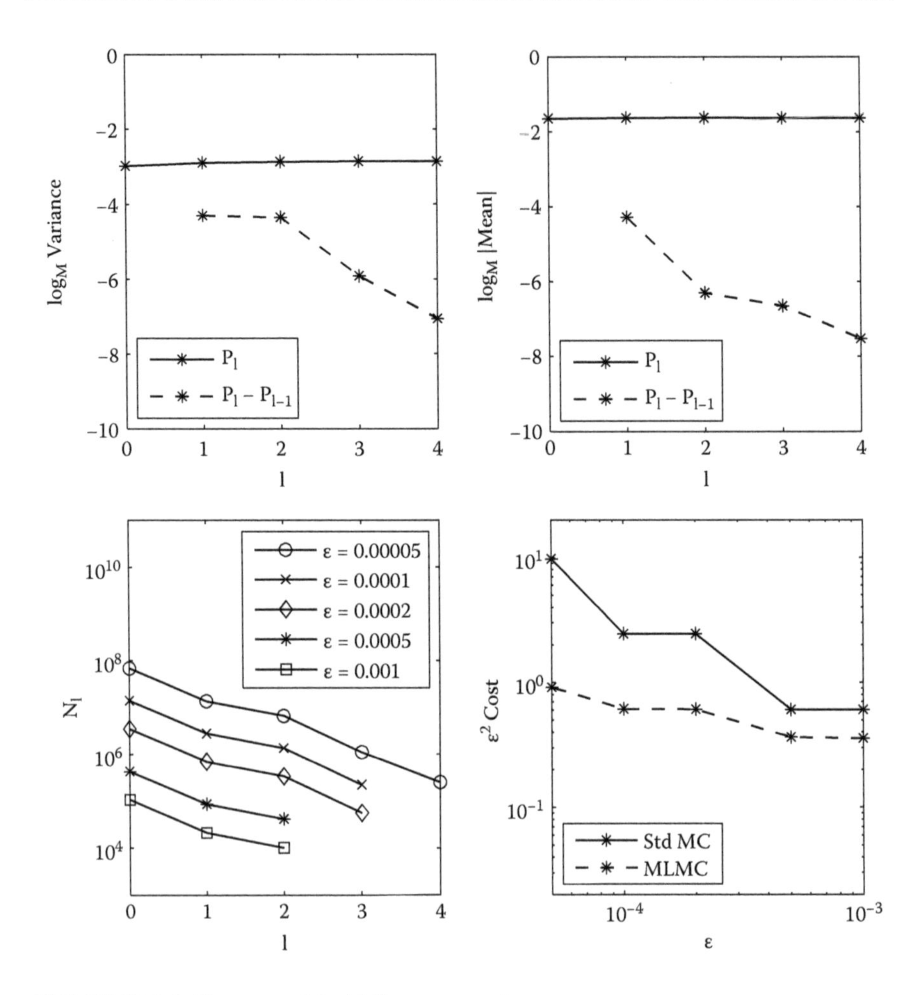

FIGURE 12.3.2 Heston model with European option.

geometric Brownian motion, and the multilevel method again yields computational savings that increase rapidly with the desired accuracy.

12.4 CONCLUDING REMARKS

In this chapter we have shown that a multilevel approach, using a geometric sequence of time steps, can reduce the order of complexity of Monte Carlo path simulations. Although the initial results look very promising, further work is needed. In particular, current research is directed toward:

- Use of the Milstein scheme with first-order strong convergence [4] to reduce the complexity to $O(\epsilon^{-2})$

- Adaptive sampling to treat discontinuous payoffs and pathwise derivatives for Greeks
- Use of quasi–Monte Carlo methods [8,9]
- Additional variance-reduction techniques

Once these features are added to the basic multilevel approach, it is hoped that this method could offer significant benefits over current Monte Carlo pricing engines.

ACKNOWLEDGMENTS

I am very grateful to Mark Broadie, Paul Glasserman, and Terry Lyons for discussions on this research, which was funded in part by a grant from Microsoft Corporation.

REFERENCES

1. V. Bally and D. Talay. The law of the Euler scheme for stochastic differential equations, I: convergence rate of the distribution function. *Probability Theory and Related Fields*, 104(1): 43–60, 1995.
2. D. Duffie and P. Glynn. Efficient Monte Carlo simulation of security prices. *Annals of Applied Probability*, 5(4): 897–905, 1995.
3. M.B. Giles. Multi-level Monte Carlo path simulation. Technical report NA06/03, Oxford University Computing Laboratory, 2006.
4. P. Glasserman. *Monte Carlo Methods in Financial Engineering*. Springer-Verlag, New York, 2004.
5. S.I. Heston. A closed-form solution for options with stochastic volatility with applications to bond and currency options. *Review of Financial Studies*, 6: 327–343, 1993.
6. A. Kebaier. Statistical Romberg extrapolation: a new variance reduction method and applications to options pricing. *Annals of Applied Probability*, 14(4): 2681–2705, 2005.
7. P.E. Kloeden and E. Platen. *Numerical Solution of Stochastic Differential Equations*. Springer-Verlag, Berlin, 1992.
8. F.Y. Kuo and I.H. Sloan. Lifting the curse of dimensionality. *Notices of the AMS*, 52(11): 1320–1328, 2005.
9. P. L'Ecuyer. Quasi-Monte Carlo methods in finance. In R.G. Ingalls, M.D. Rossetti, J.S. Smith, and B.A. Peters, editors, *Proceedings of the 2004 Winter Simulation Conference*. IEEE Press, 2004, pp. 1645–1655.
10. D. Talay and L. Tubaro. Expansion of the global error for numerical schemes solving stochastic differential equations. *Stochastic Analysis and Applications*, 8: 483–509, 1990.

CHAPTER 13

Value at Risk and Self-Similarity

Olaf Menkens
Dublin City University, Dublin, Ireland

Contents

Abstract: The concept of value at risk (VaR) measures the "risk" of a portfolio and is a statement of the following form: With probability q, the potential loss will not exceed the value-at-risk figure. It is in widespread use within the banking industry.

It is common to derive the VaR figure of d days from that of 1 day by multiplying with $\sqrt{d}$. Obviously, this formula is right if the changes in the value of the portfolio are normally distributed with stationary and independent increments. However, this formula is no longer valid if arbitrary distributions are assumed. For example, if the distributions of the changes in the value of the portfolio are self-similar with Hurst coefficient H, the VaR figure of 1 day has to be multiplied by d^H to get the VaR at Risk figure for d days.

This chapter investigates to what extent this formula (of multiplying by $\sqrt{d}$) can be applied for all financial time series. Moreover, it will be studied how much the risk can be over- or underestimated if the above formula is used. The scaling-law coefficient and the Hurst exponent are calculated for various financial time series for several quantiles.

JEL classification: C13, C14, G10, G21.

Keywords: square-root-of-time rule, time-scaling of risk, scaling law, value at risk, self-similarity, order statistics, Hurst exponent estimation in the quantiles

13.1 INTRODUCTION

There are several methods of estimating the risk of an investment in capital markets. A method in widespread use is the value-at-risk approach. The concept of value at risk (VaR) measures the "risk" of a portfolio. More precisely, it is a statement of the following form: With probability q, the potential loss will not exceed the VaR figure.

Although this concept has several disadvantages (e.g., it is not subadditive and thus not a so-called coherent risk measure; see Artzner et al. [1] and see also Daníelsson et al. [4]), it is in widespread use within the banking industry. It is common to derive the VaR figure of d days from that of 1 day by multiplying the VaR figure of 1 day with $\sqrt{d}$. Even banking supervisors recommend this procedure (see the Basel Committee on Banking Supervision [2]).

Obviously, this formula is right if the changes in the value of the portfolio are normally distributed with stationary and independent increments (namely a Brownian motion). However, this formula is no longer valid if arbitrary distributions are assumed. For example, if the distributions of the changes in the value of the portfolio are self-similar with Hurst coefficient H, the VaR figure of 1 day has to be multiplied by d^H to get the VaR figure for d days.

In the following, it will be investigated to determine to what extent this scaling law (of multiplying with $\sqrt{d}$) can be applied for financial time series. Moreover, it will be studied how much the risk can be over- or underestimated if the above formula is used. The relationship between the scaling law of the VaR and the self-similarity of the underlying process will be scrutinized.

The outline of this chapter is as follows: the considered problem will be set up in a mathematical framework in Section 13.2. In section 13.3, it will be investigated how much the risk can be over- or underestimated if the formula in equation (13.2.2) is used. Section 13.4 deals with the estimation of the Hurst coefficient via quantiles, and section 13.5 describes the used techniques. Section 13.6 considers the scaling law for some DAX stocks and for the DJI and its 30 stocks. In section 13.7 the Hurst exponents are estimated for the above financial time series. Possible interpretations in finance of the Hurst exponent are given in section 13.8. Section 13.9 concludes the chapter and gives an outlook.

13.2 THE SETUP

Speaking in mathematical terms, the VaR is simply the q-quantile of the distribution of the change of value for a given portfolio P. More specifically,

$$\mathrm{VaR}_{1-q}(P^d) = -F_{P^d}^{-1}(q), \tag{13.2.1}$$

where P^d is the change of value for a given portfolio over d days (the d-day return) and F_{P^d} is the distribution function of P^d. With this definition, this chapter considers the commercial return

$$P_c^d(t) := \frac{P(t) - P(t-d)}{P(t-d)}$$

as well as the logarithmic return

$$P_l^d(t) := \ln(P(t)) - \ln(P(t-d)),$$

where $P(t)$ is the value of the portfolio at time t. Moreover, the quantile function F^{-1} is a “generalized inverse” function

$$F^{-1}(q) = \inf\{x : F(x) \geq q\}, \quad \text{for } 0 < q < 1.$$

Notice also that it is common in the financial sector to speak of the q-quantile as the $1 - q$ VaR figure. Furthermore, it is common in practice to calculate the overnight VaR figure and derive from this the dth day VaR figure with the following formula

$$\mathrm{VaR}_{1-q}(P^d) = \sqrt{d} \cdot \mathrm{VaR}_{1-q}(P^1). \tag{13.2.2}$$

This is true if the changes of value of the considered portfolio for d days P^d are normally distributed with stationary and independent increments and with standard deviation $\sqrt{d}$ (i.e., $P^d \sim \mathcal{N}(0, d)$). To simplify the notation, the variance $\sigma^2 \cdot d$ has been set to d, meaning $\sigma^2 = 1$. However, the following calculation is also valid for $P^d \sim \mathcal{N}(0, \sigma^2 \cdot d)$.

$$\begin{aligned}
F_{P^d}(x) &= \int_{-\infty}^{x} \frac{1}{\sqrt{2\pi d}} \exp\left(-\frac{z^2}{2d}\right) dz \\
&= \int_{-\infty}^{\frac{x}{\sqrt{d}}} \frac{1}{\sqrt{2\pi d}} \exp\left(-\frac{w^2}{2}\right) \sqrt{d}\, dw \\
&= \int_{-\infty}^{\frac{x}{\sqrt{d}}} \frac{1}{\sqrt{2\pi}} \exp\left(-\frac{w^2}{2}\right) dw \\
&= F_{P^1}\left(d^{-\frac{1}{2}} x\right),
\end{aligned}$$

where the substitution $z = \sqrt{d} \cdot w$ was used. Applying this to $F_{P^d}^{-1}$ yields

$$\begin{aligned}
F_{P^d}^{-1}(q) &= \inf\{x : F_{P^d}(x) \geq q\} \\
&= \inf\{x : F_{P^1}(d^{-\frac{1}{2}} x) \geq q\} \\
&= \inf\{\sqrt{d} \cdot w : F_{P^1}(w) \geq q\} \\
&= \sqrt{d} \cdot F_{P^1}^{-1}(q).
\end{aligned}$$

On the other hand, if the changes of the value of the portfolio P are self-similar with Hurst coefficient H, equation (13.2.2) has to be modified in the following way:

$$\mathrm{VaR}_{1-q}(P^d) = d^H \cdot \mathrm{VaR}_{1-q}(P^1). \tag{13.2.3}$$

To verify this equation, let us first recall the definition of self-similarity (see, for example, Samorodnitsky and Taqqu [16], p. 311; compare also with Embrechts and Maejima [10]).

DEFINITION 13.2.1
A real-valued process $(X(t))_{t\in\mathbb{R}}$ *is* self-similar with index $H > 0$ *(H-ss) if, for all* $a > 0$*, the finite-dimensional distributions of* $(X(at))_{t\in\mathbb{R}}$ *are identical to the finite-dimensional distributions of* $(a^H X(t))_{t\in\mathbb{R}}$*, i.e., if for any* $a > 0$

$$(X(at))_{t\in\mathbb{R}} \stackrel{d}{=} (a^H X(t))_{t\in\mathbb{R}}.$$

This implies

$$\begin{aligned} F_{X(at)}(x) &= F_{a^H X(t)}(x) \quad \text{for all } a > 0 \text{ and } t \in \mathbb{R} \\ &= P(a^H X(t) < x) \\ &= P(X(t) < a^{-H} x) \\ &= F_{X(t)}(a^{-H} x). \end{aligned}$$

Thus, the assertion in equation (13.2.3) has been verified. So far, there are just three papers known to the author that also deal with the scaling behavior of VaR (see Diebold et al. [7] or [6], Dowd et al. [8], and Daníelsson and Zigrand [5]).

For calculating the VaR figure, there exist several possibilities, such as the historical simulation, the variance–covariance approach, and the Monte Carlo simulation. Most recently, the extreme-value theory has also been considered in estimating the VaR figure. In the variance–covariance approach, the assumption is made that the time series (P^d) of an underlying financial asset is normally distributed with independent increments and with drift μ and variance σ^2, which are estimated from the time series. Because this case assumes a normal distribution with stationary and independent increments, equation (13.2.2) obviously holds. Therefore, this case will not be considered here. Furthermore, the Monte Carlo simulation will not be considered either, as a particular stochastic model is chosen for the simulation. Thus the self-similarity holds for the Monte Carlo simulation if the chosen underlying stochastic model is self-similar. The extreme-value-theory approach is a semiparametric model, where the tail thickness is estimated by empirical methods (see, for example, Daníelsson and de Vries [3] or Embrechts et al. [9]). However, this tail-index estimator already determines the scaling-law coefficient.

There exists a great deal of literature on VaR that covers both the variance–covariance approach and the Monte Carlo simulation. Just to name the most popular, see for example Jorion [14] or Wilmott [20]. For further references, see also the references therein.

However, in practice banks often estimate the VaR via order statistics, which is the focus of this chapter. Let $G_{j:n}(x)$ be the distribution function of the jth order statistics. Because the probability that exactly j observations (of a total of n observations) are less than or equal to x is given by (see for example Reiss [15] or Stuart and Ord [19])

$$\frac{n!}{j! \cdot (n-j)!} F(x)^j \, (1 - F(x))^{n-j},$$

it can be verified that

$$G_{j:n}(x) = \sum_{k=j}^{n} \frac{n!}{k! \cdot (n-k)!} F(x)^k \, (1 - F(x))^{n-k}. \tag{13.2.4}$$

This is the probability that at least j observations are less than or equal to x given a total of n observations.

Equation (13.2.4) implies that the self-similarity holds also for the distribution function of the jth-order statistics of a self-similar random variable. In this case, one has

$$G_{j:n,P^d}(x) = G_{j:n,P^1}(d^{-H} \cdot x).$$

It is important that one has n observations for (P^1) as well as for (P^d); otherwise the equation does not hold. This shows that the jth-order statistics preserves—and therefore shows—the self-similarity of a self-similar process. Thus the jth-order statistics can be used to estimate the Hurst exponent, as will be done in this chapter.

13.3 RISK ESTIMATION FOR DIFFERENT HURST COEFFICIENTS

This section investigates how much the risk is over- or underestimated if equation (13.2.2) is used, although equation (13.2.3) is actually the right equation for $H \neq \frac{1}{2}$. In this case, the difference $d^H - \sqrt{d}$ determines how much the risk will be underestimated (or overestimated if the difference is negative). For example, for $d = 10$ days and $H = 0.6$, the underestimation

TABLE 13.3.1 Value at Risk and Self-Similarity—I

	$H = 0.55$			$H = 0.6$		
Days (d)	$d^{0.55}$	$d^{0.55} - d^{\frac{1}{2}}$	Relative Difference in Percent	$d^{0.6}$	$d^{0.6} - d^{\frac{1}{2}}$	Relative Difference in Percent
5	2.42	0.19	8.38	2.63	0.39	17.46
10	3.55	0.39	12.2	3.98	0.82	25.89
30	6.49	1.02	18.54	7.7	2.22	40.51
250	20.84	5.03	31.79	27.46	11.65	73.7

This table shows d^H, the difference between d^H and $\sqrt{d}$, and the relative difference $\frac{d^H - \sqrt{d}}{\sqrt{d}}$ for various days d and for $H = 0.55$ and $H = 0.6$.

will be of the size 0.82 or 25.89% (see Table 13.3.1). This underestimation will even extend to 73.7% if the 1-year VaR is considered (which is the case for $d = 250$). Here, the relative difference has been taken with respect to the value (namely $\sqrt{d}$) that is used by the banking industry.

Most important is the case $d = 10$ days, as banks are required to calculate not only the 1-day VaR, but also the 10-day VaR. However, the banks are allowed to derive the 10-day VaR by multiplying the 1-day VaR with $\sqrt{10}$ (see the Basel Committee on Banking Supervision [2]). Table 13.3.2 shows how much the 10-day VaR is underestimated (or overestimated) if the considered time series are self-similar with Hurst coefficient H.

TABLE 13.3.2 Value at Risk and Self-Similarity—II

H	10^H	$10^H - 10^{\frac{1}{2}}$	Relative Difference in Percent	H	10^H	$10^H - 10^{\frac{1}{2}}$	Relative Difference in Percent
0.35	2.24	−0.92	−29.21	0.4	2.51	−0.65	−20.57
0.45	2.82	−0.34	−10.87	0.46	2.88	−0.28	−8.8
0.47	2.95	−0.21	−6.67	0.48	3.02	−0.14	−4.5
0.49	3.09	−0.07	−2.28	0.5	3.16	0	0
0.51	3.24	0.07	2.33	0.52	3.31	0.15	4.71
0.53	3.39	0.23	7.15	0.54	3.47	0.31	9.65
0.55	3.55	0.39	12.2	0.56	3.63	0.47	14.82
0.57	3.72	0.55	17.49	0.58	3.8	0.64	20.23
0.59	3.89	0.73	23.03	0.6	3.98	0.82	25.89
0.61	4.07	0.91	28.82	0.62	4.17	1.01	31.83
0.63	4.27	1.1	34.9	0.64	4.37	1.2	38.04
0.65	4.47	1.3	41.25	0.66	4.57	1.41	44.54

This table shows 10^H, the difference between 10^H and $\sqrt{10}$, and the relative difference $\frac{10^H - \sqrt{10}}{\sqrt{10}}$ for various Hurst exponents H.

13.4 ESTIMATION OF THE HURST EXPONENT VIA QUANTILES

The Hurst exponent is often estimated via the pth moment with $p \in \mathbb{N}$. This can be justified with the following

PROPOSITION 13.4.1
Suppose $Y(k) = m(k)+X(k)$ with a deterministic function $m(k)$ and $X(k)$ is a stochastic process with all moments $\mathbb{E}\left[|X(k)|^p\right]$ existing for $k \in \mathbb{N}$ and distributions $F_k(x) := Prob(\{\omega \in \Omega : X(k,\omega) \leq x\})$ symmetric to the origin. Then the following are equivalent:

1. *For each $p \in \mathbb{N}$ it holds:*

$$\mathbb{E}\left[|Y(k) - \mathbb{E}\left[Y(k)\right]|^p\right] = c(p) \cdot \sigma^p |k|^{pH} \tag{13.4.1}$$

2. *For each k the following functional scaling law holds on $SymC_0^0(\mathbb{R})$:*

$$F_k(x) = F_1(k^{-H}x), \tag{13.4.2}$$

where $SymC_0^0(\mathbb{R})$ is the set of symmetric (with respect to the y-axis) continuous functions with compact support.

This has basically been shown by Singer et al. [18].

EXAMPLE 13.1
Let Y be a normal distributed random variable with variance σ. It is well known that

$$\mathbb{E}\left[(Y - \mathbb{E}\left[Y\right])^p\right] = \left\{\begin{matrix} 0 & \text{if } p \text{ is odd.} \\ \sigma^p(p-1)(p-3)\cdot\ldots\cdot3\cdot1 & \text{else} \end{matrix}\right\}.$$

Hence, in this case proposition 13.4.1 holds with $H = \frac{1}{2}$ and

$$c(p) = \left\{\begin{matrix} 0 & \text{if } p \text{ is odd.} \\ (p-1)(p-3)\cdot\ldots\cdot3\cdot1 & \text{else} \end{matrix}\right\}.$$

Proposition 13.4.1 states that the pth moment obeys a scaling law for each p given by equation (13.4.1) if a process is self-similar with Hurst coefficient H and the pth moment exists for each $p \in \mathbb{N}$. To check whether a process is actually self-similar with Hurst exponent H, it is most important that H be independent of p. However, often the Hurst exponent will be estimated just from one moment (mostly $p = 1$ or 2) (see Evertsz et al. [11] and, for more references on the Hurst coefficient, see also the references therein). This is because the higher moments might not exist (see, for example, Samorodnitsky and Taqqu [16], p. 18 and p. 316). Anyway, it is not sufficient to estimate the Hurst exponent just for one moment, because the important point is that the Hurst exponent $H = H(p)$ is equal for all moments, as the statement is for each $p \in \mathbb{N}$ in the proposition. Thus equation (13.4.1) is a necessary condition but not a sufficient one, if it is verified only for some $p \in \mathbb{N}$ but not for all $p \in \mathbb{N}$.

However, even if one has shown that equation (13.4.1) holds for each p, one has just proved that the one-dimensional marginal distribution obeys a functional scaling law. Even worse is the fact that this proves only that this functional scaling law holds just for symmetric functions. To be a self-similar process, a functional scaling law must hold for the finite-dimensional distribution of the process (see definition 13.2.1).

The following approach for estimating the Hurst coefficient is more promising, as it is possible to estimate the Hurst coefficient for various quantiles. Therefore, it is possible to observe the evolution of the estimation of the Hurst coefficient along the various quantiles. To derive an estimation of the Hurst exponent, let us recall, that

$$\mathrm{VaR}_{1-q}(P^d) = d^H \cdot \mathrm{VaR}_{1-q}(P^1),$$

if (P^d) is H-ss. Given this, it is easy to derive that

$$\log\left(\mathrm{VaR}_{1-q}(P^d)\right) = H \cdot \log(d) + \log\left(\mathrm{VaR}_{1-q}(P^1)\right). \qquad (13.4.3)$$

Thus the Hurst exponent can be derived from the gradient of a linear regression in a log–log plot.

13.4.1 Error of the Quantile Estimation

Obviously, equation (13.4.3) can only be applied if $\mathrm{VaR}_{1-q}(P^d) \neq 0$. Moreover, close to zero, a possible error in the quantile estimation will lead to an error in equation (13.4.3), which is much larger than the original error from the quantile estimation.

Let l be the number that represents the qth quantile of the order statistics with n observations. With this, x_l is the qth quantile of an ordered time series X, which consists of n observations with $q = \frac{l}{n}$. Let X be a stochastic process with a differentiable density function $f > 0$. Then Stuart and Ord [19] showed that the variance of x_l is

$$\sigma_{x_l}^2 = \frac{q \cdot (1-q)}{n \cdot (f(x_l))^2},$$

where f is the density function of X, and f must be strictly greater than zero.

The propagation of errors is calculated by the total differential. Thus, the propagation of this error in equation (13.4.3) is given by

$$\sigma_{\log(x_l)} = \frac{1}{x_l} \cdot \left(\frac{q \cdot (1-q)}{n \cdot (f(x_l))^2} \right)^{\frac{1}{2}}$$

$$= \sqrt{\frac{q \cdot (1-q)}{n}} \cdot \frac{1}{x_l \cdot f(x_l)}.$$

For example, if $X \sim \mathcal{N}(0, \sigma^2)$, the propagation of the error can be written as

$$\sigma_{\log(x)} = \sqrt{\frac{q \cdot (1-q)}{n}} \cdot \frac{\sqrt{2\pi} \cdot \sigma}{x \cdot \exp\left(-\frac{x^2}{2\sigma^2}\right)}$$

$$= \sqrt{\frac{q \cdot (1-q)}{n}} \cdot \frac{\sqrt{2\pi}}{y \cdot \exp\left(-\frac{y^2}{2}\right)},$$

where the substitution $\sigma \cdot y = x$ has been used. This shows that the error is independent of the variance of the underlying process if this underlying process is normally distributed (see also Figure 13.4.1).

Similarly, if X is Cauchy with mean zero, the propagation of the error can be shown to be

$$\sigma_{\log(x)} = \sqrt{\frac{q \cdot (1-q)}{n}} \cdot \frac{\pi \cdot (x^2 + \sigma^2)}{\sigma \cdot x}$$

$$= \sqrt{\frac{q \cdot (1-q)}{n}} \cdot \frac{\pi \cdot (y^2 + 1)}{y},$$

where the substitution $\sigma \cdot y = x$ has also been used. Once again, the error is independent of the scaling coefficient σ of the underlying Cauchy process. Because, for Levy processes with Hurst coefficient $\frac{1}{2} < H < 1$, closed forms for the density functions do not exist, the error cannot be calculated explicitly as in the normal and in the Cauchy case.

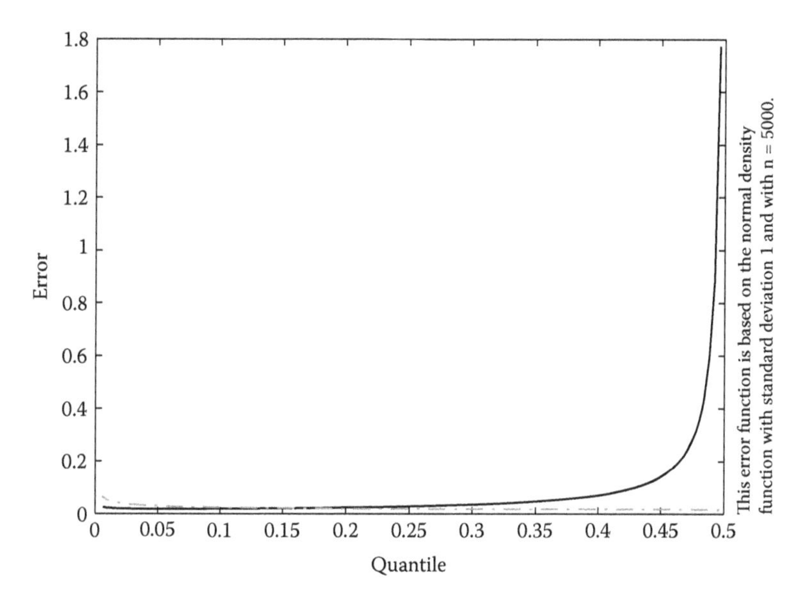

FIGURE 13.4.1 Error function for the normal distribution, left quantile.

Figure 13.4.1 shows that the error is minimal around the 5% quantile in the case of a normal distribution, but for a Cauchy distribution, the error is minimal around the 20% quantile (see Figure 13.4.2). Furthermore, the minimal error in the normal case is even less than half as large as in the Cauchy case.

This error analysis shows already the major drawback of estimating the Hurst exponent via quantiles. Because of the size of the error, it is not possible to estimate the Hurst exponent around the 50% quantile. However, it is still possible to estimate the Hurst coefficient in the (semi-) tails. Moreover, it is possible to check whether the Hurst exponent remains constant for various quantiles.

Hartung et al. [13] state that the $1-\alpha$ confidence interval for the qth quantile of an order statistic, which is based on n points, is given approximately by $[x_r; x_s]$. Here r and s are the next higher natural numbers of

$$r^* = n \cdot q - u_{1-\alpha/2}\sqrt{n \cdot q(1-q)}$$

and

$$s^* = n \cdot q + u_{1-\alpha/2}\sqrt{n \cdot q(1-q)}, \text{ respectively.}$$

The notation u_α has been used for the α-quantile of the $N(0, 1)$ distribution. Moreover, Hartung et al. [13] say that this approximation can be used if $q \cdot (1-q) \cdot n > 9$. Therefore this approximation can be used up to $q = 0.01$

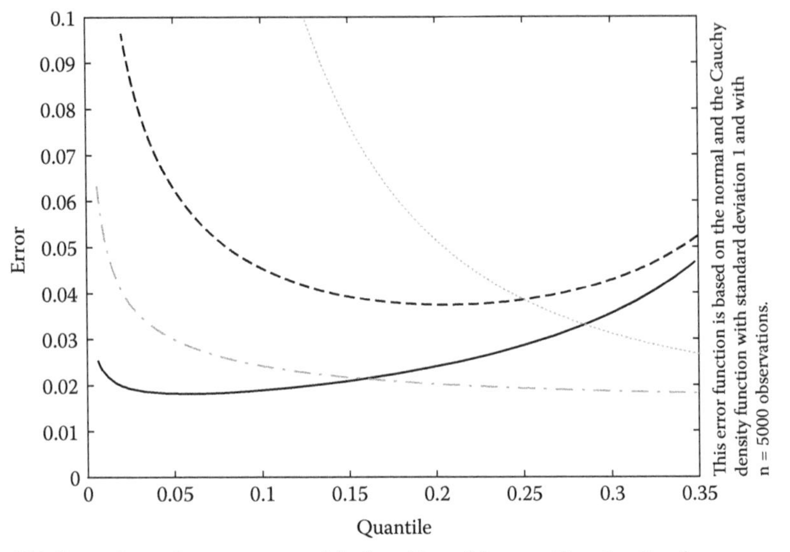

This figure shows the error curves of the logarithm of the quantile estimation for the normal distribution (solid line) and for the Cauchy distribution (dashed line). Moreover, the dash-dotted line depicts the error curve of the quantile estimation in the case of the normal distribution and the dotted line is the error curve of the quantile estimation in the case of the Cauchy distribution.

FIGURE 13.4.2 Error function for the normal and the Cauchy distribution, a zoom-in.

(which denotes the 1% quantile and will be the lowest quantile to be considered in this chapter) if $n > 910$, which will be the case in this discussion.

Obviously, these confidence intervals are not symmetric, meaning that the distribution of the error of the quantile estimation is not symmetric and therefore is not normally distributed. However, the error of the quantile estimation is asymptotically normally distributed (see, for example, Stuard and Ord [19]). Thus, for large n, the error is approximately normally distributed. Bearing this in mind, an error σ_{x_l} for the qth quantile estimation will be estimated by setting $u_\alpha = 1$ and making the approximation

$$\sigma_{x_l} \approx \frac{x_s - x_r}{2} \quad \text{with} \quad l = n \cdot q.$$

13.5 USED TECHNIQUES

Because the given financial time series do not have enough sample points to consider independent 2^j-day returns for $j = 1, \ldots, 4$, this chapter uses overlapping data to get more sample points.

13.5.1 Detrending

Generally, it is assumed that financial time series have an exponential trend. This drift has been removed from a given financial time series X in the following way.

$$^{d}X_t = \exp\left(\log(X_t) - \frac{t}{T}(\log(X_T) - \log(X_0))\right), \tag{13.5.1}$$

where ^{d}X will be called the detrended financial time series associated with the original financial time series X. This expression for ^{d}X will be abbreviated by the phrase detrended financial time series. To understand the meaning of this detrending method, let us consider $Y_t := \log(X_t)$, which is the cumulative logarithmic return of the financial time series (X_t). Assume that Y_t has a drift; this means that the drift in Y has been removed in ^{dl}Y and thus that the exponential drift in X has been removed in $^{d}X_t = \exp(^{dl}Y_t)$. Observe that $^{d}X_t$ is a bridge from X_0 to X_0. Correspondingly, $^{dl}Y_t$ is a bridge from Y_0 to Y_0.

Observe that this method of detrending is not simply subtracting the exponential drift from the given financial time series, rather it is dividing the given financial time series by the exponential of the drift of the underlying logarithmic returns, as can be seen from the formula. By subtracting the drift, one could get negative stock prices, which is avoided with the previously described method of detrending.

It is easy to derive that building a bridge in this way is the same as subtracting its mean from the 1-day logarithmic return. To verify this, let us denote with $P_l^1(t)$ the 1-day logarithmic return of the given time series X (as it has been defined in section 13.1). Therefore, one has $P_l^1(t) = Y_t - Y_{t-1}$. Keeping this in mind, one gets

$$\begin{aligned}
^{dl}Y_t - {}^{dl}Y_{t-1} &= Y_t - Y_{t-1} - \frac{t}{T}(Y_T - Y_0) + \frac{t-1}{T}(Y_T - Y_0) \\
&= P_l^1(t) - \frac{1}{T}(Y_T - Y_0) \\
&= P_l^1(t) - \frac{1}{T}\sum_{n=1}^{T}(Y_n - Y_{n-1}) \\
&= P_l^1(t) - \frac{1}{T}\sum_{n=1}^{T}P_l^1(n).
\end{aligned}$$

13.5.2 Considering Autocorrelation

To calculate the autocorrelation accurately, no overlapping data have been used. The major result is that the autocorrelation function of returns of the considered financial time series is around zero. The hypothesis that the 1-day returns are white noise can be rejected for most of the time series considered in this chapter to both the 0.95 confidence interval and the 0.99 confidence interval. Considering the 10-day returns, however, this is no longer true. This indicates that the distributions of the 1-day returns are likely to be different from the distributions of the 10-day returns. Hence, it is not likely to find a scaling coefficient for the above distributions. However, it is still possible to calculate the scaling coefficients for certain quantiles, as will be done in the following discussion.

13.5.3 Test of Self-Similarity

To be self-similar, the Hurst exponent has to be constant for the different quantiles. Two different tests are introduced in the following subsections.

13.5.3.1 A First Simple Test

This first simple test tries to fit a constant for the given estimation of the Hurst coefficient on the different quantiles. The test will reject the hypothesis (that the Hurst coefficient is constant, and thus that the time series is self-similar) if the goodness of fit is rejected. Fitting a constant to a given sample is a special case of the linear regression by setting $b = 0$. Apply the goodness-of-fit test to decide if the linear regression is believable and thus if the time series might be self-similar.

However, for the goodness-of-fit test, it is of utmost importance that the estimations of the Hurst coefficient for the different quantiles be independent of each other. Obviously, this is not the case for the quantile estimation, which the estimation of the Hurst coefficient is based on.

13.5.3.2 A Second Test

The second test tries to make a second linear regression for the given estimation for the Hurst coefficient on the different quantiles. The variable y_i is the estimation of the Hurst coefficient for the given quantile, which will be x_i. Moreover, σ_i is the error of the Hurst coefficient estimation, and N is the number of considered quantiles.

The null hypothesis is then that $b = 0$. The alternative hypothesis is $b \neq 0$. Thus the hypothesis will be rejected to the error level γ if

$$\left| \frac{b}{\sigma_b} \right| > t_{N-2,1-\frac{\gamma}{2}},$$

where $t_{\nu,\gamma}$ is the γ-quantile of the t_ν distribution (see Hartung [13]).

If the hypothesis is not rejected, a is the estimation of the Hurst exponent, and σ_a is the error of this estimation. Again, this test is based on the assumption that the estimations of the Hurst coefficient for the different quantiles are independent of each other.

Both tests lead to the same phenomenon, which has been described by Granger and Newbold [12]. That is, both tests mostly reject the hypothesis of self-similarity, not only for the underlying processes of the financial time series, but also for generated self-similar processes such as the Brownian motion or Levy processes.

It remains for future research to develop some test on self-similarity on the quantiles that overcome these obstacles.

13.6 ESTIMATING THE SCALING LAW FOR SOME STOCKS

A self-similar process with Hurst exponent H cannot have a drift. Because it is recognized that financial time series do have a drift, they cannot be self-similar. Because of this, the wording *scaling law* instead of *Hurst exponent* will be used when talking about financial time series that have *not* been detrended.

Because the scaling law is more relevant in practice than in theory, only those figures are depicted that are based on commercial returns.

13.6.1 Results for Some DAX Stocks

The underlying price processes of the DAX stocks are the daily closing prices from January 2, 1979, to January 13, 2000. Each time series consists of 5434 points.

Figure 13.6.1 shows the estimated scaling laws of 24 DAX stocks in the lower quantiles. Because this figure is not that easy to analyze, Figure 13.6.2 combines the results in Figure 13.6.1 by showing the mean, the mean plus/minus the standard deviation, and the minimal and maximal estimated scaling law of the 24 DAX stocks over the various quantiles. Moreover, Figures 13.6.2 to 13.6.5 show only the quantitative characteristics of the estimation of the scaling laws

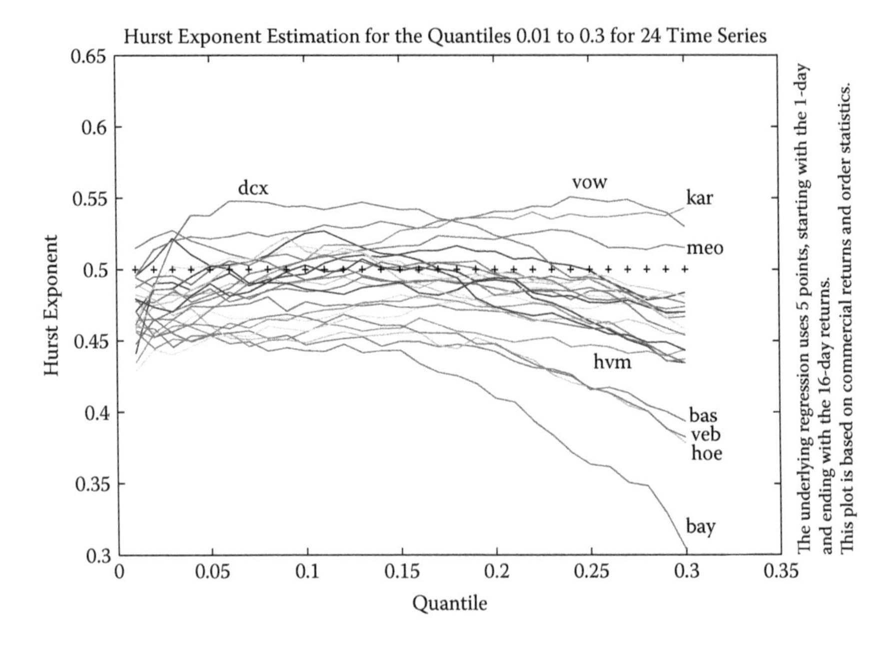

FIGURE 13.6.1 Estimation of the scaling law for 24 DAX stocks, lower (left) quantiles.

for 24 DAX stocks for the various quantiles, as these are considered to be more meaningful.

However, it is important to recognize that the 24 financial time series are not several realizations of one stochastic process. Therefore, one has to be very careful with the interpretation of the graphics in the case of financial time series. The interpretation here is that the graphics show the overall tendencies of the financial time series. Furthermore, the mean of the estimation of the scaling law is relevant for a well-diversified portfolio of these 24 stocks. The maximum of the estimation of the scaling law is the worst case possible for the considered stocks.

The estimation of the scaling law on the lower (left) quantile for 24 DAX stocks, which is based on commercial returns, shows that the shape of the mean is curved and below 0.5 (see Figure 13.6.2). The interpretation of this is that a portfolio of these 24 DAX stocks that is well diversified has a scaling law below 0.5. However, a poorly diversified portfolio of these 24 DAX-stocks can obey a scaling law as high as 0.55. This would imply an underestimation of 12.2% for the 10-day VaR.

On the upper (right) quantile, the mean curve is sloped, ranging from a scaling law of 0.66 for the 70% quantile to 0.48 for the 99% quantile (see Figure 13.6.3). The shape of the mean curve of the right quantile is totally

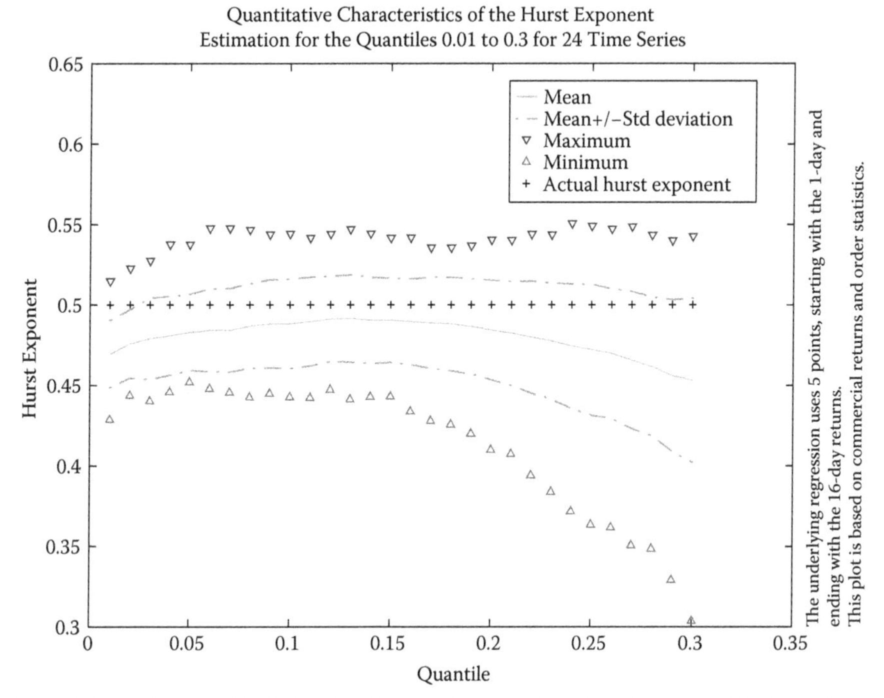

The solid line is the mean, the dash-dotted lines are the mean plus/minus the standard deviation, the triangles are the minimum, and the upside down triangles are the maximum of the estimation for the scaling law, which are based on 24 DAX-stocks. The underlying time series is a commercial return. Shown are the lower (left) quantiles.

FIGURE 13.6.2 Estimation of the scaling law for 24 DAX Stocks, lower (left) quantile.

different from the one of the left quantile, which might be due to a drift or to asymmetric distribution of the underlying process (compare Figure 13.6.2 with Figure 13.6.3). In particular, the mean is only in the 0.99 quantile slightly below 0.5. For all other upper quantiles, the mean is above 0.5.

Obviously, the right quantile is only relevant for short positions. For example, the VaR for the 0.95 quantile will be underestimated by approximately 9.6% for a well-diversified portfolio of short positions in these 24 DAX stocks and can be underestimated by up to 23% for some specific stocks.

The curves of the error of the estimation are also interesting (see Figures 13.6.4 and 13.6.5). First notice that the minimum of the mean of the error curves is in both cases below 0.1, which is substantially below the minimal error in the case of a normal distribution (compare with Figure 13.4.1) and in the case of a Cauchy distribution (see Figure 13.4.2).

However, the shape of the mean of the error curves of the left quantile is like the shape of the error curve of a normal distribution. Only the minimum

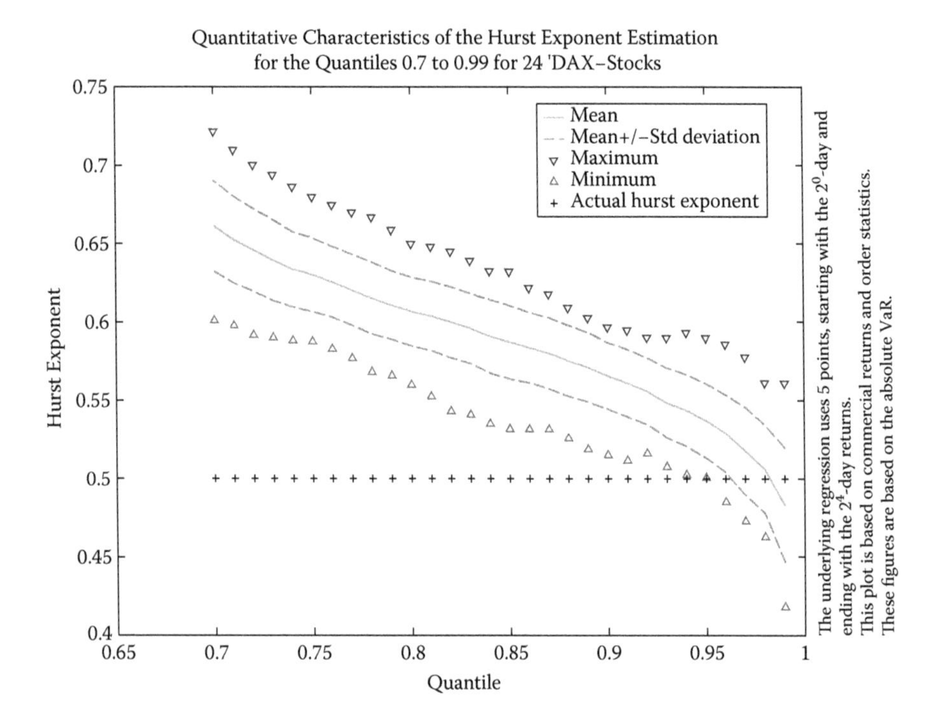

FIGURE 13.6.3 Estimation of the scaling law for 24 DAX stocks, upper (right) quantiles.

of the mean is in the 0.3 quantile and is thus even more to the left than in the case of a normal distribution. The shape of the mean of the error curves of the right quantile looks like some combination of the error curves of the normal distribution and the Cauchy distribution.

Assuming that the shape of the error curve is closely related to the Hurst exponent of the underlying process, this would imply that the scaling law for the left quantile is less than or equal to 0.5 and for the right quantile is between 0.5 and 1, as has been observed. However, the relationship between the shape of the error curve and the Hurst exponent of the underlying process still has to be verified.

The situation does not change much if logarithmic returns are considered. The mean on the lower quantile is not as curved as in the case of the commercial returns. However, it is still curved. Moreover, in both cases (of the logarithmic and the commercial returns) the mean is below 0.5 on the lower quantile. On the upper quantile, the mean curve in the case of the logarithmic returns is somewhat below that of the commercial returns, but otherwise has the same shape. Therefore, the concluding results are the same as in the case of the commercial returns.

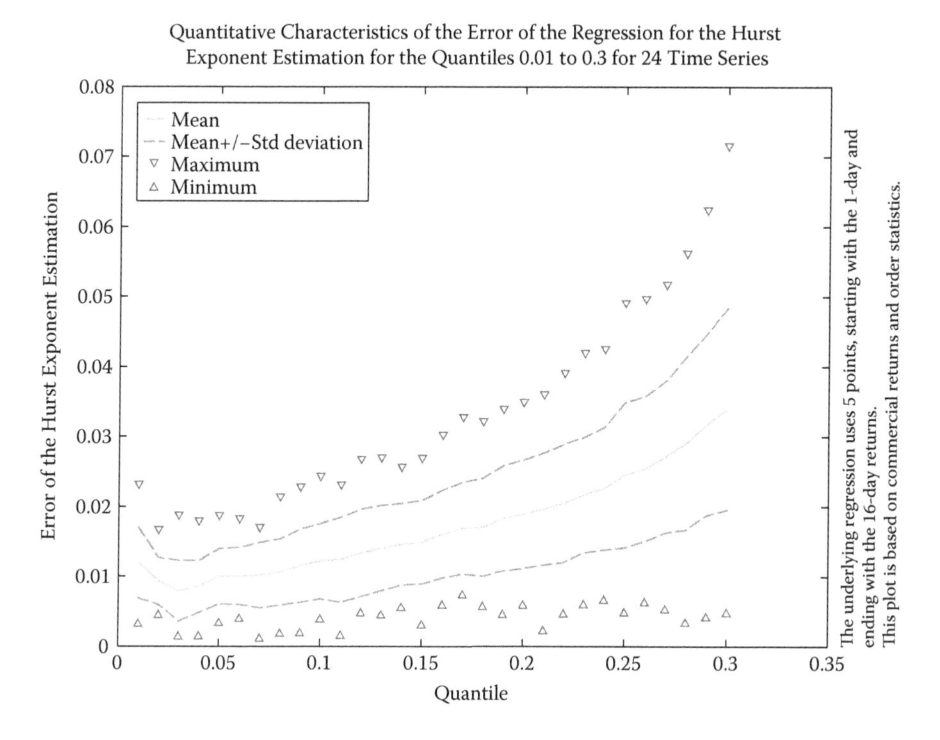

FIGURE 13.6.4 Error of the estimation of the scaling law for 24 DAX stocks, lower (left) quantiles.

13.6.2 Results for the Dow Jones Industrial Average Index and Its Stocks

The estimation of the scaling law for the Dow Jones Industrial Average Index (DJI) and its 30 stocks is based on 2241 points of the underlying price process, which dates from March 1, 1991, to January 12, 2000. As in the case of the 24 DAX stocks, the underlying price processes are the closing prices.

The results for the DJI and its 30 stocks are surprising because the mean of the estimated scaling law is substantially lower than in the case of the 24 DAX stocks for both the logarithmic returns as well as for the commercial returns. The mean of the estimation of the scaling law is also below 0.5 and has a curvature on the lower quantile (see Figure 13.6.6).

The mean of the estimation of the scaling law for the upper quantile (Figure 13.6.7) is sloped (as in the case of the 24 DAX stocks) and is below 0.5 in the quantiles that are greater than or equal to 0.95.

On the left quantile, the shape of the mean of the error curves in the case of commercial and logarithmic returns is comparable with the shape of the

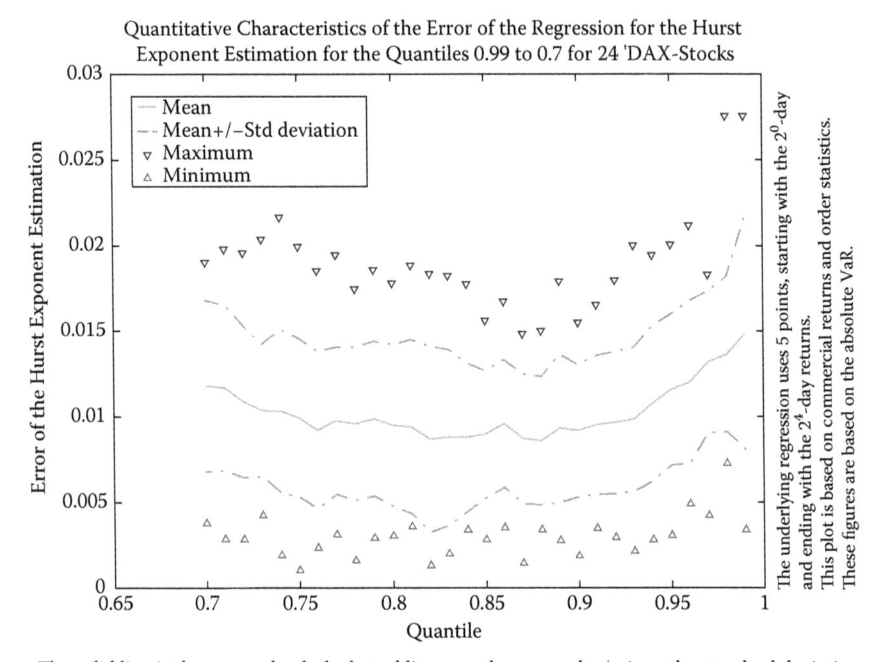

The solid line is the mean, the dash-dotted lines are the mean plus/minus the standard deviation, the triangles are the minimum, and the upside down triangles are the maximum of the error curves of the estimation for the scaling law, which are based on 24 DAX-stocks. The underlying time series is a commercial return. Shown are the lower (left) quantiles (see Figure 13.6.4) and the upper (right) quantiles (see Figure 13.6.5).

FIGURE 13.6.5 Error of the estimation of the scaling law for 24 DAX stocks, upper (right) quantiles.

mean of the error curve for the DAX stocks and therefore is comparable with the shape of the mean error curve of the normal distribution. The shape of the mean of the error curves on the right quantile is again like a combination of the error curve of the normal distribution and the error curve of the Cauchy distribution. However, the level of the mean of the error curves is the same height as the level of the error curve of the normal distribution and thus is substantially higher than the mean of the error curves of the corresponding 24 DAX stocks.

13.7 DETERMINING THE HURST EXPONENT FOR SOME STOCKS

It has already been stated, that the financial time series cannot be self-similar. However, it is possible that the detrended financial time series are self-similar with Hurst exponent H. This will be scrutinized in the following, where the

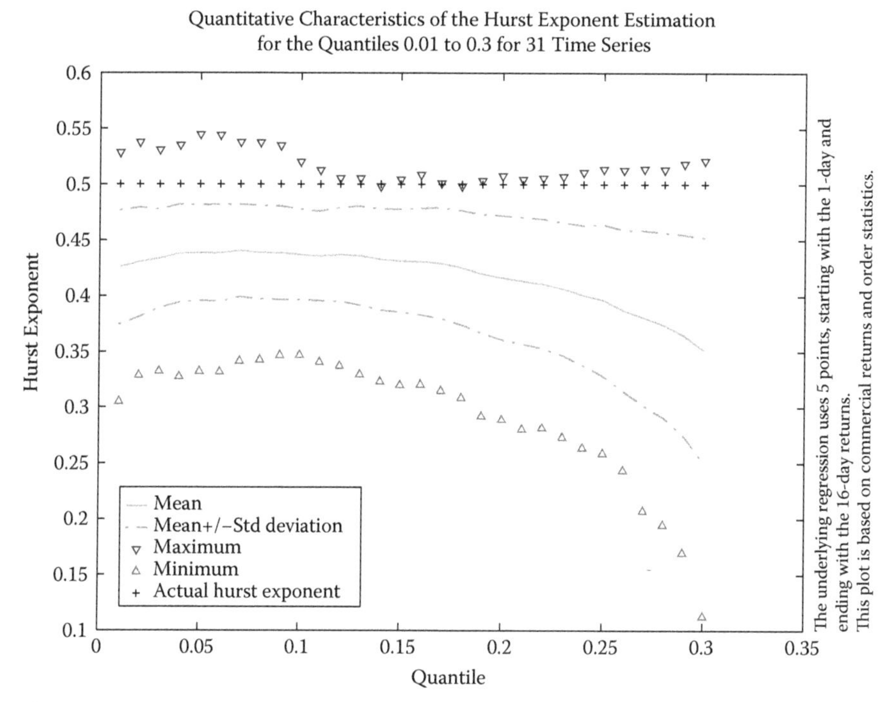

The solid line is the mean, the dash-dotted lines are the mean plus/minus the standard deviation, the triangles are the minimum, and the upside down triangles are the maximum of the estimation for the scaling law, which are based on the DJI and its 30 stocks. The underlying time series is a commercial return. Shown are the lower (left) quantiles.

FIGURE 13.6.6 Estimation of the scaling law for 30 DJI stocks, lower (left) quantiles.

financial time series have been detrended according to the method described in section 13.5.1. Because the Hurst exponent is more relevant in theory than in practice, only those figures are shown that are based on logarithmic returns.

13.7.1 Results for Some DAX Stocks

The results are shown in Figure 13.7.1 and 13.7.2. The mean of the estimation of the Hurst exponent on the left quantile is considerably higher for the detrended time series than for the time series that have not been detrended. However, the mean shows a slope for both the logarithmic as well as the commercial returns (see Figure 13.7.1). On the right quantile, the mean curve for the detrended time series has the same shape as in the case of the nondetrended time series for both the commercial return and the logarithmic return. The slope for the detrended time series is not as high as for the nondetrended time series, and

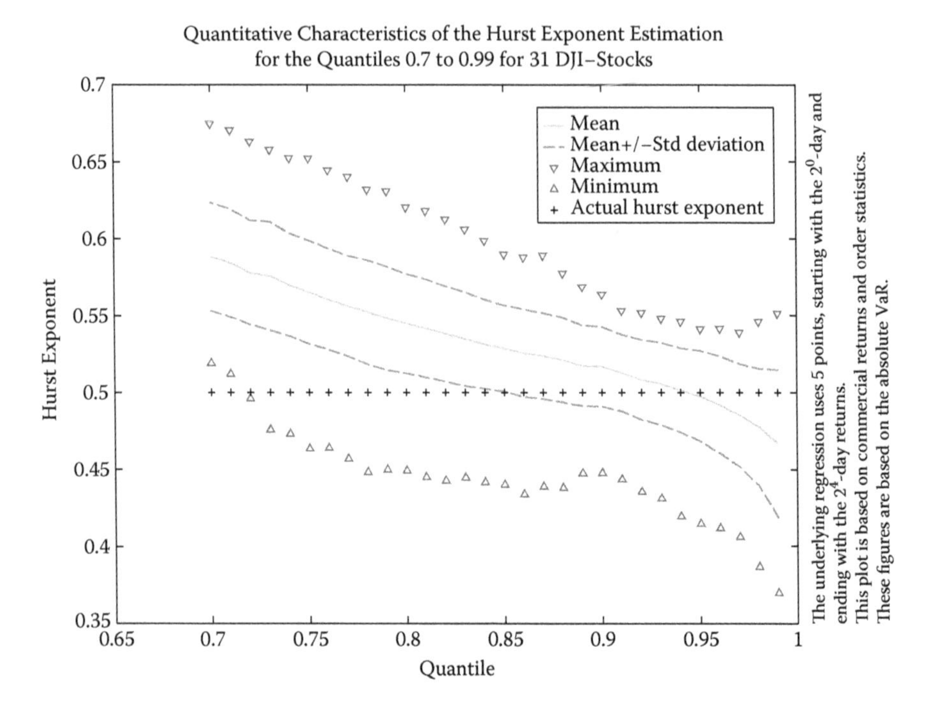

FIGURE 13.6.7 Estimation of the scaling law for 30 DJI stocks, upper (right) quantiles.

the mean curve of the detrended time series lies below the mean curve of the corresponding nondetrended time series.

The shape of the mean curve of the upper quantiles is comparable with that of the lower quantiles. This is valid for both the commercial and the logarithmic returns. However, for example in the case of the commercial return, the slope is much stronger (the mean Hurst exponent starts at about 0.6 for the 70% quantile and ends at about 0.47 for the 99% quantile compared with 0.55 for the 30% quantile and 0.48 for the 1% quantile). This indicates that the distribution of the underlying process might not be symmetric. Moreover, the wide spread of the mean Hurst exponent over the quantiles indicates, that the detrended time series are not self-similar as well.

The mean of the error curves is not much affected by the detrending. The shape of the mean of the error curves on the left quantiles are in both cases similar to the shape of the theoretical error of the normal distribution. The shape of the error curves on the upper quantiles is totally different from the ones on the lower quantiles and looks like a combination of the error curve of a normal distribution and a Cauchy distribution.

Altogether, these results indicate that the considered financial time series are not self-similar. However, the tests on self-similarity introduced in section

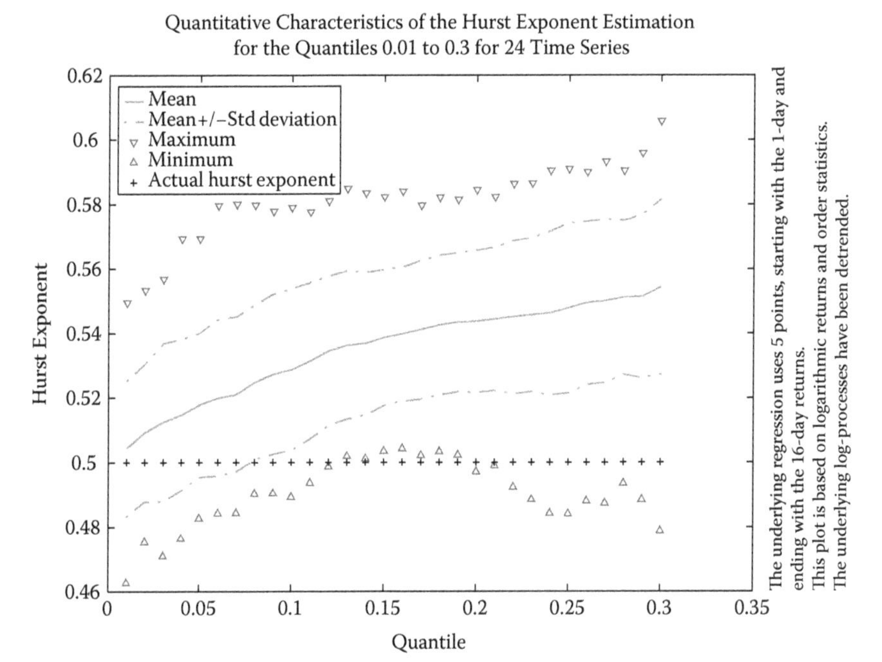

The solid line is the mean, the dash-dotted lines are the mean plus/minus the standard deviation, the triangles are the minimum, and the upside down triangles are the maximum of the estimation for the Hurst exponent, which are based on 24 DAX-stocks. The underlying time series are logarithmic returns, which have been detrended. Shown are the lower (left) quantiles.

FIGURE 13.7.1 Hurst exponent estimation for 24 DAX stocks, lower (left) quantiles.

13.5.3 are not sensitive enough to verify these findings. To verify these results, it is necessary to develop a test for self-similarity that is sufficiently sensitive.

13.7.2 Results for the Dow Jones Industrial Average Index and Its Stocks

The mean of the estimation of the Hurst exponent on the left quantile is considerably higher for the detrended time series than for the time series that have not been detrended. However, the mean shows, for both the logarithmic as well as the commercial returns, a curvature on the lowest quantiles (see Figure 13.7.3). The shape of the mean curve of the detrended time series is similar to the mean curve of the corresponding nondetrended time series.

On the right quantile, the mean curve for the detrended time series has the same shape as in the case of the nondetrended time series for both the commercial return and the logarithmic return. The slope for the detrended time series is not as high as for the nondetrended time series, and the mean curve

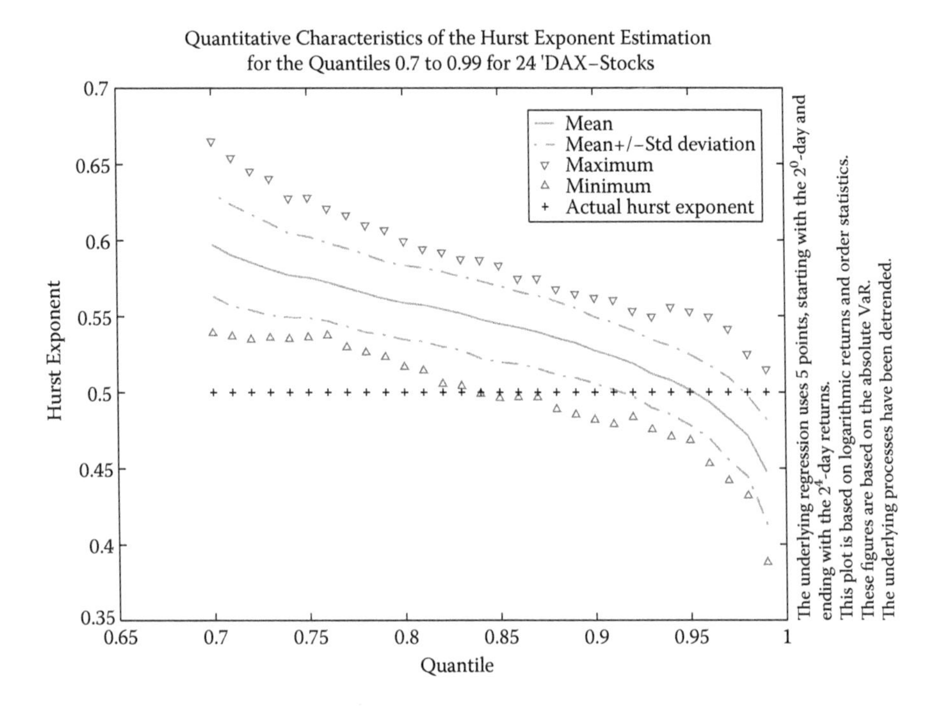

FIGURE 13.7.2 Hurst exponent estimation for 24 DAX stocks, upper (right) quantiles.

of the detrended time series lies below the mean curve of the corresponding nondetrended time series.

The shape of the mean curve of the upper quantiles is comparable with that of the lower quantiles only on the outer quantiles. This is valid for both the commercial and the logarithmic returns. Moreover, for example in the case of the commercial return, the mean Hurst exponent ranges from about 0.52 for the 70% quantile to about 0.45 for the 99% quantile compared with the range of 0.48 for the 30% quantile and 0.44 for the 1% quantile. This indicates that the distribution of the underlying process might not be symmetric. Moreover, the spread of the mean Hurst exponent over the quantiles might indicate that the detrended time series are not self-similar as well.

The mean of the error curves is not much affected by the detrending for the right quantiles. However, the mean of the error curves on the left quantiles is in both cases, about constant up to the lowest quantiles, where the curves go up. The shapes of the error curves on the upper quantiles are not as constant as those on the lower quantiles and look like some combination of the error curve of a normal distribution and a Cauchy distribution.

Altogether, these results indicate that the considered financial time series are not self-similar. However, the tests on self-similarity introduced in

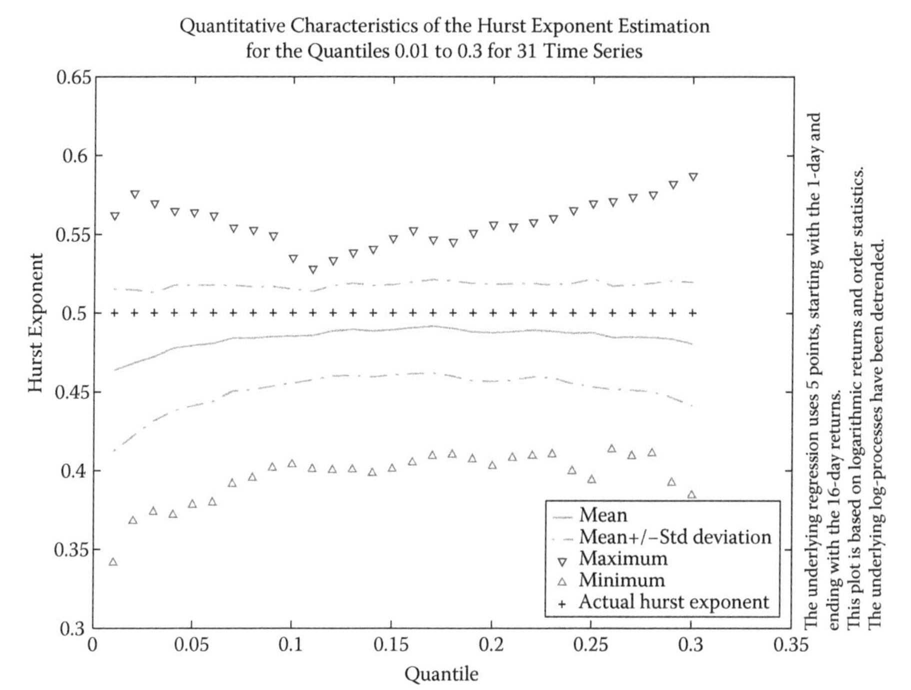

The solid line is the mean, the dash-dotted lines are the mean plus/minus the standard deviation, the triangles are the minimum, and the upside down triangles are the maximum of the estimation for the Hurst exponent, which are based on the DJI and its 30 stocks. The underlying time series are logarithmic returns, which have been detrended. Shown are the lower (left) quantiles.

FIGURE 13.7.3 Hurst exponent estimation for 30 DJI stocks, lower (left) quantiles.

section 13.5.3 are not sensitive enough to verify these findings, as has already been stated.

13.8 INTERPRETATION OF THE HURST EXPONENT FOR FINANCIAL TIME SERIES

First, let us recall the meaning of the Hurst exponent for different stochastic processes. For example, for a fractional Brownian motion with Hurst coefficient H, the Hurst exponent describes the persistence or antipersistence of the process (see, for example, Shiryaev [17]). For $1 > H > \frac{1}{2}$ the fractional Brownian motion is persistent. This means that the increments are positively correlated. For example, if an increment is positive, it is more likely that the succeeding increment is also positive than that it is negative. The higher H is, the more likely it is that the successor has the same sign as the preceding

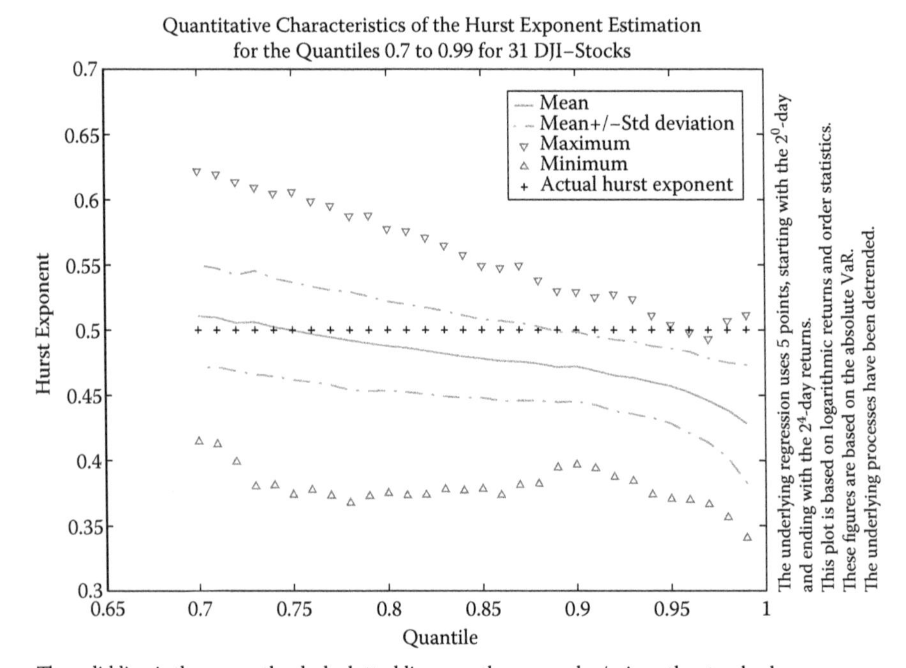

The solid line is the mean, the dash-dotted lines are the mean plus/minus the standard deviation, the triangles are the minimum, and the upside down triangles are the maximum of the estimation for the Hurst exponent, which are based on the DJI and its 30 stocks. The underlying time series are logarithmic returns, which have been detrended. Shown are the upper (right) quantiles.

FIGURE 13.7.4 Hurst exponent estimation for 30 DJI stocks, upper (right) quantiles.

increment. For $\frac{1}{2} > H > 0$ the fractional Brownian motion is antipersistent, meaning that it is more likely that the successor has a different sign than the preceding increment. The case $H = \frac{1}{2}$ is the Brownian motion, which is neither persistent nor antipersistent (see Shiryaev [17]).

This is, however, not true for Levy processes with $H > \frac{1}{2}$, where the increments are independent of each other. Therefore, the Levy processes are, like the Brownian motion, neither persistent nor antipersistent. In the case of Levy processes, the Hurst exponent H tells how much the process is heavy tailed.

Considering financial time series, the situation is not at all that clear. On the one hand, the financial time series are neither fractional Brownian motions nor Levy processes. On the other hand, the financial time series show signs of persistence and heavy tails.

Assuming the financial time series are fractional Brownian motions, then a Hurst exponent $H > \frac{1}{2}$ would mean that the time series are persistent.

The interpretation of the persistence could be that the financial markets are either rather slow to incorporate the actual given information, or this could indicate that insider trading is going on in the market. The first case would be a contradiction of the efficient-market hypothesis, and the second case would be interesting for the controlling institutions such as the SEC and the BaFin (the German analog of the SEC). A Hurst exponent of $H < \frac{1}{2}$ would mean that the financial market is constantly overreacting.

For this interpretation, compare also the findings for the DJI stocks with the results for the 24 DAX stocks. The average estimation of the Hurst coefficient of the 24 detrended DAX stocks is substantially higher than that of the detrended DJI stocks. Thus this interpretation would support the general belief that the U.S. financial market is one of the most efficient markets in the world, while the German market is not that efficient, which is often cited as the "Deutschland AG" phenomena.

Given that this interpretation is right, one could check whether a market (or an asset) has become more efficient. If its corresponding Hurst exponent gets closer to 0.5 over time, then the market (or the asset) is becoming more efficient.

Assuming, that the Hurst coefficient of a financial time series reflects persistence, the results of the detrended financial time series can be interpreted in the following way. Although the financial market is, in normal market situations, rather slow in incorporating the actual news, it tends to overreact in extreme market situations.

Not much can be said if one assumes that the Hurst coefficient of a financial time series reflects a heavy-tail property. It cannot be verified that large market movements occur more often in the German financial market than in the U.S. financial market. However, in both financial markets, big market movement does occur much more often than in the case of a Brownian motion. Therefore the financial time series are heavy tailed.

13.9 CONCLUSION AND OUTLOOK

The main results are that

- The scaling coefficient 0.5 has to be used very carefully for financial time series.
- There are substantial doubts about the self-similarity of the underlying processes of financial time series.

Concerning the scaling law, it is better to use a scaling law of 0.55 for the left quantile and a scaling law of 0.6 for the right quantile (the short positions), just to be on the safe side. It is important to keep in mind that these figures are only based on the (highly traded) DAX and Dow Jones Index stocks. Considering that infrequently traded stocks might yield even higher maximal scaling laws, these numbers should be set by market-supervision institutions such as the SEC.

However, it is possible for banks to reduce their VaR figures if they use the correct scaling law numbers. For instance, the VaR figure of a well-diversified portfolio of Dow Jones Index stocks would be reduced in this way by about 12% because it would have a scaling law of approximately 0.44.

Regarding the self-similarity, estimating the Hurst exponent via the quantiles might be a good alternative to modified R/S statistics, Q–Q plots, and calculating the Hurst exponents via the moments. However, it remains to future research to develop a test on self-similarity on the quantiles that overcomes the phenomena described by Granger and Newbold [12].

Finally, Daníelsson and Zigrand [5] mentioned that the square-root-of-time rule is also used for calculating volatilities. The presented results indicate that the appropriate scaling-law exponents for volatilities is most likely higher than the estimated scaling-law exponents for the quantiles. However, determination of specific estimates for this situation is left for future research.

ACKNOWLEDGMENTS

I would like to thank Jean-Pierre Stockis for some very valuable discussions and hints. The discussions with Peter Singer and Ralf Hendrych have always been very inspiring. Finally, I appreciate the comments and advice of Prof. Philippe Jorion, Prof. Klaus Schürger, and an anonymous referee.

REFERENCES

1. Philippe Artzner, Freddy Delbaen, Jean-Marc Eber, and David Heath. Coherent measure of risk. *Mathematical Finance*, 9(3): 203–228, 1999.
2. Basel Committee on Banking Supervision. Overview of the amendment to the capital accord to incorporate market risk. Basel Committee Publications No. 23. Bank for International Settlements, CH-4002 Basel, Switzerland, January 1996. See also http://www.bis.org/bcbs/publ.htm.
3. Jón Daníelsson and Casper G. de Vries. Value-at-risk and extreme returns. *Annales d'Economie et de Statistique*, 60(special issue): 236–269, 2000. See also http://www.riskresearch.org/.

4. Jón Daníelsson, Bjørn N. Jorgensen, Gennady Samorodnitsky, Mandira Sarma, and Casper G. de Vries. Subadditivity re-examined: the case for value-at-risk. Working paper, November 2005. See http://www.riskresearch.org/.
5. Jón Daníelsson and Jean-Pierre Zigrand. On time-scaling of risk and the square-root-of-time rule. FMG discussion paper, Financial Markets Group dp439. London School of Economics and Political Science, London, March 2003. See also http://www.riskresearch.org/.
6. Francis Diebold, Andrew Hickman, Atsushi Inoue, and Til Schuermann. Converting 1-day volatility to h-day volatility: scaling by $\sqrt{h}$ is worse than you think. Working paper, Financial Institutions Center 97–34. The Wharton School, University of Pennsylvania, Philadelphia, July 1997. See http://fic.wharton.upenn.edu/fic/papers/97/9734.pdf.
7. Francis Diebold, Atsushi Inoue, Andrew Hickman, and Til Schuermann. Scale models. *Risk Magazine*, 104–107, January 1998.
8. Kevin Dowd, David Blake, and Andrew Cairns. Long-term value at risk. Discussion paper pi-0006. The Pensions Institute. Birkbeck College, University of London, London, UK, June 2001.
9. Paul Embrechts, Claudia Klüppelberg, and Thomas Mikosch. *Modelling Extremal Events*, vol. 33 of *Applications of Mathematics*. Springer, Berlin, 3rd ed., 1997.
10. Paul Embrechts and Makoto Maejima. An introduction to the theory of self-similar stochastic processes. *International Journal of Modern Physics B*, 14(12/13): 1399–1420, 2000.
11. Carl Evertsz, Ralf Hendrych, Peter Singer, and Heinz-Otto Peitgen. *Komplexe Systeme und Nichtlineare Dynamik*, chapter Zur fraktalen Geometrie von Börsenzeitreihen. Springer, Berlin, 1999, pp. 400–419.
12. Clive W. J. Granger and Paul Newbold. Spurious regressions in econometrics. *Journal of Econometrics*, 2: 111–120, 1974.
13. Joachim Hartung, Bärbel Elpelt, and Karl-Heinz Klösener. *Statistik*. Oldenbourg, München, 10th ed., 1995.
14. Philippe Jorion. *Value at Risk: The New Benchmark for Controlling Market Risk*. McGraw-Hill, New York, 1997.
15. Rolf-Dieter Reiss. *Approximate Distributions of Order Statistics*. Springer, New York, 1989.
16. Gennady Samorodnitsky and Murad S. Taqqu. *Stable Non-Gaussian Random Processes*. Chapman & Hall, New York, 1994.
17. Albert N. Shiryaev. *Essentials of Stochastic Finance*. World Scientific, Singapore, 1999.
18. Peter Singer, Ralf Hendrych, Carl Evertsz, and Heinz-Otto Peitgen. Semistable processes in finance and option pricing. CEVIS, University of Bremen, Bremen, Germany. Preprint, 2000.
19. Alan Stuart and J. Keith Ord. Distribution theory. In *Kendall's Advanced Theory of Statistics*, vol. 1. Arnold, London, 6th ed., 1994.
20. Paul Wilmott. *Derivatives: The Theory and Practise of Financial Engineering*. John Wiley & Sons, Chichester, U.K. 1998.

CHAPTER 14

Parameter Uncertainty in Kalman-Filter Estimation of the CIR Term-Structure Model

Conall O'Sullivan
University College Dublin, Dublin, Ireland

Contents

Abstract: The Cox, Ingersoll, and Ross (1985) CIR term structure model describes the stochastic evolution of government-bond yield curves over time using a square-root Orstein–Uhlenbeck diffusion process, while imposing cross-sectional no-arbitrage restrictions between yields of different maturities. A Kalman-filter approach can be used to estimate the parameters of the

CIR model from panel data consisting of a time series of bonds of different maturities. The parameters are estimated by optimizing a quasi log-likelihood function that results from the prediction-error decomposition of the Kalman filter. The quasi log-likelihood function is usually optimized with a deterministic gradient-based optimization technique such as a quadratic hill-climbing optimizer. This chapter uses an evolutionary optimizer known as differential evolution (DE) to optimize over the parameter space. The DE optimizer is more likely to find the global maximum than a deterministic optimizer in the presence of a non-convex objective function, which may be the case in multifactor term-structure models with nonnegativity constraints and parameter constraints. The method is applied to estimate parameters from a one- and two-factor Cox, Ingersoll, and Ross (1985) model. It is shown that, in the two-factor model, the problem of local maxima arises whereby a number of different parameter vectors perform equally well in the estimation procedure. Fixed-income derivative prices are particularly sensitive to term-structure parameters such as the volatility, the rate of mean reversion, and the market price of risk of each factor. The effect of different optimal parameter vectors on fixed-income derivatives is examined and is found to be significant.

14.1 INTRODUCTION

Dynamic term-structure models describe the stochastic evolution of government bond yield curves (or swap-market yield curves) over time while imposing cross-sectional no-arbitrage restrictions between yields of different maturities. The classic dynamic term-structure models are usually assumed to be driven by a small number of latent factors motivated by the empirical results of Steeley (1990) and Litterman and Scheinkman (1991), who found that three factors can explain up to 99% of the variability of the yield curve. However modeling yield curves using a low-dimensional factor model leads to measurement error between the theoretically implied yields (or bond prices) and market yields (or bond prices). If the dynamic term-structure model (DTSM) can be formulated into a state–space representation, a Kalman-filter approach can be used, and this measurement error can be accounted for explicitly. In the Kalman-filter approach, model parameters are estimated from panel data consisting of a time series of bonds with different maturity dates. Parameter estimation is carried out by optimizing a quasi log-likelihood function that results from the prediction-error decomposition of the Kalman filter. The quasi log-likelihood function is usually optimized with a deterministic gradient-based optimization technique such as a quadratic hill-climbing optimizer. This chapter uses an evolutionary based optimizer known as differential evolution (DE) to optimize over the parameter space. The DE optimizer is more likely to find the global maximum than are the deterministic optimizers in the presence of a nonconvex objective function, which may be the case in term-structure models with nonnegativity constraints and parameter constraints such as the Cox, Ingersoll, and Ross (hereinafter CIR) model (1985).

The term-structure models used in this study are restricted to the one- and two-factor versions of the CIR (1985) model for simplicity.[1] However, DE could potentially be even more useful than deterministic optimizers in more-complex DTSMs, such as those preferred by Dai and Singleton (2000). The data set consists of Fama–Bliss U.S. government bond yields of 3-, 6-, 12-, and 60-months maturities sampled monthly from April 1964 to December 1997. Duan and Simonato (1999) have used the same data to estimate a one-factor Vasicek model and one- and two-factor CIR models using a deterministic optimizer. We find that the optimal parameter vectors using the DE optimizer agree with Duan and Simonato for the one-factor CIR case; however, for the two-factor CIR case, we get a number of different parameter estimates with the same maximum value for the log-likelihood function. The effect of the different approximately optimal parameter vectors on fixed-income derivatives is examined and is found to be more pronounced than on the underlying bonds themselves. This leads to the conclusion that fixed-income derivatives should also be considered or even included in the estimation of multifactor term-structure models, as this may reduce the problem of local maxima, whereby a number of different parameter vectors perform equally well in the estimation procedure.

The remainder of this chapter is organized as follows: first we consider the motivation for the study, and then we consider the related literature in term-structure model estimation and evolutionary computational techniques in finance. Section 14.2 introduces the one- and two-factor CIR term-structure models to be estimated. This section then outlines the use of the Kalman filter in terms of structure modeling. Section 14.3 contains a description of differential evolution. Section 14.4 contains the results and compares them

---

[1] The parameters for one- and two-factor Vasicek models were also estimated using DE. It was found that they do not exhibit local maxima. Results are not reported but are available upon request.

with the well-known paper of Duan and Simonato (1999). The effect of local maxima on derivatives prices is also examined. Section 14.5 concludes the chapter and discusses some possible future research avenues.

14.1.1 Motivation

As the models used in finance become less parsimonious and increasingly complicated to better account for the complexities of financial markets, the problem of estimating and calibrating these high-dimensional models is no longer a straightforward step, as many of these models contain local maxima. This means that a number of different parameter vectors that describe a model can result in virtually identical estimation or calibration performance. We will refer to these different parameter vectors as locally optimal parameter vectors. If the prices and hedge ratios from these locally optimal parameter vectors are virtually identical, then having a number of locally optimal parameter vectors might not be a serious problem if these models are only used in the pricing and hedging of securities similar to those used in the estimation or calibration. However, in many cases, these models are used in a wider framework, such as in the pricing of exotic securities or in asset-allocation decisions that involve many other asset classes. In these cases, the locally optimal parameter vectors may cause large differences in exotic prices or may result in a different decision-making process. Recent literature contains some examples of the occurrence of local maxima in financial models. For example, Ben Hamida and Cont (2004) retrieved a local volatility surface from a finite set of option prices and examined the multiplicity of solutions satisfing the constraints that model and market prices must be within a certain tolerance level related to the bid–ask spread. Ayache et al. (2004) calibrated a regime switching model to a set of option prices and found many possible parameter vector solutions that calibrate to the data equally well. They found that the introduction of exotic options in the calibration step helps reduce this problem. Thus in estimating or calibrating high-dimensional financial models, global optimizers may need to be used to provide information as to whether the objective function is well behaved or not, and if it is not well behaved, to reduce the probability of the optimizer returning a parameter vector that is simply a local maximum of the objective function.

Term-structure modeling is crucial in fixed-income modeling, and the proprietary trading desks of many banks and hedge funds use two- and three-factor dynamic term-structure models in their statistical arbitrage modeling of fixed-income markets. However, estimating these models based on panel data can be a difficult task for multifactor models, and these models can display locally optimal parameter vectors. The Kalman-filter approach to estimating term-structure models is by now well known and very popular (see, for example, Duan and Simonato [1999] and Babbs and Nowman [1999] among many others). Parameters are estimated by maximizing the log-likelihood function, which is usually optimized with regard to the parameter vector using a deterministic gradient-based optimizer. It is recommended that the log-likelihood function be optimized for a number of different parameter vector initializations, thereby reducing the chance of the deterministic optimizer converging to a local maximum. However this is a rather ad hoc way to proceed, especially if the model is complicated and high dimensional, such as Feldhutter and Lando's (2005) five-factor affine model with parameter constraints. The contribution of this chapter is in the use of an optimizer known as differential evolution (DE) (see Storn and Price [1997]), to estimate term-structure models. The objective function used in this chapter is the quasi log-likelihood function (CIR) resulting from the prediction-error decomposition of a Kalman filter. The objective function is calculated from the prediction-error decomposition that is obtained by running a Kalman filter on a data set based on a single choice of the parameter vector. There is no guarantee that the maximization of the objective function by appropriate choice of a parameter vector is a well-posed problem, especially as the complexity of the term-structure model increases. For example, in a two-factor CIR model the factors should always be positive. For certain choices of parameters, the factors may become negative. For example, if $2\kappa\theta < \sigma^2$, where κ is the mean reversion rate, θ is the long-run mean, and σ is the volatility of the factor, there is a positive probability that this factor will reach zero. In continuous time, the process will be reflected away from zero; however, in a discrete time setting, finite time steps are used, and the factor can become negative. To prevent this, we can ensure that the factor does not become negative by replacing a negative value of the factor with a value of zero (the approach adopted in this chapter) or by applying a constraint on the parameter space such that $2\kappa\theta \geq \sigma^2$. Both of these approaches can result in a nonconvex objective function. As the term-structure model becomes more complex, the number of such parameters and constraints increases; thus the use of a global optimizer, such as DE, becomes even more crucial.

14.1.2 Related Literature

No-arbitrage dynamic term-structure modeling began with Vasicek (1977), Dothan (1982), Courtadon (1982), and Cox, Ingersoll, and Ross (1985) (hereinafter CIR). These models posit a diffusion process for the short rate of interest under the physical probability measure, and combining this with an assumption on investors' risk preferences results in a process for the short rate under the equivalent martingale measure in which bond prices and yields can be calculated. Vasicek and CIR are affine dynamic term-structure models (DTSM), as bond yields are affine functions of the short rate, and as a result they are tractable and widely used term-structure models. Recognizing the need for the use of more than one factor to better explain the correlation structure of yields at different maturities, Brennan and Schwartz (1979), Schaefer and Schwartz (1984), Longstaff and Schwartz (1992), and Chen and Scott (1993), among others, began examining multifactor term-structure models. This culminated in the Duffie and Kan (1996) paper, where a general affine term-structure model was derived that encompasses multifactor versions of the Vasicek and CIR models. More recently, Dai and Singleton (2000) have examined maximally flexible canonical affine term-structure models. All other affine term-structure models are specific examples of their maximally flexible models, with certain parameters set to zero. These models allow for nonzero correlation between the latent factors; however, in their most general form they do not admit closed-form solutions for bond prices, so they must be solved numerically using a system of ordinary differential equations. In this chapter we focus on one- and two-factor CIR models for simplicity and clarity. However, the methodology of this chapter is applicable to any term-structure model that can be formulated into a state–space model.

Parameter estimation in DTSMs can be done using a number of different methods, including the maximum-likelihood method of Lo (1986, 1988), the generalized method of moments (GMM) of Hansen (1982), and the efficient method of moments (EMM) of Gallant and Tauchen (1996). The Kalman filter (KF) can also be used to estimate the parameters of a DTSM if it can be formulated into a state–space model (see Pennachi [1991], Duan and Simonato [1999], and Babbs and Nowman [1999], among many others). The KF is a natural approach to use when the underlying state is unobserved and follows a diffusion process. The KF allows for measurement error between the model yield and market yield, which naturally arises when using a small number of factors to model the yield curve. Thus the KF can account for model misspecification and market imperfections such as bid–ask spread and illiquidity. Parameter estimates are obtained by optimizing a log-likelihood function in Gaussian DTSMs or a quasi log-likelihood function in non-Gaussian DTSMs that results from the prediction-error decomposition of the KF. Affine DTSMs are particularly suited to estimation using the KF because of their tractability and linear nature. However, outside the Gaussian class of affine TSMs, including for example the affine CIR model, parameter estimates are inconsistent because the distribution of the shocks to the latent factors is non-Gaussian. However, the degree of inconsistency is found to be of little importance in practice according to Lund (1997), Duan and Simonato (1999), and De Jong (2000), who conducted Monte Carlo experiments to check this. Chen and Scott (2003) conducted similar Monte Carlo experiments on the multifactor CIR model and found that, while some parameter estimates were biased, those combinations of parameters that are important for asset pricing were unbiased. Extended KFs or unscented KFs can be combined with quasi-maximum likelihood to estimate models using data that are nonlinear functions of the latent factors, such as swap data. Recently Duffee and Stanton (2004) suggested that the KF is a robust method to use for parameter estimation in dynamic term-structure models, and they recommend the use of the KF when maximum-likelihood estimation is not feasible. However, many prefer the use of KF even when maximum likelihood is feasible, as is the case for Vasicek and CIR models. The KF method does not require the assumption that certain-market observable rates be observed without error, as is done in ML methods. The KF updates the first two moments of a system of latent variables, and historical information about the unobserved system is embedded in these two moments. The KF is the optimal filtering technique to use among the class of linear filters. The KF is also well suited to out-of-sample tests, given that the factors are updated in the filter with the arrival of each new observation date.

14.2 DYNAMIC TERM-STRUCTURE MODELS

In this section we introduce the two term-structure models used in this chapter and describe the Kalman filter and how it can be used to estimate parameters of dynamic term-structure models. We assume that there are n state variables, denoted $x_t \equiv (x_{1t}, \ldots, x_{nt})'$ (in this chapter we only deal with $n = 1, 2$). Uncertainty is generated by n independent Brownian motions, where we assume independence for simplicity. The independence assumption can be relaxed, and factor correlations can be crucial in modeling certain phenomena, such as the term structure of volatility (see Dai and Singleton [2000]). Under the equivalent martingale measure, these Brownian motions are denoted as $\tilde{z}_t \equiv (\tilde{z}_{1t}, \ldots, \tilde{z}_{nt})'$; the corresponding Brownian motions under the physical measure are denoted without the tildes. The instantaneous interest rate, denoted r_t, is affine in the state and given by

$$r_t = 1' \cdot x_t,$$

where 1 is the unit vector with n elements.

14.2.1 CIR Model

In the CIR model, the state dynamics under the physical measure are given by a square-root diffusion process

$$dx_{it} = \kappa_i (\theta_i - x_{it})\, dt + \sigma_i \sqrt{x_{it}} dz_{it}.$$

The equivalent martingale dynamics determine bond prices. Assume that the market price of risk of each factor λ_i is proportional to the factor. The dynamics under the equivalent martingale measure are given by

$$dx_{it} = (\kappa_i \theta_i - (\kappa_i + \lambda_i)\, x_{it})\, dt + \sigma_i \sqrt{x_{it}} d\tilde{z}_{it}.$$

Zero-coupon bonds maturing at time $t + \tau$ have prices and yields given by

$$P(x_t, \tau) = \exp[A(\tau) - B(\tau)' x_t], \tag{14.2.1}$$

$$Y(x_t, \tau) = (1/\tau)[-A(\tau) + B(\tau)' x_t], \tag{14.2.2}$$

where functions $A(\tau)$ and $B(\tau)$ are given by

$$A(\tau) = A_1(\tau) + A_2(\tau)$$

$$B(\tau) = [B_1(\tau), B_2(\tau)]'$$

$$A_i(\tau) = \frac{2\kappa_i \theta_i}{\sigma_i^2} \ln \left(\frac{2\gamma_i e^{(\kappa_i + \lambda_i + \gamma_i)\frac{\tau}{2}}}{(\kappa_i + \lambda_i + \gamma_i)\,(e^{\gamma_i \tau} - 1) + 2\gamma_i} \right)$$

$$B_i(\tau) = \frac{2\,(e^{\gamma_i \tau} - 1)}{(\kappa_i + \lambda_i + \gamma_i)\,(e^{\gamma_i \tau} - 1) + 2\gamma_i},$$

$$\gamma_i = \sqrt{(\kappa_i + \lambda_i)^2 + 2\sigma_i^2}.$$

A negative value for λ_i means that the risk premium for holding longer-term bonds is positive.

14.2.2 Kalman Filter

The Kalman filter is a powerful linear filtering technique introduced by Kalman (1960). In its original form, it is assumed that a system is driven by an unobservable state that experiences additive noise and that there are observables that are linear functions of this unobservable state. However, the observables themselves are measured with noise. The KF is a filter method that allows the estimation of the unobservable state and its covariance matrix at each point in time using only knowledge of the noisy measurements. It operates iteratively on the data, so only the current estimate of the state and its covariance matrix are needed for the prediction of the future state; thus it is computationally efficient. The reader is referred to Harvey (1989) for a thorough explanation of KFs and to Duan and Simonato (1999) and various other papers cited in the literature review for examples of KFs in interest-rate modeling. A brief explanation is outlined here for completeness.

The state transition and measurement equations are as follows

$$x_t = \Phi_0 + \Phi_1 x_{t-h} + \eta_t, \qquad \text{var}[\eta_t] = Q_t, \tag{14.2.3}$$

$$Y_t = H_0 + H_1 x_t + e_t, \qquad \text{var}[e_t] = R, \tag{14.2.4}$$

where Q_t and R are the covariance matrices of the unobservable state innovation and the measurement error, respectively. Initial conditions are $\hat{x}_0 = E(x_0)$ and $\hat{P}_0 = \text{var}(x_0)$. The KF consists of three steps. The prediction step is given by

$$\hat{x}_t^- = \Phi_0 + \Phi_1 \hat{x}_{t-h}$$

$$P_t^- = \Phi_1 P_{t-h} \Phi_1' + Q_t.$$

The predicted measurement, the prediction error, and its covariance are given by

$$\hat{Y}_t^- = H_0 + H_1 \hat{x}_t^-$$

$$u_t = Y_t - \hat{Y}_t^-$$

$$P_{yy,t} = H_1 P_t^- H_1' + R.$$

The filtered updates are given by

$$K_t = P_t^- H_1' P_{yy,t}^{-1}$$

$$\hat{x}_t = \hat{x}_t^- + K_t u_t$$

$$P_t = (I - K_t H_1) P_t^-.$$

The log-likelihood function is derived from the prediction error decomposition of the KF and is given by

$$\ln L\,(\Psi) = -\frac{Nm}{2}\ln 2\pi - \frac{1}{2}\sum_{t=0}^{N}\ln|P_{yy,t}| - \frac{1}{2}\sum_{t=0}^{N}u_t P_{yy,t}^{-1}u_t',$$

where N is the number of time steps with $t = \{h, 2h, \ldots, T-h, T\}$, $N = T/h$, m is the number of different maturity bonds used, and Ψ is the parameter vector.

To apply the KF in the one-factor CIR models, initialize the algorithm with the unconditional mean and variance of the state

$$\hat{x}_0 = \theta, \quad P_0 = \theta\frac{\sigma^2}{2\kappa},$$

and choose the elements of the KF as follows:

$$\Phi_0 = (1 - \mathrm{e}^{-\kappa h})\theta, \quad \Phi_1 = \mathrm{e}^{-\kappa h}$$

$$Q_t = \theta\frac{\sigma^2}{2\kappa}(1 - \mathrm{e}^{-\kappa h})^2 + \hat{x}_{t-h}\frac{\sigma^2}{\kappa}(\mathrm{e}^{-\kappa h} - \mathrm{e}^{-2\kappa h})$$

$$H_0 = -\frac{1}{\tau}A(\tau), \quad H_1 = \frac{1}{\tau}B(\tau), \text{ and } R = \mathrm{diag}\left(\sigma_{e_i}^2\right),$$

where $\sigma_{e_i}^2$ is the variance of the ith measurement error, and $\mathrm{diag}(\sigma_{e_i}^2)$ is a diagonal matrix with diagonal elements $\sigma_{e_i}^2$. The measurement-error covariance matrix can be nondiagonal; however, it is chosen to be diagonal in this chapter, so the covariance structure of bond yields is represented only by the model itself and not by the measurement-error covariance matrix. The application of the KF in a multifactor model is a straightforward extension of the above one-factor case.

14.3 DIFFERENTIAL EVOLUTION

Evolutionary computational techniques have been used in finance applications for quite some time now. They first came to prominence in the form of genetic algorithms and genetic programming and are used in forecasting, classification, and trading applications. However, they are also used as a complementary tool in optimization problems involving complicated high-dimensional and possibly nonconvex objective functions that may have many local maxima. Differential evolution (see Storn and Price [1997]), is a population-based search algorithm that draws inspiration from the field of evolutionary computation. DE embeds concepts of mutation, recombination, and fitness-based selection to evolve good solutions. There are a number of worthwhile gains associated with the use of DE, including: it is more likely to reach the global optimum than deterministic-based algorithms in the presence of nonconvex objective functions; DE can easily accommodate nonnegativity constraints and parameter constraints that might result in a nonconvex objective function and cause deterministic optimizers to converge to a local optimum; DE can return a population of parameter vectors that are approximately optimal, thus giving insight into parameter or model uncertainty that is perhaps more informative than standard asymptotic diagnostic tests; DE can handle large-scale optimization problems, thereby avoiding complicatious caused by the inclusion of additional parameters; and DE does not require the gradient of the objective function, which can be cumbersome to calculate for certain models. Of course in certain situations, such as the Vasicek term-structure model, DE does not offer any more insight into the problem than a deterministic-based optimizer. DE can also take longer to converge than a deterministic optimizer, and one can only use informal diagnostics in determining the convergence of the optimizer. That said, this chapter does not advocate the replacement of deterministic gradient-based optimizers with global evolutionary optimizers, but rather encourages using DE or other evolutionary optimizers in certain situations where deterministic optimizers may run into problems or even using a combination of deterministic and evolutionary optimizers.

Although several DE algorithms exist, I will describe one version of the algorithm. See Storn and Price (1997) for more detail on the various types of algorithms, Brabazon and O'Neill (2006) for a complete analysis of evolutionary algorithms in the context of financial modeling, and Ben Hamida and Cont (2005) for an example of a derivative-pricing model calibration method using evolutionary algorithms. The DE algorithm used in this chapter is based on the *DE/rand/1/bin* scheme. The different variants of the DE algorithm are described using the shorthand *DE/x/y/z*, where x specifies how the base vector is chosen, (*rand* if it is randomly selected, and *best* if the best individual in the population is selected), y is the number of difference vectors used, and z denotes the crossover scheme (*bin* for crossover based on independent binomial experiments, and *exp* for exponential crossover).

At the start of the algorithm, a population of N d-dimensional parameter vectors Ψ_j for $j = 1, \dots, N$ is randomly initialized and evaluated using a fitness function. In this chapter, the fitness function is the quasi log-likelihood function, and the parameter vector contains the structural parameters that describe the term-structure model and the parameters from the Kalman filter, namely the standard deviation of the measurement errors. For example, for a one-factor CIR model using panel data with four bonds, the parameter vector would have the following form, $\Psi_j = \left(\theta, \kappa, \sigma, \lambda, \sigma_{e_1}, \ldots, \sigma_{e_4}\right)$, where θ, κ, σ, and λ are the structural parameters of the term-structure model, and σ_{e_i}, for $i = 1, \ldots, 4$, are the four measurement-error standard deviations. During the search process, each individual (j) is iteratively refined. The modification process has three steps:

- Create a variant parameter vector using randomly selected members of the population (mutation step).
- Create a trial parameter vector by combining the variant vector with j (crossover step).
- Perform a selection process to determine whether the newly created trial vector replaced j in the population.

Under the mutation operator, for each vector $\Psi_j(t)$ a variant vector $V_j(t+1)$ is obtained

$$V_j(t+1) = \Psi_m(t) + F\left(\Psi_k(t) - \Psi_l(t)\right),$$

where $k, l, m \in 1, \ldots, N$ are mutually distinct, randomly selected indices, and all the indices $\neq j \cdot$ ($\Psi_m(t)$ is referred to the base vector, and $\Psi_k(t) - \Psi_l(t)$ is referred to as a difference vector). F is a scaling parameter, and typically $F \in (0, 2)$. The scaling parameter controls the amplification of the difference between Ψ_k and Ψ_l, and is used to avoid stagnation of the search process. Following the creation of the variant vector, a trial vector $U_j(t+1)$ is obtained

$$U_{jk}(t+1) = \begin{cases} V_{jk}(t+1), \text{ if } (rand \leq CR) \text{ or } (j = rnbr(ind)); \\ \Psi_{jk}(t), \text{ if } (rand \geq CR) \text{ or } (j \neq rnbr(ind)). \end{cases}$$

where $k = 1, 2, \ldots, d$, *rand* is a random number generated in the range $(0, 1)$, *CR* is the user-specified crossover constant in the range $(0, 1)$, and *rnbr*(*ind*) is a randomly chosen index chosen from the range $(1, 2, \ldots, d)$. The random index is used to ensure that the trial solution differs by at least one component from $\Psi_j(t)$. The resulting trial (child) solution replaces its parent if it has higher fitness; otherwise the parent survives unchanged into the next generation.

The DE algorithm has three key parameters: the population size (N), the crossover rate (CR) and the scaling factor (F). Higher values of CR tend to produce faster convergence of the population of solutions. Typical values for these parameters are in the ranges $N = 50$–100 or five to ten times the number of dimensions in a solution vector, $CR = 0.4$–0.7, and $F = 0.4$–0.9.

14.4 RESULTS

In this chapter, the data set used is the same as that used by Duan and Simonato (1999) (hereinafter DS), which is kindly posted on their Web site. It consists of four monthly yield series for the U.S. Treasury debt securities with maturities of 3, 6, 12, and 60 months taken from the Fama–Bliss data file. All interest rates are expressed on an annualized, continuously compounded basis. The data cover the period from April 1964 to December 1997, totaling 405 time-series observations. The data are out of date; however, given that DS have estimated term-structure models on this same data set, it was thought appropriate to use it for comparison purposes. As with DS, the unit of time is set to 1 year so that in the Kalman filter $h = \frac{1}{12}$. The following subsections report the results for one- and two-factor CIR models. The sensitivity of the DE optimizer solutions to the relevant DE parameters is examined in the one-factor case.

14.4.1 One-Factor CIR Model

The one-factor CIR models were estimated using the DE optimizer, and the results were very similar to those reported in DS. Table 14.4.1 reports the results for the one-factor CIR model from 100 runs of the DE optimizer. As with all stochastic optimizers, the algorithm should be run a number of

TABLE 14.4.1 Parameter Estimates for the One-Factor CIR Model

Parameter	Mean	Median	Std	Min	Max	Max LL	DS
θ	0.0611 (0.0157)	0.0610	0.0011	0.0607	0.0717	0.0610	0.0613 (0.0123)
κ	0.2247 (0.0580)	0.2251	0.0035	0.1906	0.2267	0.2251	0.2249 (0.0457)
σ	0.0702 (0.0051)	0.0702	0.0000	0.0700	0.0703	0.0702	0.0700 (0.0045)
λ	−0.1115 (0.0587)	−0.1119	0.0034	−0.1131	−0.0785	−0.1119	−0.1110 (0.0454)
σ_{e_1}	0.0028 (0.0002)	0.0028	0.0000	0.0028	0.0028	0.0028	0.0028 (0.0002)
σ_{e_2}	0.0000 (0.2883)	0.0000	0.0000	0.0000	0.0000	0.0000	0.0000 (7.6255)
σ_{e_3}	0.0030 (0.0002)	0.0030	0.0000	0.0030	0.0030	0.0030	0.0030 (0.0002)
σ_{e_4}	0.0099 (0.0003)	0.0099	0.0000	0.0099	0.0100	0.0099	0.0099 (0.0003)

Column 1 denotes the parameter. Column two is the mean parameter estimate and standard error, three is the median, four is the standard deviation, and five and six are the minimum and maximum from the 100 DE optimal parameter vectors. Column seven is the parameter vector with maximum LL value, and column eight contains DS parameter estimates and standard errors.

TABLE 14.4.2 Parameter Estimates for the One-Factor CIR Model Using Best Parameter Vectors

Parameter	Mean	Median	Std	Min	Max	Max LL
θ	0.0610	0.0610	0.0000	0.0608	0.0610	0.0610
κ	0.2252	0.2251	0.0001	0.2248	0.2258	0.2251
σ	0.0702	0.0702	0.0000	0.0701	0.0702	0.0702
λ	−0.1119	−0.1119	0.0001	−0.1126	−0.1116	−0.1119
σ_{e_1}	0.0028	0.0028	0.0000	0.0028	0.0028	0.0028
σ_{e_2}	0.0000	0.0000	0.0000	0.0000	0.0000	0.0000
σ_{e_3}	0.0030	0.0030	0.0000	0.0030	0.0030	0.0030
σ_{e_4}	0.0099	0.0099	0.0000	0.0099	0.0099	0.0099

The columns are the same as Table 14.4.1; however, only the 95 best-performing DE optimal parameter vectors are used.

times to ensure it has converged. The DE optimizer was run with the following parameters for the optimizer: $NP = 20, F = 0.8, CR = 0.8$, and setting the number of iterations to 300. The results are very similar to the results in DS that are reported in the final column of Table 14.4.1. Sensitivity analysis was conducted on the optimizer parameters. The optimizer was run 100 times for each of the following parameter settings: $\{NP, F, CR\} = \{20, 0.8, 0.8\}, \{20, 0.6, 0.8\}, \{20, 0.4, 0.8\}, \{20, 0.8, 0.6\}$, and $\{20, 0.8, 0.4\}$. The parameter estimates[2] and the parameter standard errors were not sensitive to the different settings. The convergence speed of the algorithm is not sensitive to the crossover rate, CR, but it does improve for smaller values of F. However, this faster convergence is achieved at the expense of a less extensive parameter space search. When the five worst-performing parameter vectors (with the lowest log-likelihood, [LL] values) are dropped, the dispersion of the remaining 95 runs reduces dramatically across all the different settings. Table 14.4.2 contains results using the same optimizer parameters as Table 14.4.1 but with the five worst-performing parameter vectors removed from the analysis. These results on the one-factor CIR model indicate that the DE optimizer is a useful optimizer. However, it must be run a number of times to ensure that it has not converged too quickly. The results also indicate that the one-factor CIR model does not suffer from the problem of local maxima.

14.4.2 Two-Factor CIR Model

The two-factor CIR model was also estimated, and if a factor was negative, it was replaced with zero in the optimization routine. Table 14.4.3 reports the

[2] These results are not reported to save space but are available from the author upon request.

TABLE 14.4.3 Parameter Estimates for the Two-Factor CIR Model

Parameter	Mean	Median	Std	Min	Max	Max LL	Duan and Simonato (SE)
θ_1	0.0703	0.0294	0.1080	0.0257	0.6852	0.0271	0.0303 (0.0031)
κ_1	0.9509	1.1413	0.4045	0.0412	1.3144	1.2689	1.1627 (0.1508)
σ_1	0.1146	0.1146	0.0034	0.1070	0.1221	0.1178	0.1202 (0.0079)
λ_1	−0.1632	−0.3320	0.3562	−0.4959	0.6371	−0.4395	−0.3139 (0.1222)
θ_2	0.0001	0.0000	0.0001	0.0000	0.0011	0.0000	0.0000 (0.0000)
κ_2	0.0314	0.0295	0.0171	0.0016	0.0946	0.0297	0.0007 (0.0614)
σ_2	0.0618	0.0621	0.0018	0.0558	0.0653	0.0622	0.0681 (0.0035)
λ_2	−0.0367	−0.0356	0.0176	−0.1038	−0.0075	−0.0362	−0.0266 (0.0636)
σ_{e_1}	0.0026	0.0026	0.0000	0.0025	0.0027	0.0025	0.0027 (0.0001)
σ_{e_2}	0.0000	0.0000	0.0000	0.0000	0.0002	0.0000	0.0000 (0.0985)
σ_{e_3}	0.0021	0.0021	0.0000	0.0020	0.0022	0.0021	0.0021 (0.0001)
σ_{e_4}	0.0018	0.0018	0.0001	0.0017	0.0020	0.0017	0.0012 (0.0001)

Column one denotes the parameter. Column two is the mean parameter estimate and standard error, three is the median, four is the standard deviation, and five and six are the minimum and maximum from the 100 DE optimal parameter vectors. Column seven is the parameter vector with maximum LL value, and column eight contains DS parameter estimates and standard errors.

results for the two-factor model from 100 runs of the DE optimizer. Most of the parameters estimated in this study are not very different from the parameters estimated by DS; however, the mean and median of the mean reversion for the second factor, κ_2, and the parameter estimate associated with the overall maximum-likelihood value for κ_2 are quite different from the value of 0.0007 that was estimated by DS. This suggests that the second factor might not be as close to a nonstationary process as DS have suggested. However, this study is not claiming that parameter estimates should be very different from those of DS, but rather is emphasizing the local maxima problem that arises in the estimation of the two-factor CIR term-structure model. Table 14.4.4 contains the same results as Table 14.4.3, but with the 40 worst-performing parameter vectors removed from the analysis. More parameter vectors were removed in the two-factor case due to a higher proportion of parameters that converge too quickly to suboptimal solutions.

When examined in more detail, the analysis seems to suggest that the Kalman filter is a double-edged sword when used to estimate parameters. The Kalman filter returns unobservable factors for each of the 100 optimal parameters from the DE optimizer that are almost identical from one run to another. This is quite a promising result, suggesting that the Kalman filter is capable of retrieving the underlying factors that are driving the term structure for this particular data set. Figure 14.4.1(b) shows a plot of the two factors for the 100 optimal DE parameter vectors. The factors from each

TABLE 14.4.4 Parameter Estimates for the Two-Factor CIR Model Using Best Parameter Vectors

Parameter	Mean	Median	Std	Min	Max	Max LL	Duan and Simonato (SE)
θ_1	0.0284	0.0282	0.0014	0.0257	0.0318	0.0271	0.0303 (0.0031)
κ_1	1.1993	1.2051	0.0675	1.0415	1.3144	1.2689	1.1627 (0.1508)
σ_1	0.1156	0.1154	0.0025	0.1105	0.1196	0.1178	0.1202 (0.0079)
λ_1	−0.3812	−0.3930	0.0606	−0.4959	−0.2436	−0.4395	−0.3139 (0.1222)
θ_2	0.0000	0.0000	0.0000	0.0000	0.0001	0.0000	0.0000 (0.0000)
κ_2	0.0308	0.0297	0.0152	0.0016	0.0624	0.0297	0.0007 (0.0614)
σ_2	0.0625	0.0624	0.0011	0.0610	0.0653	0.0622	0.0681 (0.0035)
λ_2	−0.0368	−0.0367	0.0156	−0.0690	−0.0077	−0.0362	−0.0266 (0.0636)
σ_{e_1}	0.0026	0.0026	0.0000	0.0025	0.0027	0.0025	0.0027 (0.0001)
σ_{e_2}	0.0000	0.0000	0.0000	0.0000	0.0001	0.0000	0.0000 (0.0985)
σ_{e_3}	0.0021	0.0021	0.0000	0.0020	0.0021	0.0021	0.0021 (0.0001)
σ_{e_4}	0.0017	0.0017	0.0000	0.0017	0.0018	0.0017	0.0012 (0.0001)

The columns are the same as Table 14.4.3; however, only the 60 best-performing DE optimal parameter vectors are used.

parameter vector are close, making it hard to tell them apart on the graph. In fact, the two unobservable factors from this analysis are > 99.9% correlated with the corresponding unobservable factors from the other runs of the DE optimizer.

However, any two parameter vectors Ψ_i and Ψ_j, for $i, j \in \{1, \ldots, 100\}$, are capable of returning almost identical log-likelihood values and unobservable factors, even though the parameter vectors themselves can differ substantially. To examine this point further, Figures 14.4.2 and 14.4.3 contain plots of the parameter values from the 100 DE optimal parameters for the first and second factors, respectively, versus the negative log-likelihood value. Figure 14.4.2(a) plots the long-run mean for factor 1, θ_1, versus the negative log-likelihood function, and it seems to be converging to its lower bound of zero. Figure 14.4.2(b) shows the mean reversion of the first factor, κ_1, and it seems to be converging to a global optimal value. Figure 14.4.2(c) suggests that there are a number of different optimal parameter vectors whose value for σ_1 can differ (from 0.11 to 0.12) yet whose log-likelihood value is almost identical. However, most of the values for σ_1 are within one standard error of its mean value. Figure 14.4.2(d) suggests that the market price of risk of the first factor, λ_1, also seems to be converging to a global optimal value. The situation for the parameters of the second factor is different, as illustrated in Figure 14.4.3. The long-run mean of the second factor, θ_2, seems to be converging to its lower limit of zero. However, there are a large number of different optimal

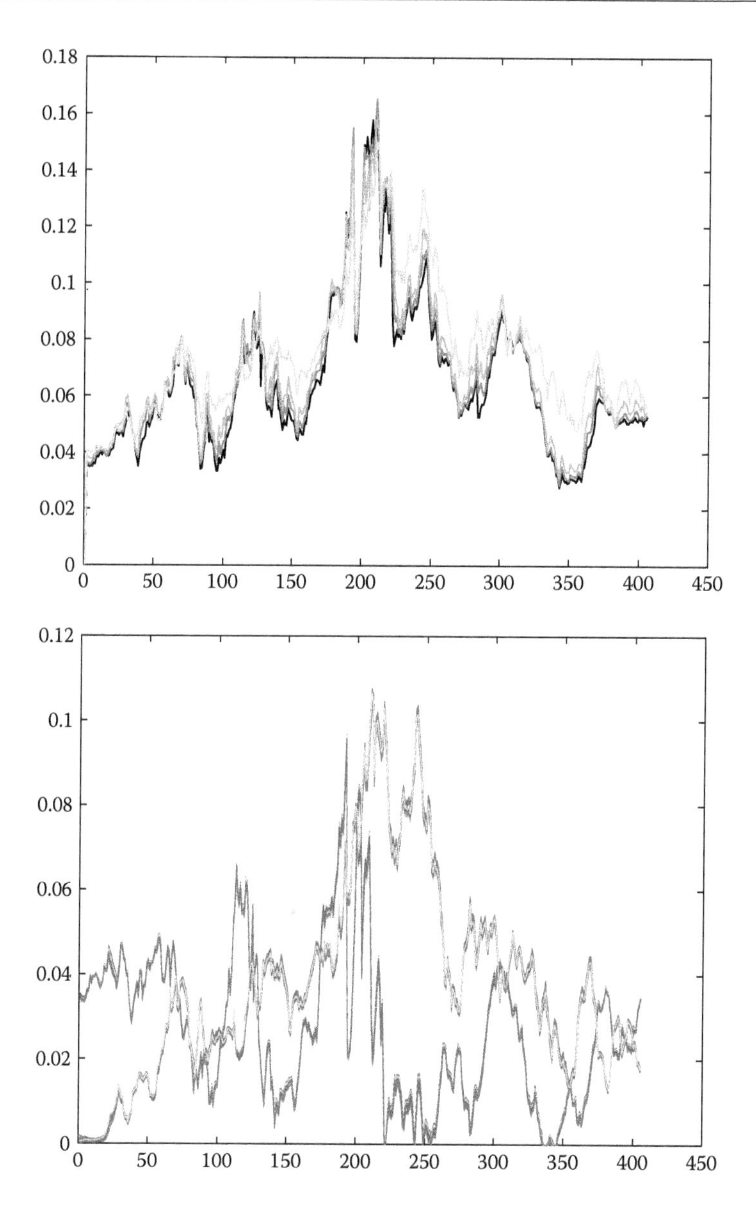

FIGURE 14.4.1 (a) Fama–Bliss data set of four U.S. Treasury yields with maturities of 3, 6, 12, and 60 months and (b) the two unobservable factors for the 100 runs of the DE optimizer.

parameter vectors whose value for κ_2 can differ substantially, yet all have similar log-likelihood values; and the same can be said for σ_2 and λ_2. This suggests that the two-factor CIR model is misspecified, a conclusion that agrees with that of DS. However, many firms still use misspecified term-structure models due to their tractability and ability to explain certain statistical features of the

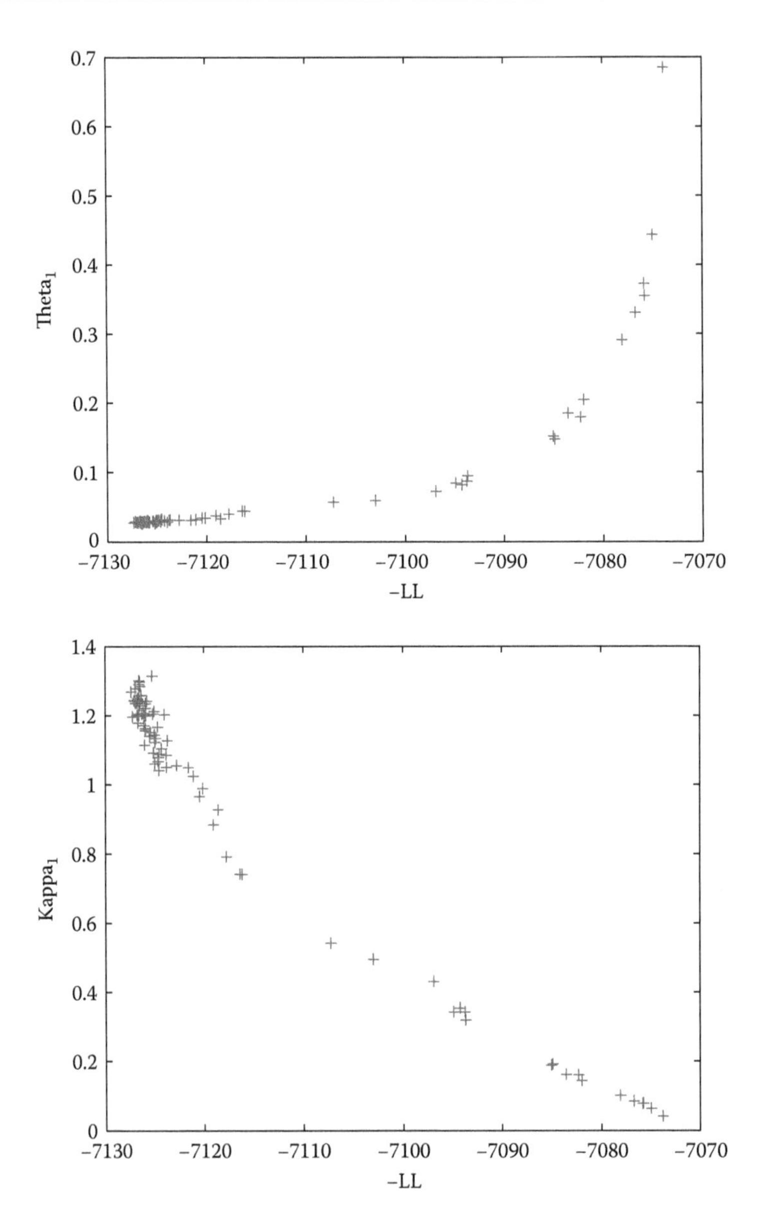

FIGURE 14.4.2 Factor 1 parameter estimates versus the (negative) log-likelihood function.

term-structure. Thus the problem of locally optimal parameter vectors may be relevant for many term structure models that are used extensively in academia and industry today.

We now examine the effect of these different approximately optimal parameter vectors on bond prices and bond derivatives. To do this we take

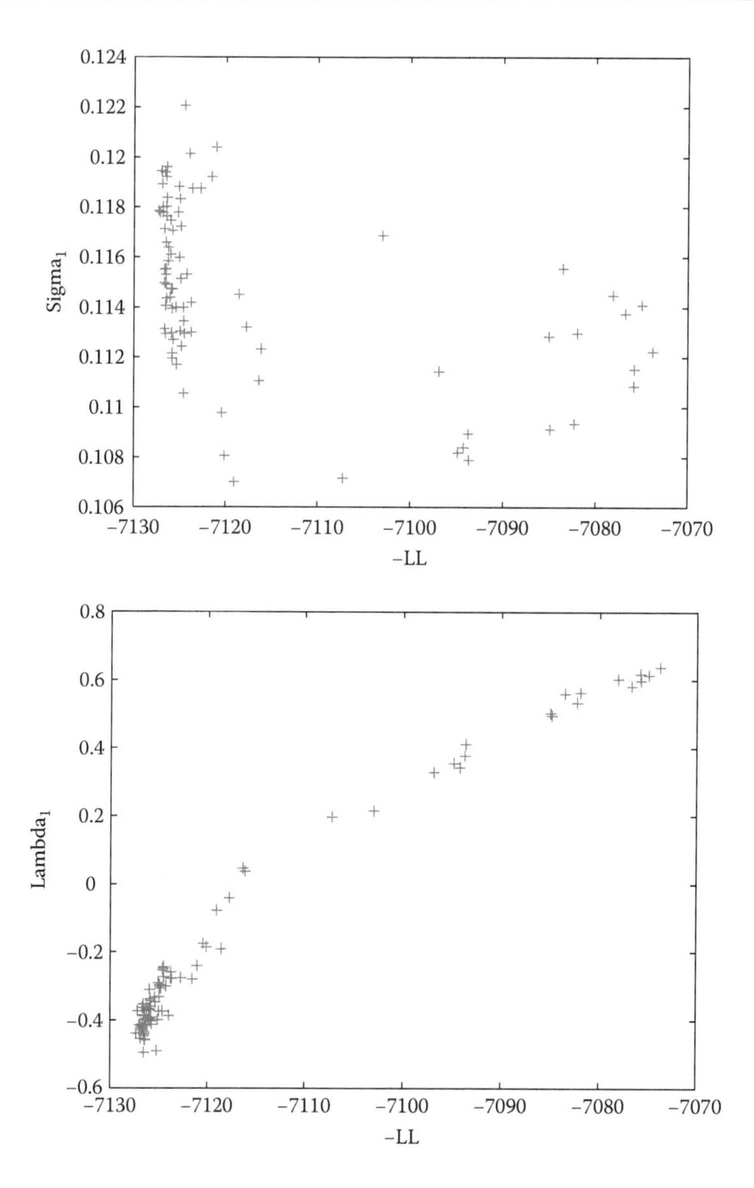

FIGURE 14.4.2 (*Continued*).

each optimal parameter vector from the 100 DE runs and the corresponding unobservable factors x_1 and x_2 on the last month of the time-series data. We then use these to price a 3-year zero-coupon bond. For a bond of face value \$100, the mean, median, and standard deviation of the bond prices are \$84.4875, \$84.4842, and \$0.0144, respectively. The locally optimal parameter vectors do not result in very different bond prices.

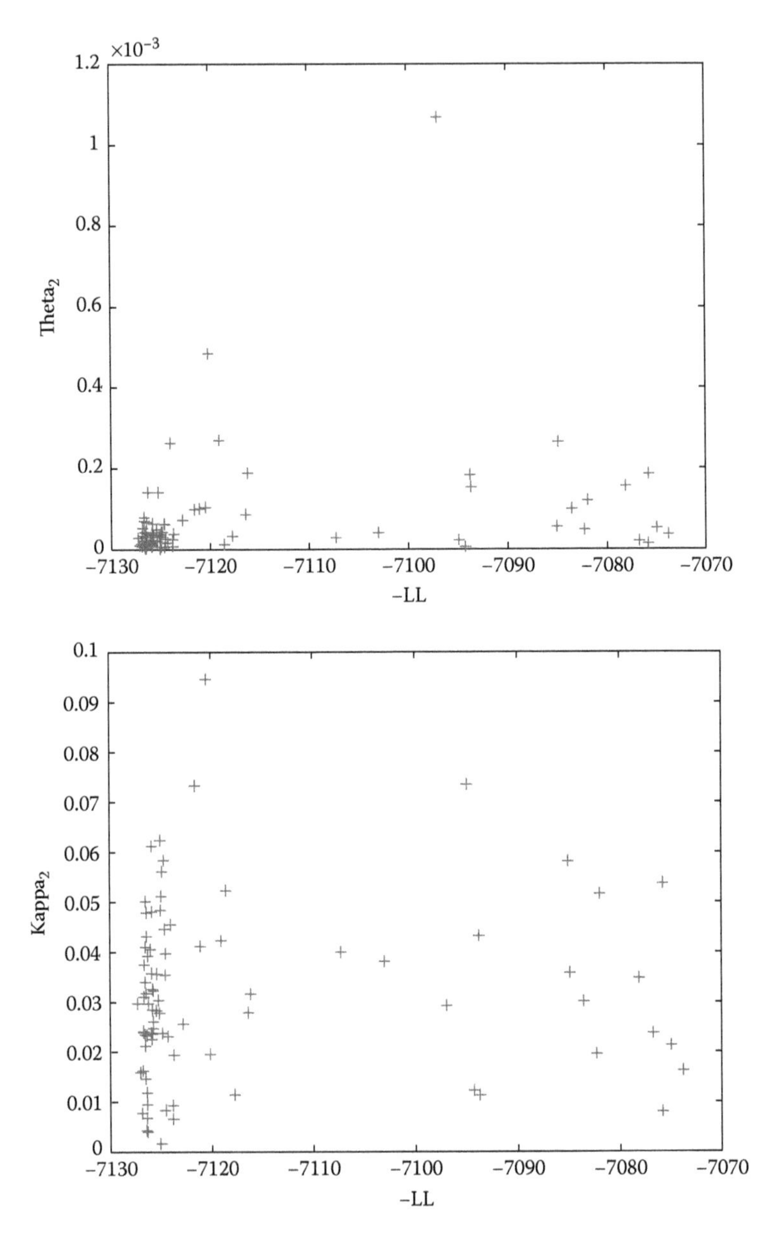

FIGURE 14.4.3 Factor 2 parameter estimates versus the (negative) log-likelihood function.

Figure 14.4.4(a) depicts a histogram of the 100 different bond prices for each parameter vector Ψ. The results show that the problem of the local maxima is very minor when taken in the context of bond pricing. However, if we price an option on the same bond using the 100 parameter vectors,

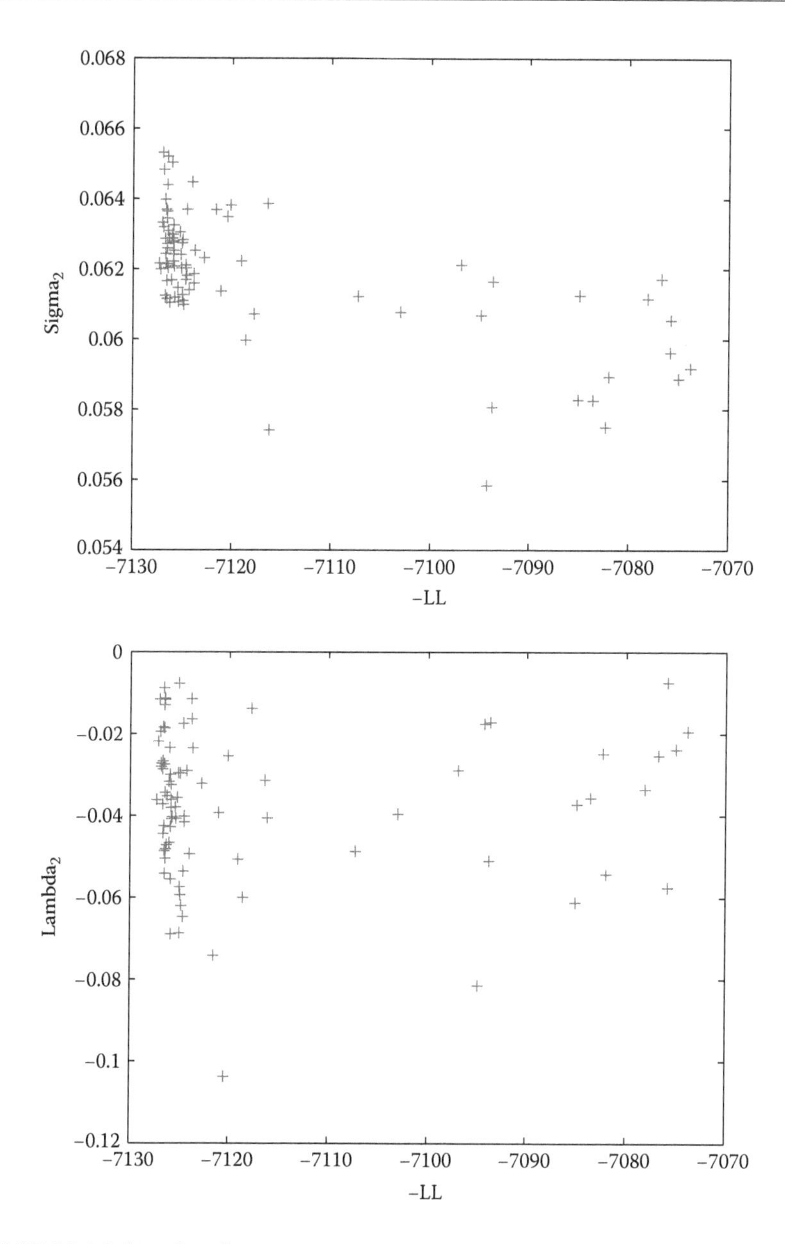

FIGURE 14.4.3 (*Continued*).

the situation becomes more problematic. Taking the same bond as above, we price an at-the-money-forward bond option with a strike price of \$89.19 and a maturity of 1 year using the two-factor CIR-option-pricing formula given in Chen and Scott (1992). The mean, median, and standard deviation of the bond option prices are \$0.6403, \$0.6371, and \$0.0188, respectively.

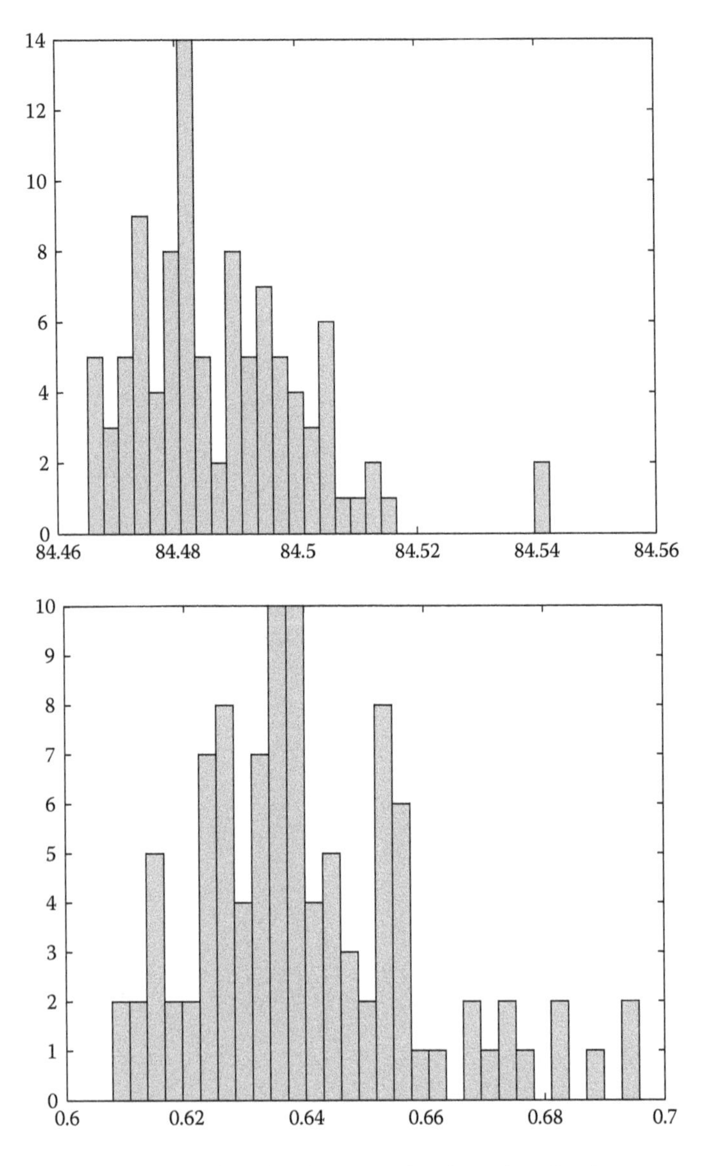

FIGURE 14.4.4 Histograms of bond and option prices for each optimal parameter vector Ψ.

Figure 14.4.4(b) depicts a histogram of the 100 option prices for each parameter vector Ψ. As can be seen, the effect on the option prices is proportionally much larger. The standard deviation of the bond prices is 0.017% of the mean bond price, whereas the standard deviation of the option prices is approximately 3% of the mean option price. This problem does not disappear as we cut out the parameter vectors with lower log likelihoods, i.e., those parameter estimates for which the optimizer converged too quickly. If we consider the 50 parameter vectors with the largest log-likelihoods, the problem still persists, with the standard deviation being 2% of the mean bond price. This is not surprising given the flatness of the log-likelihood plots, where we can see that many different parameter vectors produce the same maximum log-likelihood value. These differences in option prices are magnified when considering out-of-the money options and longer maturity options, which are typical of the fixed-income market. All though a 2–3% difference may not seem that large relative to the bid–ask spread in these markets, when these options are combined to form a typical fixed-income instrument such as a cap (given that a cap can be decomposed into a portfolio of bond options), they will quickly accumulate and result in large option-price differences.

The suggested remedies are to include more than four cross-sectional bond maturities in the estimation and perhaps even include a very liquid benchmark bond derivative in the estimation. If nonlinear securities such as derivatives are used in the estimation, it means that nonlinear Kalman-filter techniques, such as the extended or unscented Kalman filter, will to have to be used in place of the standard Kalman filter. This research is currently being undertaken by the author.

14.5 CONCLUSION

In this chapter, the problem of local maxima in the context of parameter estimation for dynamic term-structure models was highlighted using an evolutionary optimizer known as differential evolution. The effect of the local maxima on bond pricing was shown to be relatively minor; however, the effect on bond derivatives was shown to be a reasonable proportion of the derivative price. The suggested remedies are (a) to use a better-specified dynamic term-structure model than the CIR for the data set in this study or (b) to use more cross-sectional bond maturities and even a liquid benchmark derivative in the estimation procedure. Future work should include simulation studies to determine whether the local maxima are a result of the dynamic term-structure model itself or a result of some specific structure in the data set, such as near-unit root behavior or perhaps a combination of the two.

REFERENCES

1. Ayache, E., P. Henrotte and S. Nassar and X. Wang. (2005), "Can anyone solve the smile problem?" The Best of Wilmott 2, John Wiley and Sons, London.
2. Babbs, S. H. and K. B. Nowman. (1999), "Kalman filtering of generalised Vasicek term structure models." *Journal of Financial and Quantitative Analysis*, 34, 115–130.
3. Ben Hamida, S. and R. Cont. (2005), "Recovering volatility from option prices by evolutionary optimisation." *Journal of Computational Finance*, 8(4).
4. Brabazon, A. and M. O'Neill. (2006), *Biologically Inspired Algorithms for Financial Modelling*. Springer, Berlin.
5. Brennan, M. J. and E. S. Schwartz. (1979), "A continuous time approach to the pricing of bonds." *Journal of Banking and Finance*, 3, 133–155.
6. Chen, R. -R. and L. Scott. (1992), "Pricing interest rate options in a two-factor Cox-Ingersoll-Ross model of the term structure." *Review of Financial Studies*, 5, 613–636.
7. Chen, R. -R. and L. Scott. (1993), "Maximum likelihood estimation for a multifactor equilibrium model of the term structure of interest rates." *Journal of Fixed Income*, 3, 14–31.
8. Chen, R. -R. and L. Scott. (2003), "Multi-factor Cox-Ingersoll-Ross models of the term structure: estimates and tests from a Kalman filter model." *Journal of Real Estate Finance and Economics*, 27, 143–172.
9. Courtadon, G. (1982), "The pricing of options on default free bonds." *Journal of Financial and Quantitative Analysis*, 17, 75–100.
10. Cox, J. C., J. E. Ingersoll, and S. A. Ross. (1985), "A theory of the term structure of interest rates." *Econometrica*, 55, 385–407.
11. Dai, Q. and K. J. Singleton. (2000), "Specification analysis of affine term structure models." *Journal of Finance*, 55, 1943–1978.
12. De Jong, F. (2000), "Time series and cross-section information in affine term structure models." *Journal of Business and Economic Statistics*, 18, 300–314.
13. Dothan, M. U. (1978), "On the term structure of interest rates." *Journal of Financial Economics*, 6, 59–69.
14. Duan, J. C. and J. G. Simonato. (1999), "Estimating and testing exponential-affine term structure models by Kalman filter." *Review of Quantitative Finance and Accounting*, 13, 111–135.
15. Duffee, G. R. and R. H. Stanton. (2004), "Estimation of dynamic term structure models." *Working paper*, Haas School of Business, Berkeley, CA.
16. Duffie, D. and R. Kan. (1996), "A yield-factor model of interest rates." Mathematical Finance, 6(4), 379–406.
17. Feldhutter, P. and D. Lando. (2005), "Decomposing swap spreads." *Working paper*, Copenhagen Business School, Denmark.
18. Gallant, A. R. and G. Tauchen. (1996), "Which moments to match?" *Econometric Theory*, 12, 657–681.
19. Hansen, L. P. (1982), "Large sample properties of generalised method of moments estimators." *Econometrica*, 50, 1029–1055.

20. Harvey, A. (1991), "Forecasting structural time series models and the Kalman filter." Cambridge University Press, Cambridge.
21. Kalman, R. E. (1960), "A new approach to linear filtering and prediction theory." *Journal of Basic Engineering, Transactions ASME, Series D*, 82, 35–45.
22. Litterman, R. and J. A. Scheinkman. (1991), "Common factors affecting bond returns." *Journal of Fixed Income*, 1, 54–61.
23. Lo, A. W. (1986), "Statistical tests of contingent claims asset pricing models." *Journal of Financial Economics*, 17, 143–173.
24. Lo, A. W. (1988), "Maximum likelihood estimation of generalised Ito processes with discretely sampled data." *Econometric Theory*, 4, 231–247.
25. Longstaff, F. A. and E. S. Schwartz. (1992), "Interest rate volatility and the term structure: a two-factor general equilibrium model." *Journal of Finance*, 47, 1259–1282.
26. Lund, J. (1997), "Non linear Kalman filtering techniques for term structure models." *Working paper*, University of Aarhus, Denmark.
27. Pennacchi, G. G. (1991), "Identifying the dynamics of real interest rates and inflation: evidence using survey data." *Review of Financial Studies*, 4, 53–86.
28. Schaefer, S. M. and E. S. Schwartz. (1984), "A two factor model of the term structure: an approximate analytical solution. The Journal of Financial and Quantitative Analysis" 19 (4), 413–424.
29. Steeley, J. M. (1990), "Modelling the dynamics of the term structure of interest rates." *The Economic and Social Review*, 21, 337–361.
30. Storn, R. and K. Price. (1997), "Differential evolution—a simple and efficient heuristic for global optimization over continuous spaces." *Journal of Global Optimization*, 11, 341–359.
31. Vasicek, O. (1977), "An equilibrium characterisation of the term structure." *Journal of Financial Economics*, 5, 177–188.

CHAPTER 15

EDDIE for Discovering Arbitrage Opportunities

Edward Tsang, Sheri Markose, and Alma Garcia
University of Essex, Colchester, United Kingdom
Hakan Er
Akdeniz University, Antalya, Turkey

Contents

The prices of the options and futures of a stock both reflect the market's expectation of future trends of the stock's price. Their prices normally align with each other within a limited window. When they do not, arbitrage opportunities arise: an investor who spots the misalignment will be able to buy (sell) options on one hand, and sell (buy) futures on the other and make risk-free profits. In this chapter, we focus on put-call-futures parity arbitrage opportunities. The upper bound of a futures bid price, denoted by F_{bt}, is given by

$$F_{bt}e^{-ra(T-t)} \leq C_{at} - P_{bt} + Xe^{rb(T-t)} + TC. \tag{15.0.1}$$

Here, T is the expiration date and t is today, i.e., $T - t$ is the remaining time to maturity; C_{at} is the option's call premium at the ask; P_{bt} is the option's put premium at the bid; X is the exercise price for the option; TC is the transaction cost; ra is the interest rate on the borrowing to finance the futures; and rb is the interest rate to lend. If equation (15.0.1) is violated, then the arbitrageur will be able to make a risk-free profit equal to

$$F_{bt}e^{ra(T-t)} - [C_{at}P_{bt} + Xe^{rb(T-t)} + TC] > 0. \tag{15.0.2}$$

When equation (15.0.1) is violated, a short arbitrage profit can be realized by shorting futures and then protecting it by a synthetic long futures position by (a) buying a call option, (b) shorting a futures option, and (c) borrowing the present discounted value of the futures price and lending the same for the exercise price. Historical data suggest that option and futures prices on the LIFFE market (London) occasionally do not satisfy equation (15.0.1). In the LIFFE tick trade data from January 1991 to June 1998, we identified 8073 profitable short arbitrage and 7410 profitable long arbitrage opportunities when no transaction cost is considered. If we assume a transaction cost of £60 per put-call-futures arbitrage operation, then 2345 (or 29%) of the 8037 triplets would still be profitable. The profits in equation (15.0.2) are those that accrue if the arbitrageur could have obtained as quoted the trade prices recorded at these points in time. In reality, due to delay, the arbitrageur may not be able to obtain the quoted prices. Therefore, an arbitrageur may not be able to exploit all the profitable arbitrage opportunities, especially if it reacts passively. Besides, price misalignments are corrected rapidly by the market, so reacting ahead of the others is crucial to securing the risk-free profits. Therefore, the challenge is not only to spot such opportunities, but to discover them ahead of other arbitrageurs. This motivated us to turn our attention to our previous work on forecasting. EDDIE is a genetic programming tool for forecasting. A specialization of EDDIE, which we called EDDIE-ARB, was implemented for forecasting arbitrage opportunities. EDDIE uses constraints to focus its search in promising areas of the space. The task that we gave EDDIE-ARB was to predict arbitrage opportunities five minutes ahead of time.

As a tool, EDDIE enables economists and computer scientists to work together to identify relevant independent variables. The usefulness of EDDIE-ARB as a tool is fully demonstrated in this project. When data was first fed into EDDIE-ARB, no patterns were found. The economists and computer scientists in this project together noticed that certain subcomponents were repeatedly generated by the program. In response to that, data was further prepared to help EDDIE-ARB to succeed. For example, "moneyness" (spot price divided by strike price) was introduced, as (a) it is meaningful in economics and (b) this pattern was found by EDDIE-ARB repeatedly. Similarly, "basis" (futures price minus spot price) was introduced to capture mispricing in the futures leg of the arbitrage. Scaling was applied to certain variables to avoid the precision problem (which computer scientists are more sensitive to than economists).

The above preparation alone was not enough to help EDDIE-ARB find patterns reliably. The difficulty of this forecasting problem is that a large percentage of the cases were negative instances. Only about 3% of the instances in the training data represented opportunities. This meant that a program that made no positive recommendations (i.e., classifying all cases to be nonprofitable) would achieve an accuracy of 97%, even though it had 0% recall. Such forecasts would not help us to spot any arbitrage opportunities, and therefore would have no commercial value. To tackle this problem, we removed certain negative training instances to rebalance the database (we removed those instances that showed no follow-up in the market). When the data set contained around 25% positive instances, EDDIE-ARB started to pick up repeated patterns.

We trained and tested EDDIE on intraday historical tick data on the FTSE-100 European style index option traded on LIFFE from March 1, 1991, to June 18, 1998, and verified it on out-of-sample data from June 30, 1998, to March 30, 1999. The constraints in EDDIE-ARB enabled us to trade precision against recall. For example, the final data set used (i.e., after heavy preprocessing) allowed us to find rules with 99% precision and 53% recall.

Results by EDDIE-ARB were compared with those obtained by a naïve ex ante rule, which only reacted when misalignments were detected. If we assume an operational delay of one minute after opportunities are identified, then expected profit may not be realized by the naïve rule (as explained above). Under this assumption, EDDIE-ARB outperformed the naïve rule on average profit per operation in the test data. However, EDDIE-ARB only picked up a very small percentage of the profitable arbitrage opportunities. As a result, the total amount of profit made by the naïve rule was comparable with EDDIE-ARB's. Our next challenge was therefore to improve EDDIE-ARB's recall rate. Two general methods, namely the scenario method and the repository method, have been developed. Early results suggest that one can collect and combine rules from multiple decision trees to improve precision and recall with these methods.

This work falls into the research area of chance discovery, the discovery of chances through innovation. In chance discovery, the innovation part often involves human input. A "chance" here refers to an event or a situation with significant impact on human decision making—a new event/situation that can be conceived either as an opportunity (e.g., in business) or as a risk (such as an earthquake). Chance discovery extends data mining, which is often limited to pattern recognition in a given data set. For example, in the EDDIE-ARB project, we identified new attributes that were not present in the original data set. We have in this project established EDDIE-ARB as a promising tool for bringing human users and a computer program together to discover arbitrage opportunities. We have also demonstrated how economists and computer scientists could work together to achieve results that neither party alone was capable of achieving.

REFERENCES

1. A. Abe and Y. Ohsawa (eds.), Special Issue on Chance Discovery. *New Generation Computing*, 21(1), Berlin: Springer and Tokyo: Ohmsha, 2003.
2. Chance Discovery Consortium, http://www.chancediscovery.com/english/.
3. A.L. Garcia-Almanza and E.P.K. Tsang, Simplifying Decision Trees Learned by Genetic Algorithms, Proceedings, Congress on Evolutionary Computation (CEC), 2006, 7906–7912.
4. A. Garcia and E.P.K. Tsang, The Repository Method for Chance Discovery in Financial Forecasting, Proceedings, 10th International Conference on Knowledge-Based & Intelligent Information & Engineering Systems (KES2006), Bournemouth, UK, 9–11 October 2006.
5. J. Li, FGP: A Genetic Programming Based Tool for Financial Forecasting, Ph.D. thesis, University of Essex, Colchester, U.K., 2001.
6. Y. Ohsawa and P. McBurney, *Chance Discovery*, Springer Publishers, Berlin, 2003.
7. E.P.K. Tsang, S. Markose, and H. Er, Chance discovery in stock index option and future arbitrage, *New Mathematics and Natural Computation, World Scientific*, 1, (3), 2005, 435–447.
8. E.P.K. Tsang, P. Yung, and J. Li, EDDIE-Automation, a decision support tool for financial forecasting, *Journal of Decision Support Systems*, Special issue on data mining for financial decision making, 37, (4), 2004, 559–565.
9. A.L. Tucker, *Financial Futures, Options and Swaps*, West Publishing Co., St. Paul, MN, 2001.

Index

F

G

H

M

N

O

P

Q

R

S

W

Y

Z

For Product Safety Concerns and Information please contact our EU representative GPSR@taylorandfrancis.com
Taylor & Francis Verlag GmbH, Kaufingerstraße 24, 80331 München, Germany

www.ingramcontent.com/pod-product-compliance
Ingram Content Group UK Ltd.
Pitfield, Milton Keynes, MK11 3LW, UK
UKHW021620240425
457818UK00018B/659

* 9 7 8 0 3 6 7 3 8 8 5 9 1 *